Life in the Shallows

Life in the Shallows

The wetlands of Aotearoa New Zealand

KAREN DENYER & MONICA PETERS

MASSEY UNIVERSITY PRESS

Contents

Foreword

WHEN I READ BOOKS like this one, I immediately lose a lot of time: hours, days or even more. I just can't help myself. Wetlands are places of endless fascination to me, and reading these stories instantly takes me on a virtual, dry-footed wetland tour. But this book isn't just for people who love nature and science, it's also for people who love people. It cleverly blends the weird and wonderful (and often hidden) features of our bogs, swamps and estuaries with stories about the very people who work to reveal their secrets.

When we hear the word 'scientist' it's easy to defer to the old stereotype — white lab coats and thick-rimmed glasses. This book will utterly dispel that image! You will read about people venturing into the wilderness in waders or kayaks, grappling with slippery apex predators or ornery giant herons, and coming up with clever ideas to make a fish feel at home. You will also read about researchers connecting with local hapū and community volunteers, pooling their knowledge and drawing on mātauranga Māori to learn about our wetlands and how to protect or restore them.

The scientists profiled here are a mere sample of New Zealand's scientific talent, but nonetheless impressive. The chapter themes weave and meander like a braided river, but blend together perfectly to accurately sketch the facets of 'a day in the life of a wetland scientist'. It paints the picture of a place where I can crawl on hands and knees, looking at life in all its glorious forms: orders, families, genera, species with associated ecosystem connections and jobs to do; organisms ranging in size from trees to microscopic and sometimes as small as e-DNA. (No wonder my 'field-trips' are usually no longer than 50 metres!)

When it comes to wetlands, New Zealand is not doing too well at all. In fact, just 10 per cent of our original wetlands are still hanging in there. Our European settlers, a product of their time, set out to cultivate the wild lands and to turn them into productive fields for pasture or crops. Thankfully we now appreciate that wetlands — those small springs, soaks, swamps, and the margins of rivers and streams and lakes — are nature's vital tools for maintaining water tables, filtering and recycling waste, pollutants, nutrients and sediment washed from the land. We also now know that functioning wetlands are right up there when it comes to sequestering carbon!

This book exposes the reason we have a National Wetland Trust in Aotearoa: modern-day *Homo sapiens* appear to have lost the operating manual for Planet Earth, and the passionate scientists of these wetland stories are doing their darndest to point us all in the direction of conservation, or better still, restoration of these life-sustaining habitats, and recognition of the traditional knowledge that can help us get there.

Entomologist, media personality and passionate educator Ruud Kleinpaste is the patron of the National Wetland Trust.

This storytelling is of crucial importance to the way we teach young New Zealanders to understand that we live on a biological planet, which is run by aquatic life forms, botanical giants, fungi, birds and bats, grazers, predators, algae, snails, parasites and pathogens, lichens, barnacles, bogs and bugs.

Just think for a moment about all that life in a wetland. It started about 3.8 billion years ago and through evolution the resulting assemblies of organisms are our most senior, sacred, inspirational and perfectly adapted elders from which we can learn how to behave on this watery planet.

Let's listen to what they have to say!

Ruud Kleinpaste

Preface

Getting into Wetlands

MY ALL-TIME FAVOURITE MOVIE is Taika Waititi's *What We Do in the Shadows*. Not just because of its fabulous portrayal of our uniquely Kiwi sense of humour, but also because it almost perfectly encapsulates the livelihood of many of my colleagues featured in this book. What We Do in the Shallows was my personal working title for this book. Wading into amphibious ecosystems — sometimes wet, sometimes dry — or paddling around the shallow margins of lakes, ponds, rivers and estuaries is what wetland scientists do. But how many people know who they are or what their work entails?

New Zealanders are notoriously humble. It's almost a national crime to stick your head above the parapet and say, 'Look at me, look what I've done'. It's not part of the Kiwi psyche to blow your own trumpet, so to speak. Which is why most of the people featured in this book, irrespective of the incredibly significant contributions they have made to New Zealand science and wetland conservation, may be little known outside their science circle, family and friends. We certainly have scientists who are household names — Ashley Bloomfield, Nanogirl, Ernest Rutherford — but I challenge the average New Zealander to name one wetland scientist, living or passed on.

Perhaps science in general is underappreciated in Aotearoa New Zealand, or perhaps that's just true of wetlands, still shackled with negative connotations in Western society. Even in the cancel culture of the 2020s, there are still lingering images of the cartoon-like Creature from the Black Lagoon, enduring miserable European names for wetlands (Dismal Swamp, Boggy Lagoon), and Donald Trump vowing to 'drain the swamp'.

The National Wetland Trust, which championed this book, felt it was not only time to stop 'body shaming' wetlands but also to celebrate the humble, hard-working and incredibly dedicated scientists who are willing to pull on a pair of waders and head into the shallows. That is the purpose of this book: to bring wetlands, and wetland scientists, out of the shadows. It's a smorgasbord of science; a book that readers can dip in and out of and read in any order. Each chapter features a New Zealand scientist and one or more aspects of their work — new species they have discovered, novel applications of emerging technology, a better appreciation of the important role of wetlands in our landscape, and guidance on how to restore wetlands or protect their wildlife. Side boxes offer fascinating facts, and a featured wetland to visit brings the chapter to life by allowing the reader to experience some aspect of that person's research.

Karen Denyer has a life-time fascination with New Zealand's natural places. Wetlands are a particular passion of hers.

But who to include? Who to leave out? It was an incredibly hard decision, and Monica and I could fill another two books with those we left out. But, in reality, very few scientists in New Zealand work only on wetlands. Many are specialists in a particular field, and they apply that expertise variously to wetlands and other ecosystems. Dr Corinne Watts is an example; her expertise in invertebrate ecology is applied to forests, river margins, wetlands and other ecosystems. Many others, when approached with an invitation to appear in the book, hung back . . . 'Well, I'm not really a *wetland* scientist.'

It's a fair point. What do we mean by wetland? Even for our aquatic scientists, the default tends to be: wetlands = bogs + swamps + fens; in other words, soggy but well-vegetated areas. The legal definition in New Zealand's Resource Management Act 1991 and indeed the international definition under the Ramsar Convention are much broader. Wetlands are places where water sits long enough, and often enough, to alter the soil, vegetation and animal communities that live there. They include permanently or intermittently wet areas, shallow water and land–water margins that support a natural ecosystem of plants and animals adapted to wet conditions. They are diverse ecosystems, with the different types determined by their water — amount, depth, permanence, temperature, the chemicals found there — and its source, whether groundwater, surface water, rainwater or a combination of these. The book reflects this wider definition by including scientists working on a wide range of wetland types: bogs, swamps, gumlands, estuaries, lakes, rivers and even semi-frozen pools. Each scientist has fascinating stories to tell, but often most intriguing was how they got into wetlands in the first place.

I FORGET HOW OLD I was when I had my first memorable encounter with a wetland. Probably around seven or eight. I didn't actually know it was a wetland at the time. It was just a muddy, mucky mangrove swamp that stood between me and a challenge.

With my childhood buddy and co-conspirator Michael — a fellow fearless adventurer — we slopped our way through the mangroves on the banks of the Whau River in West Auckland. Our goal? To swim to the other side. We didn't make it across, but we did survive to write about it.

I don't remember how we got out of the current, but I do recall slinking back home, covered in mud, legs bloodied from barnacle-sharpened mangrove roots, faces streaked with frightened and guilty tears, sneaking round to the outdoor hose at Michael's house to wash away the evidence. As far as I know, neither of us ever told our parents.

My Whau River adventure was the sort of experience that could scar a person for life (and not just from those sharp barnacles). Wetlands didn't really factor in my life again until high school science expeditions, back to the muddy mangrove forests, but with a better appreciation for their positive values. Even with an intense interest in ecology and conservation, wetlands still weren't really front of mind for me when I studied science at university. Marine biology, as ever, was totally cool, and definitely my career path, until the cold reality hit me — it's your *job*, you *have* to go scuba diving when the weather is rubbish, or when you're tired, or hungover, or in murky, uninviting water. Plan B was a career in something on slightly more solid ground. There was one paper on limnology (lakes) where I learnt the name *Vallisneria australis* on a field trip to Auckland's Lake Pupuke, but next to nothing about wetlands. I came out with a double major in botany and zoology, and went straight into forest ecology.

I got into wetlands, in both senses of the word, when I took a job at what was then called Environment Waikato (now the Waikato Regional Council). The Waikato is 'wetland central' for New Zealand. Just about every type of wetland can be found here, including geothermal and plutonic (underground) wetlands. It has three of our seven internationally recognised Ramsar wetlands, and one of the nation's largest proportions of remaining wetlands. Little wonder then that many wetland experts are based in the Waikato or have conducted research here, which is perhaps why it features a little heavily in this book.

This was where I was inducted into the world of wetlands. It was where I bought my first pair of waders, and where I first got lost in a wetland (with two other women, and thank goodness we had chocolate with us!). It was where I first met incredible Māori wetland scientists and learnt to think beyond biology. It was where I learnt that Aotearoa

New Zealand has fish that can breathe through their skin, plants that eat insects, aquatic micro-organisms that can withstand 120°C. It's also where I first heard the word 'repo', despite leaving high school with a Sixth Form Certificate in te reo Māori.

AT THE TIME OF PUBLICATION, New Zealand universities still don't have any dedicated degrees for would-be wetland scientists. You might graduate with a BSc in ecology, and a few aquatic papers under your belt; but while you can get a degree in marine biology, even today few New Zealand universities offer a wetland-centric undergraduate degree.

To encourage more students to consider researching wetlands for their Master's degree, the National Wetland Trust established the Golden Plover Wetland Research Award, a small financial incentive, recently doubled in value through a partnership with the New Zealand Freshwater Sciences Society. Despite being open to post-graduate researchers across the country, there are still some years with no applications.

Things are slowly changing. Wetlands were given strong legal protection in a new environmental law in 2020, and the University of Waikato is developing bilingual multimedia educational resources based on New Zealand wetland research. But there are still so many things we have yet to learn about our wetlands. Many wetland species are cryptic (rarely seen), or tiny. Few undamaged wetlands remain to tell us what they were once like, to guide restoration goals. Māori knowledge, built from centuries of interacting with, living in and collecting resources from wetlands, has the potential to radically improve wetland management in Aotearoa, but there is a lot of work to do to integrate mātauranga Māori into all restoration projects, while respecting intellectual property and cultural knowledge.

It is our hope that the stories so generously shared by the scientists featured in this book, including their personal insights, their highlights and even their 'oopsie' moments, will encourage a greater investment in wetland science at our universities. We also hope it will inspire a new generation of future scientists to get into their gumboots, muck in and carry on the quest for knowledge about New Zealand's wetlands.

Writing about people you know as peers, hearing their backstory, learning about their pathway into science and hoping to capture just a little bit of their personality as well as their science in each chapter, has been a thoroughly humbling and enjoyable process. We hope these stories will inspire everyday New Zealanders to explore the wetlands in their backyard, to experience what motivates our wetland scientists, and to learn more about the significant cultural, spiritual, recreational and ecological value of what the late Gordon Stephenson, founder of the National Wetland Trust, called our 'shy places'.

Karen Denyer

Preface

Engaging Communities

Ko Rangitoto te maunga te rū nei taku ngākau.
Ko Tāmaki te awa e mahea nei aku māharahara.
Nō Tāmaki Makaurau ahau.
E mihi ana ki ngā tohu o nehe, o Kirikiroa e noho nei au.
Ko Monica Peters ahau.

Rangitoto is the mountain that speaks to my heart.
Tāmaki is the river that alleviates my worries.
I'm from Auckland.
I recognise the ancestral and spiritual landmarks of Hamilton where I live.
My name is Monica Peters.

MANY OF THE SCIENTISTS in this book honed their research skills and shaped their careers from a young age. My pathway has been much more eclectic, with years spent zigzagging between fine arts, conservation and international development. I eventually did study science, but not before volunteering in ever more remote places. From Te Hauturu-o-Toi Little Barrier Island to Raoul Island in the Kermadecs, to Malaysian Borneo and then Mongolia, my roles ranged from habitat-control volunteer to expedition artist. I've always seen art and science as different lenses through which to explore and understand the environment. I knew I wasn't cut out to specialise in one tiny area of either, but not because delving into detail isn't my thing. For science, in particular, I saw that to really make a difference it needs to be positioned within a sociocultural, economic and political setting. Unless it's relevant to society and decision-makers, science won't be picked up and applied. I also saw that an increasingly big part of doing science is how well it can be communicated to a broad and varied audience. And creativity is a big part of communication.

In my work — which sits at the interface between science, the public and conservation — the word 'community' crops up a lot. It's a convenient catch-all term that describes a group of people who have something in common. Community could easily be applied to the scientists in this book — they're united by a shared passion for learning about wetlands. Collectively, they're unpacking the mysteries and complexities of these extraordinarily diverse ecosystems: how wetlands work; the fascinating life within them; how we understand, value

Monica Peters applies her skills in conservation, research, international development and fine arts to understand our environment and our relationship to it.

and manage them; and how to right the wrongs of their past mismanagement. Most of the scientists profiled have worked together in one way or another. But perhaps what's not fully appreciated is how many of these scientists encourage, advise and support community groups all around Aotearoa New Zealand to restore their local wetlands and surrounds — and vice versa. How many of these scientists may have benefitted from community environmental groups with decades of hands-on wetland restoration experience?

Community groups and citizen scientists are adopting wetland monitoring techniques used by professionals to track the state of wetland health.

But it's not just scientists who study wetlands. Our understanding of these soggy sites and their inhabitants has been built on a much wider body of knowledge. This includes people with no formal science training but who are naturally curious, have a keen eye and appreciate the need to study, document and learn from nature. When people without formal science qualifications or who don't work in a science-based profession do science, they're commonly called citizen scientists. This includes backyard bird surveys, keen amateur botanists uploading their observations to iNaturalist,* and those who volunteer to regularly monitor stream water quality.

Citizen science, a particular passion of mine, is an umbrella term that covers community-based environmental monitoring. There are synergies between mātauranga Māori (traditional knowledge) and citizen science with data collection, analysis and the co-production of new knowledge commonly featuring in both. However, an important difference lies in provenance: mātauranga Māori is embedded in the Indigenous worldview of Te Ao Māori, while citizen science is built on the foundations of Western thinking. As both a social movement and a collaborative research method, citizen science often takes a more holistic approach to science-based studies, incorporating community engagement, education and outreach alongside data collection. It also complements the work of scientists: there simply aren't enough scientists to carry out the research needed to better understand and manage our environment.

We don't really know how many community groups are scattered throughout Aotearoa working to restore wetlands. We do know that it takes a phenomenal effort to haul degraded wetlands back from being unloved and uncared for, to being places that are a safe haven for our native species and amenable to visit. Sure, years of dedicated weeding, planting and controlling introduced predators (and sometimes putting the actual 'wet' back into the wetlands) typically produces a profound visual change — more water, cleaner water, lush native plants and secretive wetland birds slowly reappearing. But have the years of hard work resulted in a net gain in native species biodiversity? Nationally, the trend for our unique, quirky and remarkable indigenous species still shows a steady downward trajectory.

For my PhD I asked 296 community groups to describe their restoration objectives.[1] Many were very general — 'restore biodiversity', 'bring back native birds'. To better understand how they measured progress towards achieving their restoration objectives, I also asked them about their monitoring programmes. Did they use any science-based methods? If no, why not? If yes, did they measure the direct outcomes of their efforts, such as the numbers of predators trapped (rats, possums, stoats, cats, hedgehogs)? Or did they measure the indirect outcomes of their work, such as changes in birdlife or lizards resulting from their predator control? The difference is important: one paints a picture of what and how many introduced predators there are in the landscape; the other reveals whether reducing predator numbers is actually making a difference to our native species.

I'm still passionate about developing a more nuanced understanding of the institutional, bureaucratic and social barriers that hinder or block groups' monitoring efforts and

---

* See https://inaturalist.nz.

encouraging a culture where their data are valued more widely. Their data could, and should, be used to help flesh out how we, as a nation, are faring with respect to enhancing our indigenous biodiversity. Quantifying what community groups have achieved is important, because I (and many others working in the conservation sector) know through years of observation that their work is significant. Many of the wetlands we have left would be much worse off if it weren't for community groups and those organisations and individuals that support them: the National Wetland Trust and other environmental NGOs, funders and the science/technical advisors in councils, the Department of Conservation, universities and consultancies.

CREATING RESOURCES TO HELP COMMUNITY GROUPS manage wetlands means sweeping away desiccated formal language, jargon and tight stylistic conventions to create content that can be understood and then applied. I've had plenty of opportunities to put this philosophy into practice with my colleagues.

Working with Dr Beverly Clarkson (Manaaki Whenua — Landcare Research) to bring together *Wetland Restoration: A Handbook for New Zealand Freshwater Systems* (2010) was a matter of fortuitous timing.[2] We were already working on an experimental wetland restoration project translocating rare restiads (rushes) to new sites in the Waikato,[3] when the opportunity arose to revamp a much earlier version of the handbook. The content was due for a major makeover to reflect new developments in the field and, importantly, to expand readership from technical experts to community wetland restoration practitioners.

In the updated version, experts each contributed a chapter on one aspect of wetland restoration, and each also included a case study of a community group's project(s) and achievements. A rich array of photos, diagrams and illustrations were also added. Several years later, colleagues at Manaaki Whenua used the design template of the handbook for two further publications, *Te Reo o Te Repo: The Voice of the Wetland* (2017)[4] and *Te Reo o Te Repo: Kei Konei Tonu Au* (2021). Both are grounded in Māori culture and values, and expand on the mahi (work) of restoring wetlands.

I've previously joined forces with Karen Denyer on projects where we've turned techno-speak into everyday language. It started with the 2014 Wetland Monitoring and Assessment Kit (WETMAK).[5] With more community groups restoring wetlands, an ecosystem-specific guide was desperately needed to help them measure change over time — both big picture (for example, carrying out a 'wetland warrant of fitness') and fine grain (such as surveying for weeds). Karen, as an ecologist skilled at working with communities, drove the science component of the project, developing modules written in everyday language. I wrangled funding and designed two 'how to' videos (vegetation plots and photopoints). We carried out WETMAK training for community groups around Aotearoa and taught groups' agency supporters how to use each module. In the meantime, technology has marched ahead, so the paper templates for data entry seem quaint. It's a project that's waiting to be modernised: an app to speed data entry and instantly visualise and share results would be ideal. That's one we'd still love to tackle!

Broadening the citizen science theme, we again collaborated (this time with Peter Handford of Groundtruth*) to develop the grandly titled: *Auckland Community Ecological Monitoring Guide* (2018).[6] It's a framework designed to help environmental restoration groups choose science-based monitoring protocols. Monitoring can be a complicated business for non-scientists: What do we monitor? Why? Which protocol should we use? Why this one and not that one? We researched these questions with groups as well as council staff who support community groups, and brought together 42 established environmental monitoring protocols. They cover all ecosystem types, but none need specialist expertise, expensive equipment or iron-clad health and safety plans. So, electrofishing didn't make the cut, but gauging fish diversity and abundance by spotlighting at night did. We developed a set of 11 criteria that groups and council staff could systematically work through to find the protocol that best matched their project objective, skill level, time commitment of group volunteers and, of course, budget.

Developing resources and sharing information are integral to broadening and strengthening the network of people who research, restore, manage and protect our remaining wetlands. But so is meeting face to face. The National Wetland Trust symposia have become a who's who of wetland restoration — the biannual event draws together not only scientists but also citizen scientists, mātauranga Māori specialists, landowners, policy makers and dedicated community group members.

I'd like to mihi (give thanks) to not only the scientists Karen and I are proud to have cajoled into sharing their career highlights, specialist knowledge and field anecdotes with us to create this book, but also those restoring their local wetlands and surroundings — the farmers, landowners, iwi, community groups, trusts, schools and others. As much as we need science to underpin the work on the ground, we need the people on the ground doing the hard yards — restoring wetlands takes persistence, dedication and experimentation. We need the research of both scientists and citizens, we need communities that are supported by scientists, and we need scientists who are willing to listen and to learn from communities.

Monica Peters

---

* See https://trap.nz.

Introduction

The Wetlands of Aotearoa New Zealand

LET'S BE HONEST, it rains a lot in Aotearoa New Zealand. According to NIWA (National Institute of Water and Atmospheric Research Taihoro Nukurangi), most areas of the country receive between 600 and 1600 millimetres of rainfall each year. That's an average. Te Tai Poutini, the West Coast of Te Waipounamu South Island, can have more than 4 metres of precipitation per year. All that rain needs somewhere to go. Much of it flows out to sea, but in low-lying areas, basins, and land blocked off by dunes or gravel bars, it forms freshwater or semi-salty wetlands.

Just how you define a wetland appears to depend on who you are talking to. The authors of the latest calculations of the amount of wetland area in New Zealand, who titled their paper 'Revised Extent of Wetlands in New Zealand', define them as 'terrestrial ecosystems that are permanently or intermittently wet' and exclude saline wetlands such as estuaries.[1]

The International Convention on Wetlands* takes a broader view, defining wetlands as 'all lakes and rivers, underground aquifers, swamps and marshes, wet grasslands, peatlands, oases, estuaries, deltas and tidal flats, mangroves and other coastal areas, coral reefs, and all human-made sites such as fish ponds, rice paddies, reservoirs and salt pans'.

New Zealand law is relatively vague. The Resource Management Act 1991 defines wetlands as 'permanently or intermittently wet areas, shallow water, and land water margins that support a natural ecosystem of plants and animals that are adapted to wet conditions'. This definition could potentially apply to a poorly maintained gutter depending on how you define 'natural ecosystem', which the Act does not do.

A national wetland classification developed for New Zealand identifies nine hydrosystems (including saline) and nine wetland classes, differentiated by water source, substrate and nutrient status.[2] The compilers of the directory of New Zealand wetlands link these hydrosystems and classes to processes that formed them: 'rivers and bogs from

* The full name of the convention is the Ramsar Convention on Wetlands of International Importance Especially as Waterfowl Habitat. See www.ramsar.org.

From the mountains to the coast, Aotearoa has a diverse range of wetland types. The alpine tarns pictured here are at Key Summit, on the Routeburn Track in Fiordland National Park.

frequent rain; lakes from glaciation and volcanic actions; swamps from the deposition of erosion products by rivers and the sea; estuaries and lagoons from tidal flooding of old Pleistocene valleys'.[3] Less common, or naturally rare, wetland types include ephemeral kettle holes formed on glacial moraines, and geothermal wetlands.

Most of the freshwater wetlands found in Aotearoa today formed at or after the end of the last glaciation, about 18,000 years ago.[4] As the world warmed up, the glaciers melted, forming herbaceous fens.[5] By the time humans arrived about 800 years ago, they had developed into either wooded swamps dominated by conifers including kahikatea (white pine, *Dacrycarpus dacrydioides*) and manoao (silver pine, *Manoao colensoi*), or extensive bogs dominated by rush-like restiads (*Sporadanthus* and *Empodisma* species) and *Sphagnum* mosses.

Prior to the arrival of humans, about 10 per cent of the landscape was freshwater wetlands.[6] In addition, the nation has mapped 425,000 kilometres of rivers and streams,[7] and more than 50,000 lakes. Today, wetlands occupy only a small part of New Zealand — less than 1 per cent of the total land area. Extensive and repeated burning of the woody vegetation drastically altered catchment hydrology, resulting in wetter, herbaceous wetlands of harakeke (flax, *Phormium tenax*), raupō (bulrush, *Typha orientalis*) and rush and sedge species (for example, *Carex*, *Juncus*, *Cyperus* and *Machaerina*). Drainage later caused the total loss or fragmentation of these wetlands, with many of the remnants invaded by introduced species, including tree willows (*Salix* species). The latest figure for the extent of freshwater wetlands with emergent vegetation (technically termed 'palustrine') in New Zealand is now calculated at 249,214 hectares, or 10 per cent of the pre-human extent.[8] In addition, many of the lakes, lowland rivers and estuaries are affected by pollution, introduced plants or animals, channelisation, infilling, impounding or water diversion.

LEFT: ***Kāmana (crested grebes) are a diving bird that inhabit South Island alpine and sub-alpine lakes.*** **RIGHT:** ***The red percher dragonfly,*** **Diplacodes bipunctata,** ***is the smallest of New Zealand's 14 species of dragonfly.***

As to what animals live in New Zealand wetlands, it's mostly birds, invertebrates and fish. Long-tailed bats are known to forage along the edge of wetlands, and some skinks and geckos inhabit wooded wetland areas, but native frogs, mammals and reptiles are not generally considered wetland species in New Zealand.

Around 55 native bird species regularly inhabit or use freshwater or coastal wetlands.[9] These include the Australasian crested grebe (*Podiceps cristatus australis*); matuku (Australasian bittern, *Botaurus poiciloptilus*); herons (*Ardeidae* species); royal spoonbill (*Platalea regia*); ducks, including pūtangitangi (paradise shelduck, *Tadorna variegata*) and *Anas* species; a rail — mioweka (banded rail, *Gallirallus philippensis assimilis*); crakes (*Porzana* species); and pūkeko (purple swamp hen, *Porphyrio melanotus*). Large numbers of migratory bird species, including Pacific golden plovers (*Pluvialis fulva*), red knots (*Calidris canutus*) and bar-tailed godwits (*Limosa lapponica*), seasonally visit New Zealand wetlands to feed on intertidal areas.

A similar number of freshwater fish (53, with others yet to be formally described)[10] reside in lakes, rivers and swampy areas, and many also spend time in estuaries or the ocean as part of their migratory lifestyle. One group, waikaka (mudfish, *Neochanna* species), are adapted to living in dried-up pools and tributaries, and are capable of burrowing into mud for extended periods and 'breathing' through their scale-less skin.

The third major group to inhabit aquatic ecosystems is the invertebrates. In lowland wetlands, midges (Chironomidae), aquatic mites (Acarina), copepods, nematode worms and ostracods are abundant.[11] Other species commonly found include flies (Diptera), wasps, bees and ants (Hymenoptera), beetles (Coleoptera) and bugs (Hemiptera).[12] Many are new to science, including the stem-boring moth *Houdinia flexilissima*, which is host-specific to the nationally threatened giant cane rush *Sporadanthus ferrugineus*.

While healthy wetlands are undoubtedly important habitat for native species, they

SINKING CARBON

Wetlands play a crucial role in climate change. Globally, peatlands store around twice as much carbon as all the forests of the world.

In Aotearoa, Dr David Campbell (right, holding a peat core) and his students at the University of Waikato have been measuring carbon dioxide fluxes, to determine whether the 10,000-hectare Kopuatai peatland in the Waikato region is a carbon sink or source. The research team found that for more than 10,000 years, plants have been continuously taking up carbon dioxide from the atmosphere and converting it into organic matter, stored away in perpetuity as peat. Kopuatai accumulates carbon at four to six times the rate of northern hemisphere bogs — in recent years, as much as 2 tonnes per hectare.[13]

While healthy Waikato bogs are net carbon sinks, drained peatlands are a significant carbon source. Campbell's team estimate that in a dry year, the total area of drained Waikato peatlands might emit around 2 megatonnes of carbon dioxide per year — equivalent to about 25 per cent of the region's agricultural emissions.[14] It is expected that globally many peatlands will release carbon if the climate warms and dries.

perform many additional crucial roles in the landscape. The things they do for us free of charge is staggering — stripping nutrients and silt, storing floodwaters and recharging aquifers, to name a few. They are also, finally, now being recognised for their potential to act as carbon sinks. The dollar value of the 'ecosystem services' that wetlands provide is ten times more than either farmland or forest.[15] Addressing the legacy of wetland loss in Aotearoa will require increased understanding of their ecosystem services and their cultural and ecological value, and novel approaches to reconnect people with aquatic habitats. Many of the wetland scientists featured in this book are doing just that, but it doesn't stop with the research. Applying their findings to proactive wetland protection and restoration, and investigating sustainable practices such as paludiculture, will become increasingly important to support the many iwi-, agency- and community-led initiatives aiming to restore and protect our vulnerable and valuable wetlands.

POLLUTE OR PALUDE?

Paludiculture is the practice of farming land that is wet — cultivating species that are adapted to a waterlogged environment. Thousands of hectares of peatland in Aotearoa have been drained for farmland, releasing large amounts of carbon, and polluting waterways with livestock effluent and fertiliser run-off. Instead of constantly pumping water from former peatland, Dr Brian Sorrell (Aarhus University, Denmark, formerly NIWA), shown in the field above, believes we should look at research and trials to re-wet New Zealand's drained lowlands and transform them to sustainable paludiculture.

This isn't fanciful. Māori have a long history of sustainably harvesting natural materials from wetlands, putting them to a wide array of uses, and commercial-scale paludiculture is already starting to happen overseas. Sorrell says the only way to stop massive loss of carbon from farmed peatland is to reflood it. The carbon dioxide released from drained farmland is not included in New Zealand's greenhouse gas budget at the moment — but Sorrell thinks it should be, and says there will be more international and domestic pressure for this to happen. He says, 'There's a choice after flooding: the site could be returned "to nature" and [you can] compensate the farmers by buying their land — that's the approach used in Denmark . . . but larger countries including Germany, Holland and Poland are going down the paludiculture track as well.'[16]

He believes there is great potential in areas like the Hauraki and Southland plains to re-wet drained peatlands that currently rely on expensive floodgates and pumps, or ever-deepened artificial drains, to support farmland. 'Then we plant commercial crops of native wetland plants — raupō, purua grass, kuta — and sustainably harvest them. You can use them for a whole range of things . . . biofuels, particle board, paper . . . [There are] lots of possibilities to produce valuable biodegradable commodities to replace plastics, synthetic rubber and toxic insulation. Paludiculture would sit well with the Aotearoa clean green ethos.'

New Zealand Wetland Systems*

A national classification of New Zealand wetlands includes the full range of wetland types covered by the Ramsar Convention on Wetlands, and is presented in the book *Wetland Types in New Zealand* (2004).[17] The classification is a family-tree in which hydrosystem, the broadest grouping sits at the top. New Zealand hydrosystems include palustrine (freshwater wetlands with emergent vegetation); lacustrine wetlands (lakes, tarns and their margins); riverine wetlands (rivers and streams); estuaries; geothermal wetlands; and plutonic (underground) wetlands.[18]

Palustrine wetlands

Bogs, fens, swamps and freshwater marshes are types of palustrine wetland. Bogs receive all their water from rain, so they are not enriched by nutrients from adjacent land or underlying mineral soils. Being almost constantly waterlogged, they are poorly aerated and usually acidic. They support a unique community of plants, including mosses, lichens, cushion plants, jointed rushes, wire rush, giant cane rush, orchids, sundews, pink pine and manoao (silver pine). They also host mātātā (fernbirds), native fish, and several unique species of moth. Fens also have a peaty substrate but being on slight slopes or at the edges of raised bogs, they receive groundwater and some nutrients from adjacent mineral soils and have a slightly higher nutrient status than bogs. Fen vegetation includes tangle fern, sedges, tall herbs, tussock grasses and mānuka.

Swamps are fertile wetlands that occupy basins, valley floors, deltas and plains. They are fed by both groundwater and abundant surface run-off, so receive a regular supply of nutrients and sediments from adjacent land. Typical plants include sedges, rushes, reeds, harakeke, tall herbs and mānuka, often intermingled, and also forest trees such as kahikatea, pukatea, maire tawake and tī kōuka. Freshwater marshes are better drained than swamps, and occur on valley margins and the edges of water bodies. They typically support rushes, grasses, sedges and herbs.

Lacustrine wetlands

Lacustrine wetlands are associated with lakes and ponds and their margins. Vegetation includes submerged macrophytes (large-leafed plants), suspended diatoms and algae, and floating ferns or herbs. Edge vegetation includes tall reedlike plants such as kūwāwā (lake clubrush, *Schoenoplectus tabernaemontani*), kuta (*Eleocharis sphacelata*), raupō and wīwī (jointed twig rush, *Machaerina articulata*). Lakes can be associated with peat (for example, Rotopiko in the Waikato), sand dunes (such as Lake Wainamu in West Auckland), lagoons (such as Te Waihora Lake Ellesmere in Canterbury), impounded rivers (including the Waikato hydroelectricity lakes), volcanoes (such as the Rotorua Lakes) or glaciers (such as Te Manahuna Mackenzie Basin and Nelson Lakes National Park).

---

* Derived with permission from the National Wetland Trust website, www.wetlandtrust.org.nz/what-we-do/resources/types-of-wetlands.

Wetland types in New Zealand. **CLOCKWISE FROM TOP LEFT:** *Riverine scroll plain, palustrine swamp, lacustrine tarn, estuarine mangrove forest.*

Riverine wetlands

Riverine wetlands are associated with flowing water in rivers or streams. They include riverbeds that support algae and submerged aquatic plants; temporary wet areas of braided rivers, where kakī (black stilts, *Himantopus novaezelandiae*) feed; and swampy backwater wetlands behind river levees. The cool, clear water of the South Island's Te Waikoropupū Springs, which feed the river of the same name, support abundant submerged aquatic plants.

Estuaries

Estuaries are embayments where freshwater flows into intertidal areas, resulting in a mix of fresh, salt and brackish water. In northern New Zealand, a single species of mangrove (*Avicennia marina* subsp. *australasica*) is the dominant plant of this habitat, but where frost restricts its distribution wīwī (sea rush, *Juncus kraussii*) and oioi (jointed wire rush, *Apodasmia similis*) form extensive swards. At estuary edges, or within the influence of windborne salt spray, wave splash or groundwater intrusion, are salt meadows with tiny herbs, including shore primrose (*Samolus repens* var. *repens*), glasswort (*Salicornia quinqueflora*), remuremu (*Selliera radicans*), bachelor's button (*Cotula coronopifolia*) and the rare Māori musk (*Thyridia repens*), edged by taller mānuka, harakeke and marsh ribbonwood (*Plagianthus divaricatus*). Estuarine mudflats are vital habitat for thousands of migratory and resident wader birds.

Geothermal wetlands

Geothermal wetlands are influenced by superheated water and/or mud, hot springs, thermal streams and geyser run-off sites. Two frost-tender fern species, *Cyclosorus interruptus* and *Thelypteris confluens*, survive in the warm, steamy environment of New Zealand's geothermal wetlands in the Waikato and Bay of Plenty regions.

Plutonic wetlands

Plutonic wetlands are underground waterways. Being too dark for plants to grow, they are inhabited by fungi, microbes, insect larvae and some fish species. They occur mainly as caves and underground streams in limestone country, but also as caves in lava tubes and as aquifers. Such wetlands are best viewed on guided cave tours such as in Waitomo and Tākaka.

There is also a range of smaller-scale specialised wetland types throughout New Zealand, including ephemeral (temporary) wetlands and nival (snow-derived wetlands).

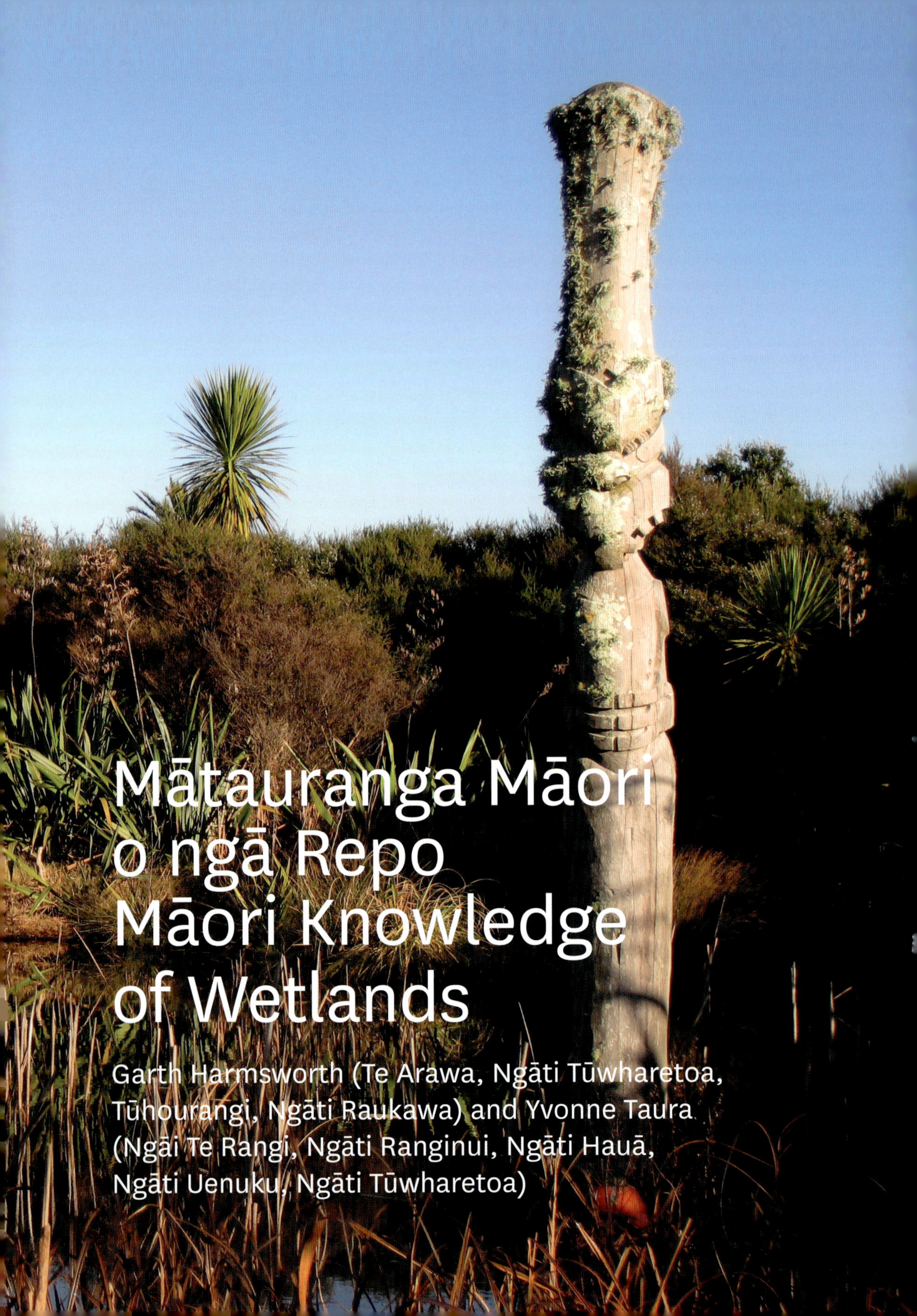

Mātauranga Māori o ngā Repo Māori Knowledge of Wetlands

Garth Harmsworth (Te Arawa, Ngāti Tūwharetoa, Tūhourangi, Ngāti Raukawa) and Yvonne Taura (Ngāi Te Rangi, Ngāti Ranginui, Ngāti Hauā, Ngāti Uenuku, Ngāti Tūwharetoa)

INDIGENOUS MĀORI HAVE BEEN observing and interacting with their environment for centuries. Te Ao Māori, the traditional Māori worldview and belief system, acknowledged a natural order to the universe, built around the living and the non-living, with the central belief that all parts of the environment were interrelated or interdependent through the domains of ngā atua or departmental gods. Traditionally, Māori believed that small shifts in the mauri or life force of any part of the environment — for example, through use or misuse — would cause shifts in the mauri of immediately related components, which could eventually affect the whole system.

All activities and relationships with the environment were governed by mythology, religion and Māori values. Within this framework spiritual qualities guided resource use through an elaborate system of ritenga or rules, with goals to regulate and sustain the well-being of people, communities and natural resources. Guiding values and concepts included kaitiakitanga (to care for, guardianship), tapu (sacred), mauri (life force), rāhui (temporary ban), mana (having status), noa (without restriction) and wairuatanga (the spiritual dimension).[1]

Water is fundamental to all life on Earth. For Māori, water is the essence of life, well-being and identity: Ko wai ko au, ko au ko wai (I am water, water is me). This important whakataukī (proverb) asks not only who am I, but also who are my waters. It stresses an interconnection with water and the importance water gives to Māori cultural identity and survival. However, only about 2.5 per cent of the total water in the world is freshwater, and 99 per cent of all freshwater is frozen or underground.[2] This places in perspective the vital role wetlands play in our planet.

Besides life itself, freshwater wetlands provide significant ecosystem services, customary resources and biodiversity values.[3] Wetlands are regarded by Māori as taonga (treasures), with historical, cultural, political, economic and spiritual significance. Many words are used to describe a wetland, one of the most common being repo. Repo are reservoirs of mātauranga Māori (Māori knowledge), well-being, resources and technology. They are sites of ceremonial activity (for example, wai tohi; see Table 1) and customary use, and support a range of activities, such as kaukau and kautāhoe (swimming and bathing), and waka ama (canoe regatta). As well as being mahinga kai (traditional food-gathering sites) used by local communities and marae, including whānau, hapū and iwi, they provide significant habitats for a large range of taonga plants and animals — fish, birds, reptiles, insects and micro-organisms. They therefore contain a variety of cultural resources for customary practice and technology, plants for rongoā (Māori medicines), implements and tools, and highly valued resources for raranga (weaving) and waihanga (building and construction).

In the past 150 years, more than 90 per cent of repo in Aotearoa New Zealand have been destroyed, and remaining repo are under threat from further land modification and other human activities. Māori have had to adjust to rapid changes in their cultural landscape, and the effects of colonisation and oppression on their culture, and they now look to a future in which they can have a greater say in the management of precious, and degrading, resources.

This chapter introduces Te Ao Māori to give a glimpse into mātauranga Māori, providing examples of this knowledge and showing how it has been applied, particularly to environmental research and resource management in Aotearoa.

Mātauranga Māori

Mātau means 'to know how', 'acquainted with' or 'to understand'. Mātauranga Māori or Māori knowledge systems are specific to Indigenous Māori, and the term encompasses knowledge and values that have been developed, retained and transferred through Māori for centuries — since at least 1300 CE in Aotearoa.[4] This knowledge has developed within the natural environment, reflecting the interrelationship and interdependency between Māori and living ecosystems, and can be seen as dynamic and evolving. It therefore refers to the observations, experience, study and understanding of the world from an Indigenous cultural perspective, and is often equated with wisdom. It has its origins in Polynesian settlement, custom, activity and migration, and in Aotearoa was linked to hapū and iwi before important dialectal and specific elements became more generic, such as te reo Māori (the Māori language).

Mātauranga Māori has many definitions, covering traditional belief systems, cosmology, philosophy, custom, values, experience, observation and practice, all of which build an epistemology* that reflects ancient, traditional, historical and contemporary periods of time.[5] It can be defined as the knowledge, comprehension or understanding of everything visible and invisible existing in the universe.[6] It encompasses everything from the physical — such as values, including mahinga kai — through to the metaphysical — such as principles, including whakaaroaro (logic), tikanga (ethics), whakaponotanga (epistemology), kaitiakitanga (resource management) and wairuatanga (spirituality). Under te Titriti o Waitangi (Treaty of Waitangi) and government statutes, mātauranga Māori is recognised in legislation. The Wai 262 claim defined mātauranga Māori as 'the unique Māori way of viewing the world, encompassing both traditional knowledge and culture'.[7]

A consistent theme from much of the literature is the multifaceted nature of mātauranga Māori. As with Western knowledge (knowledge brought to Aotearoa by the settler European cultures), mātauranga Māori has both qualitative aspects, including kōrero (narratives), pūrākau (ancient stories), observation, experience, interrelationships, learning, verbal data and evaluation, and quantitative aspects, such as measurement, assessment, metrics and numerical data.

---

* Epistemology is the branch of philosophy concerned with knowledge. Epistemologists ask such questions as 'What is knowledge?', 'How is knowledge acquired?', 'What do people know?' and 'How do we know what we know?'

Waihora Lagoon in Pureora Forest Park is an ephemeral pool surrounded by podocarp forest. Mātauranga Māori, traditional Māori knowledge systems, are a vital part of the care and conservation of these special places.

Te Ao Māori

Modern Māori use mātauranga Māori alongside Western science and other knowledge systems to comprehend and make sense of the world today. Te Ao Māori draws on traditional mātauranga Māori, and the traditional Māori belief system is based on whakapapa (ancestral lineage and descent).

The creation stories start with the beginning of time, the beginning of the universe and the beginning of life. They begin with darkness, then move through a series of progressive events to reach the emergence of the ancestral primordial parents, Ranginui (Sky Father) and Papatūānuku (Earth Mother). The couple preside over their children, a large number of ngā atua (founding elements, domains or deities) who rule over every aspect of nature and phenomena, including the stars, the sun, the moon, the rain, the winds, the seasons, the tides, the forests, the land and the waterways, and over all forms of energy, the underworld and the spiritual realm. The parents are prised apart in dramatic fashion by one of their sons, Tāne Mahuta (atua of forests and birds), to create light (te ao marama). With the emergence of light, we see the flourishing of flora and fauna, and through a series of genealogical webs whakapapa connects humans within the realms of the environment and all ecosystems. Survival and well-being for Māori are therefore inherently part of te taiao (nature).

Through this intricate genealogical web and tradition, iwi have their own kōrero, pūrākau, pakiwaitara (stories), waiata (songs), karakia (prayers), mōteatea (chants, laments) and oriori (lullabies) to pass on knowledge, connection and understanding. All connect back to ancestors and ancestral teachings, and link humans with different parts of the environment, such as repo, explaining their origins, values and meanings.[8]

Mātauranga Māori gives ancestral lineage and connection to both male and female elements. Many examples start with Tangaroa (atua of the seas and oceans) and Tutewehiwehi (atua of inland waters and reptiles). Tutewehiwehi is a brother of Ikatere (atua of fishes) and son of Punga, whose father was Tangaroa. In another example from the East Coast, Wai-nui-ātea (Clear Mighty Waters) is the second wife of Ranginui and gives rise to Te Moana-nui-a-Kiwa (the Great Ocean of Kiwa, Pacific Ocean), and through successive generations to primary waterbodies. In yet another version, Te Moana-nui-a-Kiwa and Para-whenua-mea (Muddy Soil of Mother Earth) create the moana (oceans), and Kiwa is described as the father of Onepū (Sand) and Te Parahua (Fresh Alluvial Deposit).

From the Waikato region, the union of Tangaroa and Hine-i-te-huhi gives rise to repo, while the union between Tangaroa and Hine-moana gives rise to roto (lakes) and awa (rivers).[9] Hine-o-te-repo, a sister of Māui, is a kaitiaki (guardian) of swampland. From Te Waipounamu (the South Island), tribal stories describe Hine-i-te-huhi and Hine-i-te-repo (goddesses of the swamp) as the personification of swamps.[10]

For Indigenous Māori, repo are significant in tradition, ancestral teachings and stories, and also in modern life. They reflect a deep history and set of narratives, through whakapapa, songs, stories, beliefs, origins, dependencies, uses and activities, which are an essential and integral part of understanding the whenua or Papatūānuku (cultural landscape) and te ira tangata (the human dimension).

Māori values

Indigenous Māori values regarding freshwater have existed since time immemorial. Kōrero tuku iho o ngā tūpuna (stories passed down from the ancestors) pre-date the time of Ranginui and Papatūānuku.

Māori values are derived from the traditional belief system based on mātauranga Māori.[11] They can be defined as instruments through which Māori make sense of, experience and interpret their environment.[12] Values can be used to shape Te Ao Māori and provide the concepts, principles and lore Māori use to varying degrees in everyday life to guide behaviour and carry out responsibilities. They therefore form the basis for developing ethics and principles, such as kaupapa (Māori research). They can govern responsibilities and the relationship Māori have with the environment and the way they make decisions. Māori values can therefore provide a basis to identify and locate what is treasured by Māori, and give substance and information to establish their significance and level of priority. These taonga include natural resources, wetlands, soils, significant cultural sites, significant biodiversity habitats and significant or iconic plant and animal species. Many core Māori values therefore guide the planning, management and monitoring of repo, including:

- **tikanga** — customary practice, values, protocol
- **rangatiratanga** — chieftainship, right to exercise authority, self-determination
- **mana whenua** — authority over land and resources
- **manaakitanga** — hospitality, acts of giving and looking after
- **whakakotahitanga** — reaching consensus, unity, respect for individual differences and participatory inclusion for decision-making
- **arohatanga** — expression of compassion for others, the notion of care, respect, love
- **whanaungatanga** — relationships, sense of family connection
- **tūrangawaewae** — sense of belonging, place of home and connection
- **hau kāinga** and **ahi kā** — home people, Indigenous people with primary rights and responsibilities over a defined area.

Key Māori concepts

Māori concepts derive from an understanding of mātauranga Māori and Māori values.[13] Although they begin at the local level with whānau, hapū, iwi, and mana whenua and tangata whenua, they are being used increasingly in a generic way. They form the basis for Māori perspectives when seeking to understand Te Ao Māori, mātauranga Māori, wetland ecosystems, relationships or connections and environmental management. A brief outline of key generic Māori concepts is given below:[14]

- **kaitiakitanga** — stewardship or guardianship of the environment, an active rather than a passive relationship
- **ki uta ki tai** — a whole-of-landscape approach, understanding and managing interconnected resources and ecosystems from the mountains to the sea (the Māori concept of integrated catchment management)

- **mana** — having status, authority or control over the management of natural resources
- **mauri** — an internal energy or life force derived from whakapapa, an essential essence or element sustaining all forms of life. Mauri provides life and energy to all living things, and is the binding force that links the physical world to the spiritual worlds (for example, wairua). It denotes a health and spirit that permeates through all living and non-living things. All plants, animals, water and soil possess mauri. Damage or contamination to the environment is therefore damage to, or loss of, mauri
- **taonga tuku iho** — intergenerational protection of highly valued taonga, passed on from one generation to the next, in a caring and respectful manner
- **ritenga** — the area of customs, protocols and laws that regulate actions and behaviour related to the physical environment and people. Ritenga includes concepts such as tapu, rāhui and noa, practical rules to sustain the well-being of people, communities and natural resources. Everything was balanced between regulated and deregulated states, where tapu was sacred, rāhui was restricted and noa was relaxed or unrestricted access
- **tau utu utu** — giving back what you take; giving benefit to the resource
- **te ao turoa** — an intergenerational concept of resource sustainability
- **te mana o te wai** — giving elevated status, power and prestige to the resource and reasserting ancestral connection
- **wairua, wairuatanga** — a spiritual energy and dimension as a concept for Māori well-being
- **whakapapa** — ancestry, genealogical connections, relationships. The connection, lineage or genealogy between humans, ecosystems and all flora and fauna. Māori seek to understand the total environment or whole system and its connections through whakapapa, not just individual parts of these systems, and their perspective today is holistic and integrated.

Conceptually in Te Ao Māori wetlands are often personified as ngā whatukuhu o Papatūānuku — the kidneys of the Earth — acting like giant sponges or filters across the whenua. They have the ability to cleanse the water of excess nutrients and sediment, control floodwater and pollutants and act as important carbon sinks and reservoirs, removing carbon dioxide from the atmosphere. Reflecting a range of integral values, they sustain life, support well-being and reinforce cultural identity.

The mauri of repo is linked to the overall concept of ecological health and human well-being and prosperity. The continuation and strengthening of whakapapa between the ecosystem and tangata whenua is an important demonstration of this continued existence, connection and interdependency that helps sustain the mauri. Whakapapa can be explained through generations, layers and interconnections in the use of many important kupu (words) and kōrero te reo Māori (speaking the Māori language). Repo are reservoirs for mātauranga, hauora (health and wellness) and utilisation, and are regarded by Māori as taonga with historical, social, environmental, cultural, economic and spiritual significance.

Wetland taonga species such as tī kōuka and harakeke can be seen at Bushy Point, a reconstructed pond and wetland on private land near Invercargill, open to visitors by arrangement.

Kupu Māori mo te repo — Māori terms for wetlands

The Resource Management Act 1991 defines wetlands as 'permanently or intermittently wet areas, shallow water, and land water margins that support a natural ecosystem of plants and animals that are adapted to wet conditions'. They are manifest as a variety of forms, including lakes, rivers, streams, dune lands, estuaries, swamps, bogs and marshes.

For Māori, many names and classifications are used for wetlands, including repo, reporepo, kōrepo, wairepo, poharu, kūkūwai and roto. These have been handed down through centuries from tūpuna (ancestors) and knowledgeable experts such as tohunga, mātanga, kaumātua, kuia, koroua, kairangahau and kaiako. Repo are typically wet areas, including pīpīwai and wāhi mākū and are important as part of an overall land system that regulates and accommodates water levels and flooding.

Māori have developed terms, frameworks and classifications to understand, communicate knowledge about, regulate, restrict and manage parts of their natural and spiritual environment. These have been represented through an understanding of cultural values and the interconnection between natural resources and human beings. In respect to wai (water), Māori fully understood that water was a fundamental requirement for survival and were conscious of the links between water and health.[15] Several degrees of water purity were recognised by Māori and separate sources of water were used for different purposes (for example, mahinga kai). Many local tangata whenua, hapū, iwi and kaitiaki groups have up to 70 or 80 names for water, often related to description, properties, āhuatanga (form), character, āhua or momo (type), activities, use, regulation and the values associated with water.[16] A generic Māori classification of water types is given in Table 1.

Table 1. A classification of water and its connection to mauri (from hapū and iwi)[17]

Māori expression	Description
Wai kino	Literally means 'bad water' or 'impure water' (for example, stagnant pools). Often associated with past events, polluted or contaminated water, and includes water that is dangerous, such as rapids.
Wai māori	Freshwater, for normal consumption. Water becomes wai māori when it comes into unprotected contact with people, as in the case of running streams and lakes. It therefore becomes normal, usual or ordinary, and no longer has any particularly sacred associations. Wai māori is often used to describe water that is running and unrestrained, or water that is clear or lucid. Wai māori has a mauri (which is generally benevolent) and was controlled by ritual.
Wai mātaitai	Significant estuarine or brackish waters

Māori expression	Description
Wai mate	Water that has lost mauri, is degraded, and is no longer able to sustain life. Mate is associated with death, and wai mate may have been used in places of contamination and tapu; historical battles; dead, damaged or polluted water; and where water has lost the power to rejuvenate itself or other living things. Wai mate, like wai kino, has the potential to cause ill fortune, contamination or distress to the mauri of other living things or spiritual things, including people, their kaimoana (seafood) or their agriculture. The subtle difference between wai kino and wai mate seem to be based on a continued existence of mauri (albeit damaged) in the former and its total loss in the latter. Wai mate also has geographical meaning: to denote sluggish water, a backwater to a mainstream or tidal area; in this sense, the wai mate retains its mauri.
Wai ora	The purest form of water, such as rainwater, wai ora is the spiritual and physical expression of the long desire of Ranginui (Sky Father) to be reunited with Papatūānuku (Earth Mother). Pure water is termed te wai ora a tāne, and to Māori it contains the source of life and well-being. Contact with Papatūānuku gives it purity as water for human consumption and for ritual. Traditional water could remain pure only without being mixed and was protected by ritual prayer. Traditionally, wai ora had the potential to give life, sustain well-being and counteract evil.
Wai puna	Generally pure spring water that comes from the ground, such as hillside or underground springs.
Wai tai	Seawater, saltwater and the surf or the tide — used to describe any water that is tidal, influenced or related to the sea (the domain of Tangaroa), including waves, surf, estuaries, tidal channels and river mouths. It is used to distinguish seawater from freshwater (wai māori, wai ora). Wai tai is water that is returned to Tangaroa (traditionally, Māori often thought in cycles and processes of generation, degradation and rejuvenation). Wai tai had uses for kaimoana, bathing and healing.
Wai tapu	Sacred waters used in rituals. Running water, sometimes termed wai matua o Taupapa (virgin water as it flows from the earth), was used in rituals and applied using certain plants, not human-made vessels.
Wai tohi	Areas of pure water, ceremony and baptism, used to remove tapu (whakanoa)
Wai whakaika	Ritual waters, pools and ceremonial waters
Wai whakahe-ketūpāpaku	Water burial sites

Many wetland terms and definitions are provided in the cultural wetland handbook *Te Reo o Te Repo — Kei Konei Tonu Au: The Voice of the Wetland — I Am Still Here*. This list of terms was expanded following research into national land-cover databases and remote sensing technology to explore Māori terms and expressions that align with scientific wetland classifications and terms.

Table 2. General Māori terms for describing wetlands and wetland features[18]

Term	Description	Term	Description
Hawai	Shallow of a lagoon or swamp	**Pīpīwai**	Damp, humid, swampy
Hīwai	Open water in a swamp	**Po(w)haru**	Soft, boggy, sodden
Hū	Mud, swamp, mire, quagmire	**Poharu/ pōharuharu**	Bog, mud, quagmire
Kōrepo	Shallow swamp	**Pūrekireki**	Tufts of sedge in a swamp
Kōreporepo	Swampy	**Rawa, rarawa**	Swamp
Kūkūwai	Wet, swampy land	**Rei**	Swampy ground, peat
Mākū	Wet, wet place	**Pukaha**	Spongy or swampy
Manga	Tributary, drain, creek	**Pukaki**	Head of a creek, or where a stream meets tidal waters (pu means 'heaped up' or 'bunched up', and kaki means 'neck')
Mangareporepo	Muddy creek	**Puna**	Spring of water
Mātā, mātātā	Deep swamp	**Repo**	Swamp, bog, marsh
Mato	Deep swamp, deep valley	**Reporepo**	Swamp, marsh, to be swampy, marshy
Mimi	Stream or creek	**Putarepo**	A place at the end of a swamp that can be crossed
Monoku	Wet	**Roto**	Lake, bog, swamp
Ngae, ngaengae, ngahengahe	Swamp	**Tāpokopoko**	Bogginess, bog

Term	Description	Term	Description
Ngaeki	Swamp	**U(w)hi**	Covering (applied to vegetation in a swamp, etc.)
Oaro	Bog (aro means 'bog', 'boggy place')	**Repo, ngaere**	Swamp
Opara	Muddy place	**Reporoa**	Long swamp
Oru	Boggy	**Wairepo**	Swamp waters, bog, marsh
Pākihi	Opening, clearing, flat land that is dried up and poor quality, swampy, low fertility	**Te ngae**	Swamp
Papawheki	Firm, solid (of a swamp with a good bottom)	**Waro**	Deep swamp, deep hole, pit
Para	Muddy, sediment	**Wai**	Water
Paretai	Bank of a river	**Waipara**	Stream across a plain, or a river with thick, muddy sediment
Paru	Mud	**Wharu**	Mud, quagmire, bog, mire
Pīpīwai	Damp or swampy	**Waipuke**	Flood
Puaha, ngutu awa	Mouth of a river	**Whakatahe(a)**	Drained

Table 3. Māori terms for wetland hydrosystems[19]

Wetland classification — Level 1 Hydrosystem[20]	Māori equivalent terminology examples	Associated taonga features and/or taonga species
Marine (saline, coastal, subtidal)	Tai, ākau, tahatai waitai, nō te moana, mātaitai, taimoana, taiwaitai, matāwhanga, rohe-a -tangaroa, paetai, ara o Hinekirikiri, tahatai, taimoana matāwhanga	Kaimoana, ika-a-moana, mahinga mātaitai, rimurimu, kōura, ika, tiora, pāoraora
Estuarine (estuaries, lagoons, etc.), e.g. mudflats	Pūahatanga, mātaitai, wahapū, hāpua, pūwaha, ngutuawa, koraha, ngutuawa	Kaimoana, pipī, toheroa, kuku, pūkeko, kōpūngāwhā, patiki, kanae raukura, wīwī, nana, kina
Geothermal (warm to hot sub-surface and surface water)	Waiariki, wai puia, ngāwhā, waiwera, waipuna (springs)	Ngāwhā, waiwera, waipuna (springs)
Lacustrine (lakes and ponds and their margins)	Roto (inland lake), waimāori (freshwater), moana (large lake)	Kuta, raupō, harakeke, kōura, pūkeko, weka, ika-a-roto, pāteke, kawau
Palustrine (emergent plants over freshwater, swamps, bogs and marshes)	Repo, wairepo, kōrepo (shallow swamp), rohe kōreporepo, kōrepo, kūkūwai, ngae (swamps), kūkūwai, ngae, ngaere, hūhi, mātātā, nowaiwai, wharu (bog, mire)	Harakeke, kahikatea, tī kōuka, wīwī, poniu, pūkeko, weka
Plutonic (underground water, from springs, possibly limestone or karst terrains)	Rarowhenua, manawa whenua, puna manawa whenua, ana-a-pākeho, waipuna	Waipuna (culturally significant springs are the surface expression), wētā
Riverine (rivers, streams, and creeks and their margins)	Awa, ngutu awa, manga, wai pekapeka (rivers, streams), wai māori (freshwater), parenga (riverbank), kāuru (head of river, stream)	Tuna, kōkopu, īnanga, kōaro, *Galaxias* fish species, kākahi, ika-a-awa, whio

Table 4. Māori terms for wetland classes based on a New Zealand wetland classification[21]

Wetland class (a level based on substrate, water regime, nutrients, pH) — giving the wetland type[22]	Māori equivalent term
Bog	• Repo, kōrepo, wairepo, rohe kōreporepo, oru (boggy, marshy, quagmire), wharu, pōharuharu, kūkūwai, ngae, ngaere, tāpokopoko (bogginess, bogged, bogged, soft, sink in mud), hūhi, mātātā, nowaiwai, rei — swampy ground, peat, mire • Monoku — damp, wet, moist (boggy) • Onekupuru — an organic soil found in wet situations
Ephemeral wetland	Wairangatahi, rangatahirepo, rangitahiawa
Fen	Repo, kōrepo, wairepo, kūkūwai, ngaere
Marsh	Repo, kūkūwai, ngaere
Pākihi and gumland	Pākihi (to be dried up), pākōkō (dried up, such as a spring, to be infertile), rake (barren land), tuakau, koraha, hahore (sterile land, wasteland, barren land), whenua akeake
Saltmarsh	Toterepo, waitote, ngaere (examples include mākaka, nana, rimurehia), mātaitai, totetote
Seepage	Papītanga
Shallow water	Pāti wai, kāraraha wai, pāpaku awa, pāpaku roto, kōrepo (shallow swamp), pākihikihi (to be shallow water)
Swamp	• Repo, kōrepo, wairepo, rohe kōreporepo, ngaere, kūkūwai, mātātā, rarawa, nowaiwai, hūhi • Pipī, kūkūwai, tihau, tokakawa — swampy

Traditional uses for oneone (soils)

In terms of mātauranga Māori, oneone have a wide range of uses.[23] Food was sometimes cooked by enveloping it in clay mixed with water and placing it in a hot fire.[24]

Another important use of oneone was for organic dyes, which were applied to a range of materials, including clothing made from fibres, wood products, instruments, whare interiors and art designs. Many dyes were sourced from clays, muds, organic soils, peats and plants typically found in wetter areas such as the muds or para of a swamp. They included:

- **kōkōwai, kura, tākou** — a red ochre found in clay, smeared on people's faces and bodies as a sign of chiefly status. It was also used on carved items such as waka or whare, and even on the bones of the dead. It was rolled into balls, baked in fire or hot ashes, and mixed with shark oil to form dyes
- **paraharaha** — a black mud used for dying flax fibre
- **parapara** — a soft mud used for dying flax fibre
- **paru** — a mud dye rich in iron salts; unfortunately, it is acidic and in later years led to progressive breakdown and damage of cloaks and garments
- **pukepoto** — dark cobalt-blue earth found in clay in swamps and used as a pigment
- **taioma** — a white paint, created by burning and pulverising clay and then mixing it with oil
- **tareha** — an ochre
- **te uku** — a white clay
- **uku** — a white or bluish clay with a soapy consistency and feel used for washing.

Different colours were derived from muds mixed with plant materials and soils. These included:

- black — **hīnau** (*Elaeocarpus dentatus*) bark soaked in dark-coloured mud, and charcoal or soot often mixed with fish oil; harakeke fibre was often dyed black
- blue-black — **tūpākihi** (*Coriaria* species), **whāwhākou** (*Syzygium maire*) and **mako** (*Aristotelia* species)
- brownish yellow — **pūriri** (*Vitex lucens*) mixed with **tānekaha** (*Phyllocladus trichmanoides*)
- red — **kōkōwai** (iron-oxide-stained earth) mixed with shark-liver oil
- reddish brown — **tānekaha** (*Phyllocladus trichomanoides*) bark, pounded and soaked in water; the fibre being dyed was then immersed in the water
- yellow — **raurēkau** (*Coprosma* species), and **tānekaha** (*Phyllocladus trichomanoides*) bark
- white — halloysite and kaolinite white clays.

Oneone (soils) have a range of traditional uses for Māori, ranging from crop cultivation and cooking to dyeing and preparing pigments.

Taonga species

Taonga means 'property of value', 'what is treasured' and 'what is significant'. Repo contain significant habitats for a large number of taonga species, including plants, animals, fish, birds, reptiles, insects and micro-organisms. The term taonga is used generically in te Tiriti o Waitangi claims and settlements. The Waitangi Tribunal, the commission of inquiry of all Treaty claims, states that the term 'taonga species' refers to species of flora and fauna that are significant to the culture or identity of Māori iwi or hapū, based on a large body of inherited knowledge, through whakapapa, and to fulfil responsibilities of kaitiakitanga.

Taonga species are therefore significant cultural species that were also sources of technology (for example, used for food, clothing, shelter and medicine) and give cultural identity and well-being to iwi and hapū. According to the Wai 262 claim, taonga species are:

> subject to considerable inherited knowledge relating to their characteristics and properties (such as habitats, growth cycles, sensitivity to environmental change, and requirements for their care) . . . The exercise of kaitiaki responsibilities towards those species is a fundamental aspect of Māori culture, and kaitiaki relationships are important sources of identity.[25]

Wetlands provide habitat for a large range of customary resources, including taonga species such as endemic plants, fish, birds and invertebrates — including mātātā, kōkopu and kōaro, tuna, kōura, raupō and harakeke, many of which live only in wetlands. Many native wetland species are under threat from land-use changes. As a result, some harvesters whose traditional harvesting sites have been lost need to travel to access taonga material and may intrude on sites traditionally harvested by others. This is a significant issue for hapū with limited traditional harvesting sites. To ensure that the mana and knowledge of species like kuta and that their traditional uses continues, it is important to nurture existing harvest sites, and to preserve the connection with these and celebrate the mana and wairua of finished items.

TE TIRITI O WAITANGI, the Treaty of Waitangi, is the founding constitutional agreement in Aotearoa New Zealand. It is a partnership that was signed between the British Crown and Indigenous Māori rangatira (chiefs) in 1840. Prior to this agreement, in 1835, rangatira had collectively developed and signed He Whakaputanga, the Māori Declaration of Independence, to ensure that their mana and rangatiratanga were maintained in the face of colonisation. The Treaty provides the essential foundation for forming meaningful relationships and partnerships between government and hapū and iwi. Colonisation has had devastating effects on Māori culture and their ability to practise water governance and management based on Māori values and principles, and their own tikanga and mātauranga. The extent to which Māori values are considered and given effect in water management today remains inequitable when compared with that of Western values and approaches to water management.

Taonga as customary food resources

Fish

Īnanga, kōaro kōkopu, kōpakopako, kōriwhariwha, kōawheawhe, kōkopuruao, taiwhara, moruru, para, parakeke, ruao, ruwao — species of galaxiid fish, including īnanga (*Galaxias maculatus*), kōaro (*G. brevipinnis*), giant kōkopu (*G. argenteus*), banded kōkopu (*G. fasciatus*) and shortjaw kōkopu (*G. postvectis*). Many are capable of climbing moist vertical faces to inhabit forested streams and inland wetlands. The juvenile form of the five species is harvested as whitebait.

Kanakana, korokoro, nganangana, piharau, pipiharau, pihapiharau — lamprey (*Geotria australis*), ancient cartilaginous fish that parasitise larger marine animals. Adults were considered a delicacy and traditionally caught in traps during their upstream migration to spawn in riverbeds.

Tuna — the native longfin eel (*Anguilla dieffenbachii*) and shortfin eel (*A. australis*).

Waikaka, hauhau, kōwaro — mudfish (*Neochanna* species). All mudfish except the Chatham and Canterbury mudfish lack pelvic fins, which distinguishes them from *Galaxias* species.

ABOVE: *Adult banded kōkopu were traditionally harvested at night (when they are most active) using a torch and a scoop net.* **BELOW:** *Tuna were an important source of protein, fat and oil in the Māori diet. This hardy species could be stored alive until needed, eaten fresh, and preserved by smoking or drying in the sun.*

Birds

Kawau — a general term for several varieties of shag, including the black shag or great cormorant (*Phalacrocorax carbo*).

Pāteke — brown teal (*Anas chlorotis*).

Pohowera, piopio, tūturiwhatu — banded dotterel (*Charadrius bicinctus*).

Pūkeko, Pākura — purple swamp hen (*Porphyrio melanotus*).

Invertebrates

Kākahi — freshwater mussel (*Hyridella menziesi*). A bivalve mollusc that lives in fine mud or sand in freshwater lakes, ponds, streams and rivers.

Kōura, kēwai — North Island freshwater crayfish (*Paranephrops planifrons*) and South Island freshwater crayfish (*P. zealandicus*).

Wētā, tokoriro, wētā punga, pūtangatanga — wētā is the general common name for about 70 insect species belonging to the families of a flightless locust-like insect endemic to Aotearoa.

Māori traditionally had a varied diet that included birds, freshwater invertebrates and insects, including kākahi (freshwater mussels, above left), Wellington tree wētā (*Hemideina crassidens*, above right), and pāteke (brown teal, below right).

Plants

Aruhe, rarauhe, rarauwhe, rauaruhe, rārahu, mārohi, tākaka — bracken fern (*Pteridium esculentum*). The rhizome was pounded and cooked as a staple food for starch and bread in pre-European times.

Nana, rimurehia — eelgrass (*Zostera* species). An aquatic plant forming wet carpets on muddy sandflats and saltmarshes throughout Aotearoa.

Poniu — marsh cress (*Rorippa palustris*). A creeping herb common on riverbanks and damp ground.

Pūhā — includes perennial sowthistle (*Sonchus arvensis*), sowthistle (*S. oleraceus*), prickly sowthistle (*S. asper*) and New Zealand sowthistle (*S. kirkii*), all small leafy plants with thistle-like leaves and milky juice. They are boiled and eaten as a green vegetable. *Sonchus kirkii* is the only species native to Aotearoa.

Tī kōuka — cabbage tree (*Cordyline australis*). The young inner leaves are eaten both raw and cooked.

Wātakirihi, kōwhitiwhiti — watercress (*Nasturtium officinale*) and one-rowed watercress (*N. microphyllum*). A peppery aquatic herb, introduced to Aotearoa as a culinary ingredient in 1852. As well as being a food, wātakirihi embodies cultural practices and stories unique to respective rohe.

LEFT: *Tī kōuka (cabbage tree) leaves were eaten raw or cooked.*
RIGHT: *Aruhe, the edible rhizome of the bracken fern, was a staple food for Māori and an important source of carbohydrates.*

Taonga plants for weaving (raranga, whatu, tāniko, whiriwhiri, aho), rongoā (medicines), construction (waihanga), crafts and tool-making*

Harakeke — flax (*Phormium tenax*). A tall herb with long, fibrous leaves found in lowland swamps throughout Aotearoa. Integral to customary Māori life for medicinal uses and for creating many domestic items. Today, harakeke raranga (weaving) tikanga and techniques are being revitalised and are flourishing on marae and in wānanga. Harakeke is a general name for the leaf and plant, but each variety has its own name.

Kahikatea — white pine (*Dacrycarpus dacrydioides*). Blue dye (kapara), prepared from soot obtained by burning heart of kahikatea and rimu, is used for tattooing. Māpara, the heartwood, is used for implements and weapons.

Kōpūngāwhā, kūwāwā, paopao, wāwā — lake club-rush (*Schoenoplectus tabernaemontani*). A tall, spiky sedge found in shallow freshwater and estuarine habitats. Harvested in summer, the stems are hung in bundles and dried. When completely dry, they can last for years. The dry stems must first be dampened before being used for plaiting mats and baskets.

Kuta, kūkuta, kutakuta, paopao, ngāwhā — tall spike sedge (*Eleocharis sphacelata*). Highly valued as a weaving resource. Woven gently, the kuta stem holds air, which provides warmth and softness — an important quality for tāpau (mats) used for sleeping. Kuta remains a living repository of ecological knowledge that is passed down from older generations.

Mānuka — tea tree (*Leptospermum scoparium*). Very important culturally, used for fires (hāngī and umu), palisades, sticks and weapons.

Raupō — bulrush (*Typha orientalis*). The long, narrow leaves grow to about 2 metres. Traditionally, raupō stalks were used for thatching the walls and roofs of whare and storehouses, and the down was used to stuff bedding. The leaves were used for canoe sails and kites, while bundles of the stalks made temporary rafts. The starchy rhizomes were an important food, and the yellow pollen was gathered and baked into a sweet, light cake.

Tī kōuka, tī, tī rākau, whanake — cabbage tree (*Cordyline australis*). Snares made from tī kōuka leaves were durable and kept their shape well. Because tī kōuka fibre doesn't shrink in water, the leaves were used for anchor ropes. Fishing nets made from harakeke were strengthened with runners of tī kōuka leaves. Kawe (back straps) for carrying heavy loads were often plaited from tī kōuka. Paraerae (sandals) made from dried tī kōuka leaves were a valued item for traversing rough ground and avoiding prickly plants underfoot.

Wīwī — knobby clubrush (*Isolepis nodosa*) and sea rush (*Juncus kraussii*) used for thatching the roofs of whare.

---

* For more information, see www.sciencelearn.org.nz/resources/1192-harakeke-under-the-microscope, www.sciencelearn.org.nz/resources/1271-preserving-harakeke-taonga and www.sciencelearn.org.nz/videos/2016-rongoa-and-repo.

ABOVE: *The leaves, stalks, rhizomes and flower heads of raupō* (Typha orientalis) *were used for construction and food.* **BELOW:** *Harakeke (a generic term for many varieties of swamp flax) had many domestic uses integral to customary Māori life.*

Cultural environmental monitoring and cultural indicators

Māori have monitored their environment for centuries to assess natural resources as a basis for sustenance and well-being. Using mātauranga Māori, strengthened through a close interdependent relationship with the environment, they were able to detect subtle changes in resource availability and condition to help manage resources. Traditionally, Māori used the term tohu (a sign or marker) to indicate a signal or direction of change, often relying on specialist or expert knowledge (such as that maintained by tohunga) and systems to manage resources.

Many taonga species are used as cultural indicators to assess change in the environment, in habitats, and within species and communities — for example, through their extent, abundance, quality, mauri and condition. Cheri van Schravendijk-Goodman (Te Atihaunui-a-Pāpārangi, Ngāti Apa, Ngāti Rangi) explains:

> Our species — our birds, our insects, our fish — tell us a lot about what's going on because of what they do and their roles and functions within repo and wider systems. Our plants also give us indicators of things like climate change or when different things are happening within fisheries. Our birds also tell us when there's new growth, when there are migrations. A number of them, like the kawau and kōtare (kingfisher) in particular, are really good indicators of fish movement. Because those birds, they go after fish. Pīpīwharauroa, the shining cuckoo, is a really good indicator of shifts in season. So, what I'm talking about — they become tohu or indicators. In a way, we're kind of exploiting their behaviour to tell us what's going on in the rest of the environment.[26]

Yvonne Taura adds:

> Cultural indicators are our markers of how healthy an ecosystem is and specifically for repo. Our birds, our fish, our plants, diversity — all these things play a role in showing us how healthy our wetlands are. Without them, we can see quite easily that there's something wrong within that repo; that it's not functioning the way it should.[27]

Māori have become increasingly interested in combining mātauranga Māori and locally based indicators with scientific or ecologically based indicators under the umbrella of cultural monitoring, using kaupapa Māori frameworks and methods.[29] A large range of Māori-led cultural monitoring methods have been developed over the past 20 years, which have advanced and built Māori capacity in many regions and tribal areas, particularly over the past decade (Table 5).[29] Cultural monitoring is being used not only to respond to national policy and legislative requirements,* but also increasingly by local Māori alongside active work programmes such as riparian planting and wetland

---

* For example, the Resource Management Act 1991, the National Policy Statement for Freshwater Management 2014, 2017 and 2020, and the Environmental Reporting Act 2015.

restoration to sustain and enhance freshwater resources and cultural health. The use of cultural monitoring approaches and indicators has been a natural corollary to these local, catchment and regional work programmes.

Table 5. Key Māori-led cultural monitoring approaches and tools that have been developed in Aotearoa[30]

Name of approach	Specific examples
Baselines	Cultural health assessment[31]
Cultural flow	Cultural flow preference studies[32]
Historical data and information	Mapping of Māori values, historic places, cultural resources, etc.[33]
Kaitiaki tools (guardianship tools)	Kaitiaki tools: an internet-based iwi resource management planning tool[34]
Mapping cultural habitats	Mahinga kai, cultural harvest sites[35]
Mauri assessments	• The mauriOmeter1 and Mauri Model[36] • The Mauri of Waterways Kete and Framework[37] • The Mauri Compass[38]
Monitoring, sampling and recording of significant/ key taonga species (e.g. flora and fauna)	• Kōura (freshwater crayfish)[39] • Tuna (freshwater eel)[40] • Kākahi (freshwater mussels)[41] • Kanakana/pihirau (lamprey)[42] • Īnanga, kōkopu, kōaro and other native fish species[43] • Plants such as kuta, raupō, harakeke, etc.[44]
Ngā Waihotanga Iho (estuarine assessment tools)	Ngā Waihotanga Iho: iwi estuary monitoring toolkit[45]
Report cards	2016 Pilot Waikato River Report Card: Methods and Technical Summary[46] • Framework and methods guided by river iwi
Sampling contaminants	Risk, customary resources, kaitiaki tools[47]
Science and cultural indicators	Linking cultural and science indicators[48]
Significance assessment method	Significance assessment method for determining Māori values/ tangata whenua river values[49]

Vegetation monitoring is a module in the Wetland Monitoring and Assessment Kit (WETMAK), a citizen science guide that can be combined with culturally based environmental indicators for better environmental outcomes.

Name of approach	Specific examples
The Cultural Health Index (CHI) for rivers and streams[50]	• CHI method and application: used extensively by iwi/hapū groups to inform decisions and provide knowledge to support collaborative processes[51] • Adaptations have been made for freshwater and estuarine environments[52]
Tribal/regional state of environment reporting — State of Takiwā	A 'toolbox' for iwi environmental monitoring and reporting in Te Waipounamu South Island, developed by Ngāi Tahu[53]
Wetlands	• National monitoring approaches and indicators of wetlands[54] • Wetland habitats along the Waikato west coast, e.g. Toreparu wetland assessment approach[55]

The application of cultural monitoring, and the development of indicators, can make use of both mātauranga Māori and science approaches and indicators. Table 6 shows the complementarity of approaches.

Table 6. Complementary assessment/monitoring approaches[56]

	Māori or culturally based knowledge	Citizen science	Scientific or highly technical assessments
DEFINITION	High degree of local knowledge, experience and expertise New Māori knowledge is being created from traditional knowledge	Non-professional scientific assessments/ monitoring, often low-technical, low-cost, local, rapid assessments using scientific-based approaches Community or local knowledge is generated and used	Professionally based assessments or research Scientific knowledge is created
METHODS/TOOLS	Examples: • Cultural impact assessment (CIA) • Iwi/hapū monitoring of cultural heritage sites/ wetlands/freshwater/ contaminated sites • Cultural health assessments (e.g. Cultural Health Index, Mauri Compass, Mauri Model) • Māori wetland indicators • Culturally based environmental indicators • Taonga lists, knowledge of values, uses and preparation of taonga	Examples: • Stream Health Monitoring and Assessment Kit (SHMAK) • Waterway Self-Assessment Form • Wetland Monitoring Kit (WETMAK) and other community-based environmental performance indicators • Wetland restoration output and outcome monitoring (e.g. trap catch numbers, bird surveys) • Amateur surveys/school monitoring programmes	Examples: • River and stream water-quality monitoring methods (chemistry, water quality, waterflows, pH, nutrients, bacterial counts) • Coastal survey and monitoring • Archaeological survey • Scientific environmental indicators • Laboratory analysis • Flora and fauna surveys/vegetation mapping
REQUIREMENTS	• In-depth Māori knowledge and understanding of particular environments and issues • Understanding of Māori values, goals, and aspirations • Utilisation of Māori indicators (tohu, tohutohu), defining values, cultural history	Moderate levels of technical input and skill but scientifically robust and part value based. Cost-effective, relatively simple and short duration	Higher levels of scientific and technical input and skill, robust sampling strategies, analysis and interpretation. Expensive and may be time-consuming

Cultural monitoring approaches based on mātauranga Māori and science are often guided by Māori values to determine the management changes and practices that need to take place for the aspirations, goals and objectives to be achieved (Table 7).

Table 7. Links between values, objectives, monitoring and actions to sustain or enhance the mauri of repo[57]

Values	Objectives	Performance measures and/or tools	Examples of management activities and/or variables
Kaitiakitanga **Mahinga kai** **Mana** **Mātauranga Māori** **Mauri** **Wairua** **Whakapapa**	Restore the mauri of the repo	Monitoring tools such as CHI, taonga inventories and mauri assessment	• Catchment management planning • Repo restoration • Identifying and mitigating point source and diffuse source pollution/ critical areas • Pest control/ pathogens • Fencing/stock exclusion • Riparian planting • Managing nutrient loads • Setting minimum flows/cultural flows • Mitigating erosion/ erosion control • Sediment control • Improving connectivity • Improving habitat extent and abundance
	Define standards/ limits/bottom lines to support cultural values	• Establish baselines (e.g. 1840, present day) • Identify change/ trends in the state of mauri	
	Sustain/enhance cultural resources, mahinga kai, taonga species	Taonga surveys, e.g. fish and bird habitat	
	Restore life-supporting capacity, ecological integrity	Plant/ecological surveys	
	Ensure human well-being	Assess condition of taonga	
	Strengthen and enhance mātauranga Māori	Assess abundance/ extent	
	Healthy and prosperous communities/reinforce connection	Habitat surveys, habitat condition	

Conclusion

This chapter provides a small glimpse into the rich and broad knowledge system that Māori draw on for understanding and enlightenment, and for valuing, managing and monitoring their environment, with particular reference to wetlands. The intricate and special relationships between human beings and ecosystems is based on whakapapa and connection. The examples present a culmination of centuries of ancestral knowledge transfer, ngā taonga tuku iho o ngā tūpuna, of teaching and learning, through to contemporary development and applications of this knowledge. It also provides background to, and highlights, Māori perspectives, goals, aspirations and desired outcomes for resource management.

All the examples show how Māori knowledge, values and concepts, through terms and expressions, give a unique voice and perspective. The chapter provides reference to the special rights, interests and values that emanate in particular from te Tiriti o Waitangi the Treaty of Waitangi, the founding document of Aotearoa New Zealand, and various settlement Acts, statutes and policies that influence modern freshwater governance and management. They present strong representations of indigeneity at many levels — locally, regionally and nationally. Over the past decade, acknowledgement and efforts to take Māori knowledge and values into account when managing water and wetlands have generally increased, as seen in the growing recognition and interest in this area, and in the inclusion of Māori rights, values and interests in plans, policies and legislation. Appreciation of cultural values has expanded from a rather limited awareness and understanding, to one of widespread interest and inclusion.

Māori are still trying to actively pursue a greater role in research, policy and planning, and to find the right mechanisms to be heard, assert their rights and values, and contribute to complex societal and environmental problems. It is hoped that a better awareness and understanding of this rich Indigenous knowledge base will lead to better freshwater outcomes for Aotearoa, especially in freshwater and wetland management and policy, and have relevance to international issues.

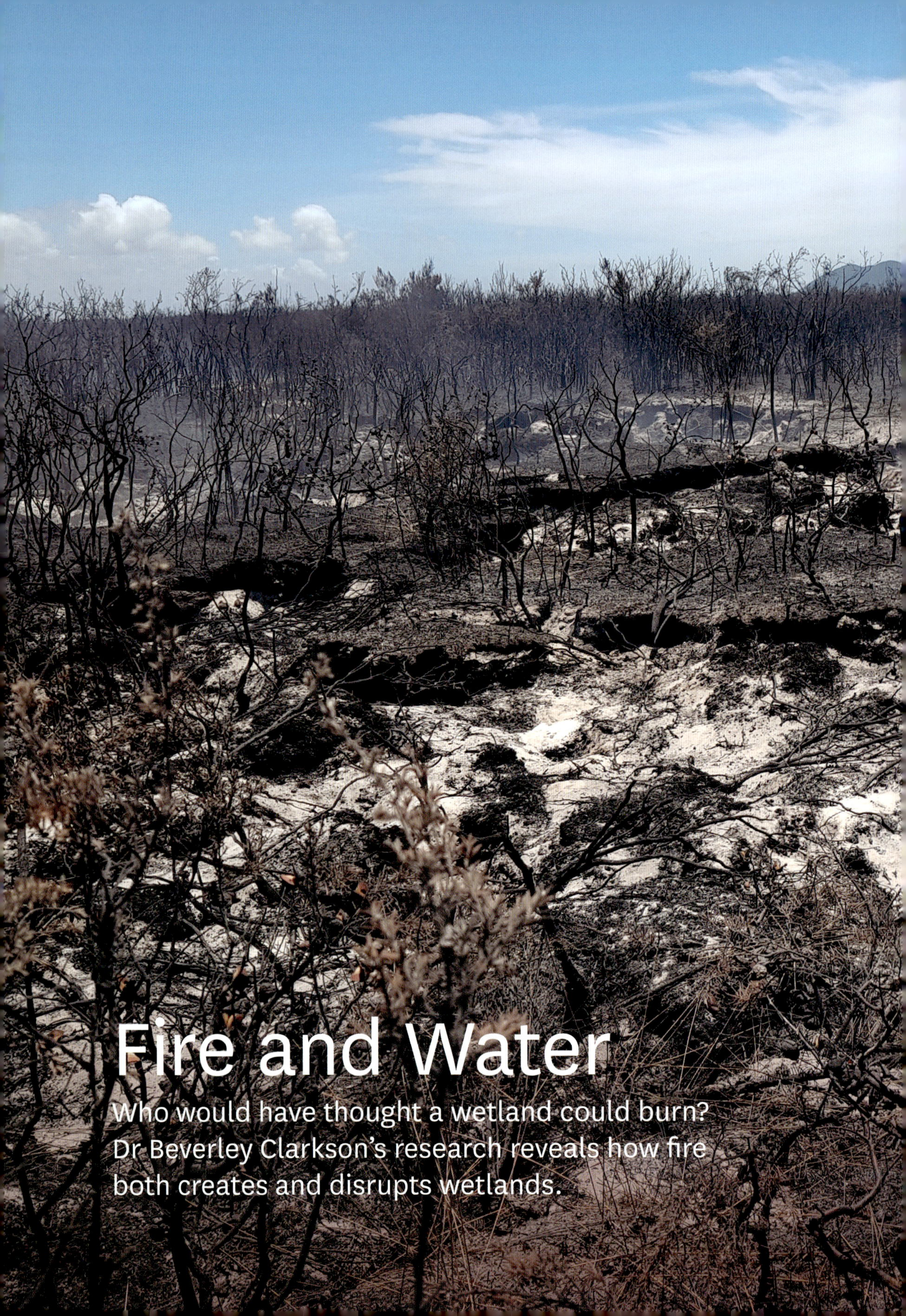

Fire and Water

Who would have thought a wetland could burn? Dr Beverley Clarkson's research reveals how fire both creates and disrupts wetlands.

Dr Beverley (Bev) Clarkson

Specialist field: Plant ecologist specialising in wetland functioning and restoration

Currently works for: Manaaki Whenua — Landcare Research, Kirikiriroa Hamilton

Bev Clarkson, MSc, PhD, is a plant ecologist specialising in the functioning and restoration of wetlands. Since 1996 she has led the national wetland research programme, funded by the Ministry of Business, Innovation and Employment (previously the Foundation for Research, Science and Technology), and she has authored more than 100 publications on various aspects of the ecology and restoration of native flora and vegetation. These include a wetland restoration best-practice handbook and national protocols for wetland monitoring and delineation.

WE KNOW FROM GEOLOGISTS that the world's most violent volcanic eruption since the end of the Stone Age took place near Taupō around 1800 years ago.[1] But it took a wetland and a botanist to work out what time of year it happened and what type of forest it obliterated.

The 232 CE Taupō eruption rocketed a plume of scorching-hot pumice and ash 50 kilometres into the air.[2] When it collapsed, it radiated outwards across the land as a searing pyroclastic flow, incinerating and burying everything in its path. The shock wave of compressed air that was pushed ahead of the flow was strong enough to flatten a 30-metre-tall rimu and tānekaha forest, laying the trees down like a giant domino chain.

This description is not based on oral history. Humans would not arrive in Aotearoa New Zealand for another thousand years. With no one to describe the pre-eruption landscape, the story is instead revealed by a wetland that consumed the forest carcass.

Some 50 kilometres northwest of the eruption, near the present-day settlement of Pureora, the pyroclastic flow cooled somewhat — enough to char but not incinerate the levelled forest. A blanket of volcanic debris settled over the devastated landscape. Pumice clogged up streams and rivulets, and ash created an impermeable surface layer, preventing rainwater from draining away and allowing a peaty fen to develop on top of the flattened forest.

From a spectacular fiery event, a wetland was born.

Unearthing secrets from the swamp

The wetland's anaerobic, waterlogged soils perfectly preserved fragments of the pre-blast forest — logs, leaves, seeds, fruit and bark — even retaining their rich colours. From these macrofossils, botanist and wetland ecologist Bev Clarkson and her colleagues were able to piece together the species composition of the buried forest and reveal the time of year the eruption took place.

Ironically, the significance of the site, and the role the wetland played, were only unearthed when the then New Zealand Forest Service set about attempting to drain the network of sedge and tangle fern fens to plant exotic pines for timber. According to Bev, in the early 1980s wetlands were often considered 'worthless, a nuisance' and regularly drained to make the land 'more productive'. In a 1988 paper co-authored with Rajni Patel and Bruce Clarkson, she describes how the Pureora buried forest was discovered:

> In November 1983, many large logs were uncovered during ditching to improve drainage of a network of small mires in Pureora Conservation Park (formerly Pureora State Forest Park), central North Island, New Zealand. Closer inspection revealed the well-preserved trunks, branches, roots, leaves and seeds of a forest burned by a pumice flow . . . The forest was apparently flattened by an initial shock wave or blast which left most of the trees lying in one direction, mainly towards the north, and then covered by the pumice and finer rhyolitic tephra from settling ash clouds.[3]

The discovery of the preserved pocket of history made headline news and, with funding from the Lottery Grants Board, Bev was employed to sift through the macrofossils and reconstruct the pre-eruption forest.

In all, Bev collected 60 litter samples from the ditches dug by the Forest Service, to examine leaves, seeds, wood and fruit, comparing the remains with reference samples from nearby forests and the Forest Research Institute herbarium. She was able to identify 41 plant species — 23 trees and shrubs, 3 climbers, 11 ferns, 2 sedges, 1 grass and 1 herb. The quality of the material was so good that she was able to determine from the toro and rimu seeds still attached to their branchlets that the blast took place in late summer or early autumn.

Bev also unearthed insect fragments, but sadly no bird remains. The naturally acidic water of the fen (with pH levels of 4.5–4.9) probably dissolved the calcium in bones, beaks and feathers.[4]

In preserving the remains of the ancient forest, the wetland inadvertently secured its own protection. Once exposed to oxygen, the macrofossils soon blackened and degraded, so the Forest Service vowed to preserve the geologically significant buried forest, which meant keeping the site waterlogged. They ditched their plans to plant the area in pines and blocked up the drains, leaving one open for people to see the preserved charred logs — a window into an ancient natural treasure trove.

Forensically examining the treasures preserved by the wetland was the start of Bev's research career linking fire and wetlands.

ANATOMY OF AN ANCIENT FOREST

The Pureora mires — an interconnected system of five small fens — formed on top of, and perfectly preserved, fragments of the podocarp forest destroyed by the Taupō eruption almost 2000 years ago. From the ancient remains (macrofossils, shown below), Bev and her colleagues have reconstructed the buried forest — at least on paper (below left). Fruit that remained attached to branchlets reveal that the eruption took place in late summer or early autumn.

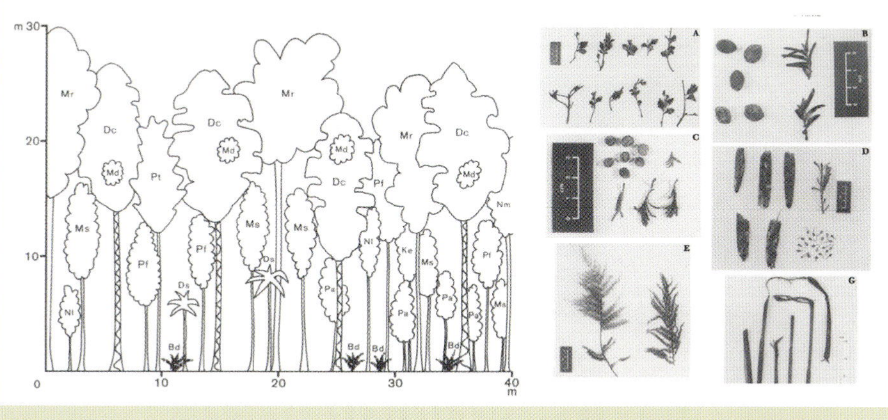

KEY

Bd — *Blechnum discolor*; **Dc** — *Dacrydium cupressinum*; **Ds** — *Dicksonia squarrosa*; **Ke** — *Knightia excelsa*; **Md** — *Metrosideros diffusa*; **Mr** — *Metrosideros robusta*; **Ms** — *Myrsine salicina*; **Nl** — *Nestegis lanceolata*; **Nm** — *Nestegis montana*; **Pa** — *Pseudowintera axillaris*; **Pf** — *Prumnopitys ferruginea*; **Pt** — *Phyllocladus trichomanoides*

Sparking an interest in wetlands

Bev has always liked unusual habitats. With an interest in botany kindled by her grandfather, she relished the opportunity to explore wetlands in the field, find new species and build up a bigger plant list. She also enjoyed indulging in the variety and uniqueness of oddball species — miniature orchids, carnivorous plants, primeval clubmosses, tiny herbs and sedges — all finding refuge in wetlands from faster-growing terrestrial plants that would overwhelm them in drier sites. She likes to refer to wetlands as 'shy places', a term coined by the late conservationist Gordon Stephenson.

Bev's fascination with wetlands was fanned at the University of Waikato, where she completed her Bachelor of Science degree. Inspirational lecturer Bruce Bulloch took his students on a field trip to nearby Moanatuatua bog — a 15,000-year-old peat wetland spread out in a triangle between Ōhaupō, Cambridge and Te Awamutu. It was there that Bev was introduced to the mysteries of the mire. She recalls standing in the broad

BUILDING A BOG

The transition from swamp to fen to bog, as depicted here for Waikato bogs, can take thousands of years. New Zealand's oldest bogs, which are on Rēkohu Chatham Island, are up to 40,000 years old and are home to many unique species.

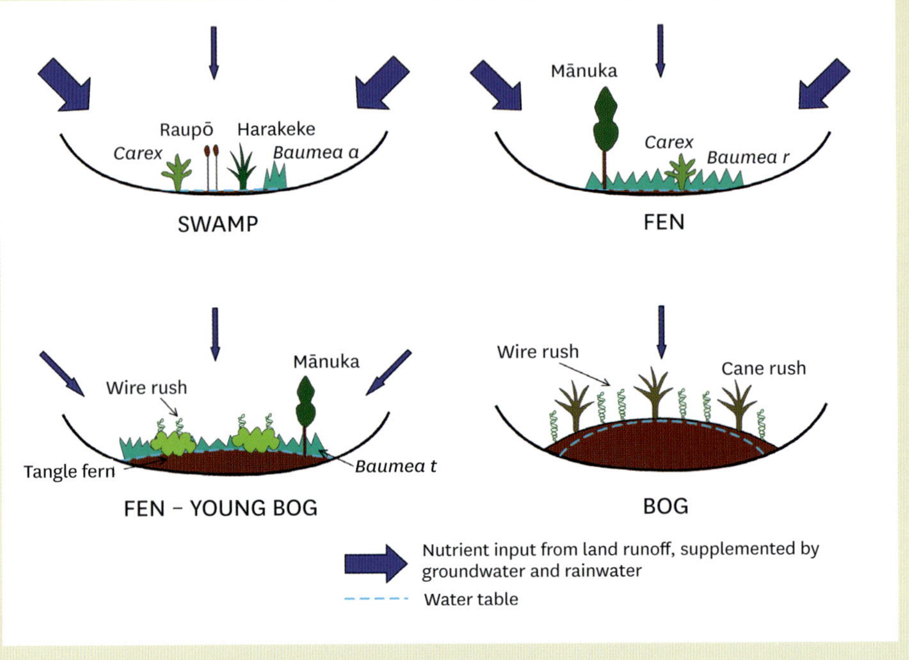

expanse of gently waving giant cane rush and thinking, 'Wow, what a neat place — it's so different!' She chose to return to Moanatuatua to survey the peat soil and vegetation patterns for her third-year BSc special topic.

From that early work, Bev went on to become one of New Zealand's leading experts in bog vegetation. Her career has helped us understand how bogs are created, how they evolve, how their characteristically low nutrient levels determine what grows there, and what happens when they burn.

Bogs are literally the ultimate wetland. They are our oldest wetlands and the end point of a long successional process, from swamps through fens to bogs.[5] Swamps develop in basins or valley bottoms, watered by nutrient-rich land run-off that is supplemented by groundwater and rainwater. They're an ideal habitat for large, lush plants like raupō, cabbage trees and harakeke. In shallow basins, away from the erosive powers of major rivers, the ground level slowly builds up with dead plant material, lifting the surface above the influence of the run-off. Still fed by groundwater and rainwater, but without the regular bath of nutrients carried in off the land, the vegetation changes to species with lower nutrient demands, such as mānuka and tangle fern. At this point the wetland is a fen, an intermediate stage between swamp and bog.

As the surface builds higher still, wire rush (*Empodisma*) takes over. This is a jointed rush in the Restionaceae family, which includes four species that occupy — and in fact create — bogs in New Zealand. Wire rush starts to form peat vigorously, especially in the centre of a wetland where the nutrient levels are at their lowest and decomposing organisms are less active. Over thousands of years, a low, broad dome of peat slowly rises above the influence of groundwater, and the area is then defined as a bog: a wetland fed by rainwater only. In northern parts of Aotearoa the final stage is yet to come — the arrival of the giant cane rush *Sporadanthus ferrugineus*.* Growing up to 2 metres tall, it towers over the wire rush. Many of the bogs that once supported giant cane rush have been drained and converted to farmland, leaving the species restricted to just four wetlands in the Waikato.

Frequent fires

Bev carried out her Moanatuatua BSc survey project two years after a fire, and her vegetation sampling lines extended from the main cover of giant cane rush, wire rush and shrubs of tamingi right into a fire zone. Even two years on, the vegetation in the burnt area was different, with abundant rush-like sedges, the extremely rare bog clubmoss (*Lycopodiella serpentina*) and the even rarer, tiny swamp helmet orchid (*Corybas carsei*). The fire had created a patch of bare, open ground — perfect habitat for such shade-intolerant, early successional plants to colonise.

Fire caused by lightning, spontaneous combustion or volcanic eruption is a natural but relatively uncommon phenomenon in New Zealand peatlands, says Bev. Ash layers and charcoal fragments in peat cores drilled from Waikato bogs indicate that fires occurred only every hundred years or so before humans arrived.[8] However, fire frequency has increased dramatically in the past 700 years, with some fires deliberately set to clear the land and some accidental. Incendiary sparks from steam trains regularly ignited wetlands adjacent to the main trunk line from the 1930s to the mid-1950s.

The concept of a wetland on fire may seem counter-intuitive, but New Zealand's northern peatlands generally have below-surface water tables in summer, and the low, uniform expanses of twiggy vegetation or dry restiad and sedge stalks offer abundant fuel and few natural firebreaks.[9] Peat fires can burn for months underground, emerging in scattered hotspots. A dramatic account is captured in a 1949 newsreel on the *Te Ara — the Encyclopedia of New Zealand* website.[10]

Twenty years after her survey, Bev revisited Moanatuatua and a much larger, but up to

---

* This is the vegetation trajectory for northern bogs from Kaitaia to Te Awamutu. South of the 'kauri line' at 38 degrees south, *Sporadanthus* species are absent from the New Zealand mainland, and the peat-forming plants are wire rush and sphagnum mosses. Bog formation on Rēkohu Chatham Island follows a different pattern.

13,000 years younger, Waikato wetland to the north — Whangamarino, an internationally listed Ramsar site. Lying alongside the main trunk line, it was frequently set alight by train sparks. Since steam trains were phased out, fires at Whangamarino have been rare, but in the 1980s two fires tore through the site. The earlier one (10 February 1984) was accidental, spread from a rubbish fire, and burnt 150 hectares of the 7000-hectare wetland. Arsonists set off the second fire, deliberately lit at several locations on 13 January 1989, and destroyed half of the area of peat bog vegetation.[11]

Bev describes the post-fire scenes at Whangamarino and Moanatuatua: 'fires destroyed virtually all the vegetation above ground; blackened standing dead stems of *Leptospermum*, *Epacris* and *Sporadanthus* (the last only at Moanatuatua), close-cropped clumps of *Baumea* (now *Machaerina)* species and *Schoenus brevifolius*, and occasional scorched hummocks of *Sphagnum* were scattered over peat strewn with burnt vegetation debris and ash.'[12] Despite the devastation, from a scientific perspective the blazes offered an opportunity to learn about the impacts of fire on peatland vegetation, and how long it takes to recover (if ever). The Department of Conservation (DOC) wasted no time in sending botanists to set up permanently marked vegetation plots after each of the fires. They visited the plots (4-metre by 4-metre squares representing bite-sized samples of the wider area) regularly to monitor vegetation recovery.

During each visit, the botanists slogged their way into the bog with only a compass and flagging tape track markers to guide them — there were no smartphones or GPS units back then. They found their plot markers (stakes in the ground) and laid out tape measures to re-create the same plot boundaries each time. Within the plot, the botanists listed every plant species and recorded their maximum and average foliage heights, along with the area of ground that each covered to give a percentage cover value — an estimate of relative dominance per species.

Rapid-fire recovery

Bev was contracted to re-measure the plots from 1991 to 1995 and turn the numerical data into a chronicle of wetland resilience. In 1997, soon after she started working at Manaaki Whenua — Landcare Research, she published a scientific paper describing how Waikato peatlands recover after fire.[13]

Within a month of the Whangamarino fire, Bev reports, a greenish hue had already appeared on the black peat surface. Stems of rush-like sedges, mainly pakihi rush (*Machaerina teretifolia*), soft twig rush (*Machaerina rubiginosa*) and the imaginatively named zig-zag bog-rush (*Schoenus brevifolius*), readily sprouted from rhizomes (underground stems) that had survived the blaze. A few harakeke (flax, *Phormium tenax*) and tangle ferns (*Gleichenia dicarpa*) also sent up new shoots from underground rhizomes or charred leaf bases. Patches of the carnivorous forked sundew (*Drosera binata*) and even rare orchids popped up here and there, thriving on the new habitat of freshly bared peat. It was a similar pattern to the burn site Bev first mapped at Moanatuatua as a university student.

TWO NEW BOG SPECIES

Peat is an organic (biologically derived) 'soil', formed by the build-up of partially decayed vegetation in waterlogged sites. In the northern hemisphere temperate regions, peat is largely composed of the remains of sphagnum moss.

In New Zealand, the peat-forming plants that build domed bogs are restiads, jointed rush-like plants in the Restionacae family. When Bev started her work in bogs, just two species were recognised: the cane rush *Sporadanthus traversii*, found in Waikato and Rēkohu bogs; and the wire rush *Empodisma minus*, found throughout mainland New Zealand.

It was on her first trip to isolated Rēkohu, 800 kilometres east of the South Island, that Bev discovered the cane rushes there were different to the plants she was familiar with in the Waikato bogs. Walking through the broad, windswept Lake Rotokawau peatland with threatened plants expert Dr David Given in 1996, Bev asked her colleague, 'Where's the *Sporadanthus*?'

'It's right here,' he replied.

'That's not *Sporadanthus*! It's spindly. Our one's huge!'

From that discourse a taxonomic review was initiated,[6] and New Zealand now had two species of cane rush — one exclusive to Rēkohu and one reduced to just a handful of bogs in the Waikato. Because the first species was described using specimens from the Chatham Islands, the Waikato giant cane rush (shown top) got the new species name *Sporadanthus ferrugineus* — after its rusty-coloured (ferric) seedheads.

Later, it was wire rush that became the focus of a taxonomic review,[7] this time triggered by a conversation about fire that Bev had with palaeoecologist Dr Matt McGlone:

> Matt used to go on about this damn fireweed — every time there's a fire, wire rush just comes back up. I said, 'No, it kills it!' So I checked in our Waikato bogs after a fire, and it did kill it. We worked out from that they were different species. The southern one does re-sprout, but not the northern North Island one. The kauri line 38 degrees south latitude generally splits the two species: the bigger *Empodisma robustum* [shown above] to the north, *Empodisma minus* to the south.

LAST CHANCE FOR TINY ORCHID

A single arm of the Whangamarino wetland is now the world's only known site of the critically endangered swamp helmet orchid (*Corybas carsei*), whose population was down to just 70 or so individuals in 2008.[16] Once present in Moanatuatua and several other Northland and Waikato bogs, this little orchid has been driven to the brink of extinction by wetland drainage, plant collectors and shading from taller plants. This 1966 account of the species by John Horsman was published in the Auckland Botanical Society newsletter.

> The rediscovery of that rare orchid *Corybas carsei*, by Bill Butcher of Hamilton on the Moanatuatua bog in 1962 was a most welcome one. On October 1st I had the pleasure of being taken to see the plants . . . [A] search in the vicinity produced several flowers of this little orchid, which . . . required almost a hands and knees search to spot . . . [W]e decided the following day to visit Carse's original locality in the Restiad peat bogs of the Waikato at Motumaoho Bog, near Tauhei . . . and there sure enough we found it, also inflower [*sic*].[17]

Horsman observed in his newsletter report that several species, including *Corybas carsei*, are quick to colonise bogs following fires. Bev's research supports the view of those 1966 botanists that the orchid benefits from fire, taking advantage of the newly bared surface and nutrient pulse from ash. DOC has been trialling controlled burning to test if it will boost the species' population. In 2020, a record number of plants was reported — 501, up from 386 the previous year[18]. There is one other hope for this plant — a very similar-looking helmet orchid in Australia could prove to be the same species, offering back-up populations should we fail to save it from extinction in New Zealand.[19]

Sedges emerge from underground rhizomes following a devastating fire in a peatland on Aotea Great Barrier Island.

The peat-forming restiad plants, wire rush and (in Moanatuatua) giant cane rush,* did not fare so well. Existing plants, including their rhizomes, were killed by the fire. Where seeds were shed or blown into the site from nearby unburnt plants they were able to recolonise, but their recovery took several years.

Occasional seedlings of windblown weeds (daisy and thistle species) were also recorded early in the recovery plots. Sun-loving weeds can take advantage of the initial flush of nutrients deposited in the ash, but nutrients return to background levels within the first couple of years, allowing native plants, adapted to the naturally low-nutrient conditions of a rain-fed bog, to take over and shade the weeds out.

Within a year of the fire at Whangamarino, the recovering sedges were a metre tall and made up most of the plot vegetation. But their reign was short-lived. Shrubs of mānuka and tamingi (*Epacris pauciflora*), and eventually the restiad plants, overshadowed them. Whangamarino was back to its pre-fire state in five years. The vegetation at Moanatuatua took almost 12 years to recover to its pre-fire condition — twice as long as at Whangamarino — in part because of the slow return of the giant cane rush, so prevalent in Moanatuatua but absent from the much younger Whangamarino.[14]

As devastating as fire can be in a wetland, particularly for less mobile birds, insects, fish and lizards, small, cool burns can provide a temporary opportunity for smaller — and in many cases rarer — plants to flourish. Such is the value of small-scale disturbance for orchids, sundews and other light-demanding species, that Bev and others have simulated fire, either via

* The giant cane rush *Sporadanthus ferrugineus* is absent from Whangamarino wetland, and is currently restricted to only four peatlands (Moanatuatua, Kopuatai, Torehape and Motumaoho, all in the Waikato region), although it has been translocated to several other Waikato wetland sites.

small, contained burns or by clipping away taller plants, to test the potential to deliberately create habitat for them in Waikato bogs. Says Bev, 'Fires reset the bog development process but generally do not alter the recovery pathways. In the absence of future major disturbance, including invasion by troublesome adventive species, the younger Whangamarino peatlands are expected to develop eventually into floristically poor, oligotrophic raised bogs similar to those at Moanatuatua and elsewhere in the Waikato district.'[15]

Changing focus

In recent years, Bev's research has shifted from learning about how peat bogs evolve, operate and cope with fire, to developing better ways to manage and restore freshwater wetlands. One major research area involved developing wetland restoration techniques at Torehape, another ancient Waikato bog, which has been systematically mined in parts for potting mix material.[14] The mine owner's resource consent required them to restore the site after removing the top metre of peat. Through various trials and experiments, Bev and her colleagues hit on a winning formula. By creating small 'islands' of milled peat in restoration zones and laying mānuka branches ripe with seed over them, they were able to get full coverage of bog plants in around two years. Even faster, albeit more expensive, recovery was achieved by directly transplanting large plugs of vegetation from areas about to be mined to the restoration sites — in effect, lifting and shifting sections of the wetland.

In 1999, Bev became a founding member of the National Wetland Trust, which has helped foster a growing interest in wetland protection and restoration among agencies, Māori, landowners and community groups. Encouraged by a series of wetland restoration conferences initiated by the trust in 2006, Bev co-edited the comprehensive *Wetland Restoration* handbook, published in 2010. This was followed in 2017 and 2021 by two further collaborations — *Te Reo o Te Repo: The Voice of the Wetland* and *Te Reo o Te Repo: Kei Konei Tonu Au*, which provides a mātauranga Māori based approach to wetland restoration.

The shift in public attitudes towards wetland protection and restoration over the past 20 years is important for Bev, who reflects on the changes she has witnessed at Moanatuatua, the bog that inspired her wetland career:

> When I was there in the 1970s it was huge, thousands of hectares. The area I worked in was the reserve, 114 hectares. When I did my Bachelor's study, they were starting to develop the paddocks. I remember diggers being in there, piles of peat alongside the drains, and thought, Oh my god, but there was still plenty of cover of *Sporadanthus* and *Empodisma*, and the water table was high — it was wet.

When she returned some 20 years later to re-measure her vegetation plots, Bev was shocked at the change. 'I couldn't believe it. They'd burnt and developed huge areas, and some of my plots were destroyed. The reserve was fine, but it's really sad. A lot's gone in my lifetime.'

At least we know from Bev Clarkson's work that we can, if we choose, bring them back.

WALKING IN A WETLAND

You don't need the lanky legs of a pied stilt to walk through a wetland — but experience and knowing where to plant your feet helps. Bev once heard her PhD supervisor go 'whump' behind her as he sank into a peat hole on his first excursion with her into Kopuatai wetland.

One of the first scientists in Aotearoa to delve deeply into freshwater wetlands (figuratively, academically and occasionally literally), Bev uses the footwear test as a handy way to distinguish various wetland types: Red Bands for bogs (such as the one shown below), tall gummies for fens, and waders for swamps. She describes moving through a bog as like walking on a set pudding — if you move quickly enough, the surface will hold you up, but there's always a risk you may break through the crust and sink into a hole. It's not quite the lethal quicksand of old cowboy movies — the ones that looked suspiciously like sodden sawdust. Generally, you'll only go in up to your knees or, at worst, your thighs. Having a field buddy to haul you out is a vital health and safety tool when you are bog-slogging for science.

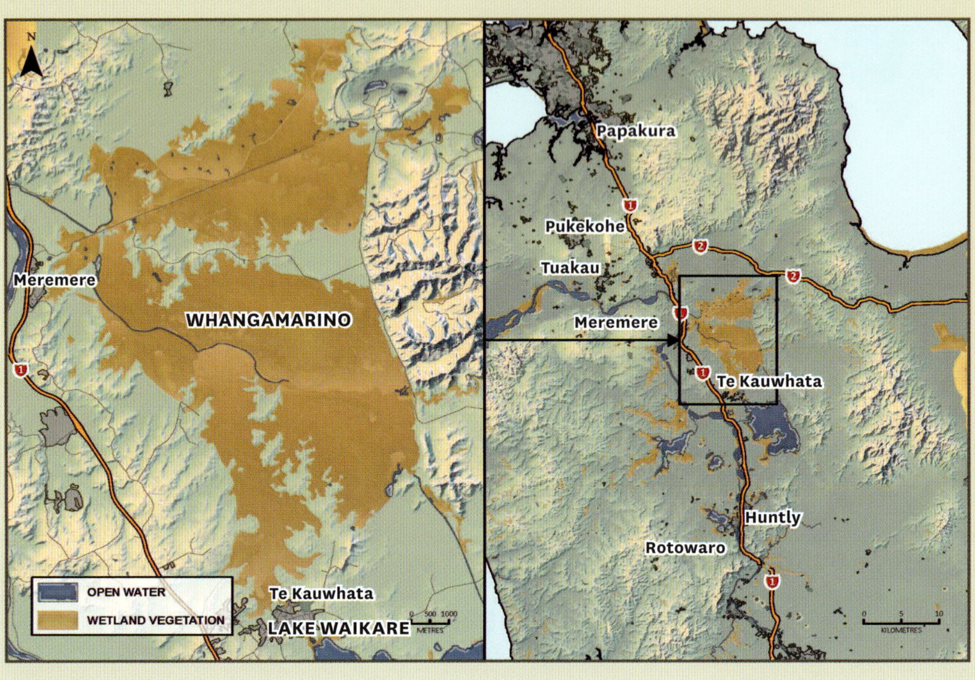

Whangamarino

Lowland swamp and bog complex

Bev has visited Whangamarino wetland many times to study vegetation recovery after fire. But while it might rate as a wetland of international importance, it's certainly not the easiest site in the world to visit. Every day, 13,000 vehicles pass by the wetland near Meremere on State Highway 1, with most southbound occupants likely unaware that the row of cabbage trees on their left marks the edge of a massive wetland.

A good example of a restiad bog can be seen on both sides of Island Block Road. In the northern hemisphere, peat bogs are formed by build-up of sphagnum moss, but in New Zealand rush-like plants from the restiad family are particularly important peat formers.

Highlights

Rare animal and plant species have found a haven in the Whangamarino area, including:

- wire rush (*Empodisma robustum*), which is the main peat former in the wetland
- the only remaining New Zealand population of the extremely rare, tiny swamp helmet orchid (*Corybas carsei*)
- black mudfish (*Neochanna diversus*), which has one of its last remaining strongholds here.

Access

Because of the wetland's size (more than 7000 hectares) and limited access, the National Wetland Trust has created a driving trail around it. The 50-kilometre loop is an easy diversion for southbound drivers on State Highway 1 with time to kill and takes about an hour to complete (add extra time for stops). It leaves the highway at Island Block Road just north of Meremere and rejoins it near Te Kauwhata. There are several options to explore the wetland:

- view the main part of the wetland from Island Block Road or near Falls Road Bridge
- walk into the wetland from Falls Road
- launch a kayak or boat at the beginning of Island Block Road and head north up the Maramarua River or east along the Whangamarino River.

www.wetlandtrust.org.nz/what-we-do/wetland-trails

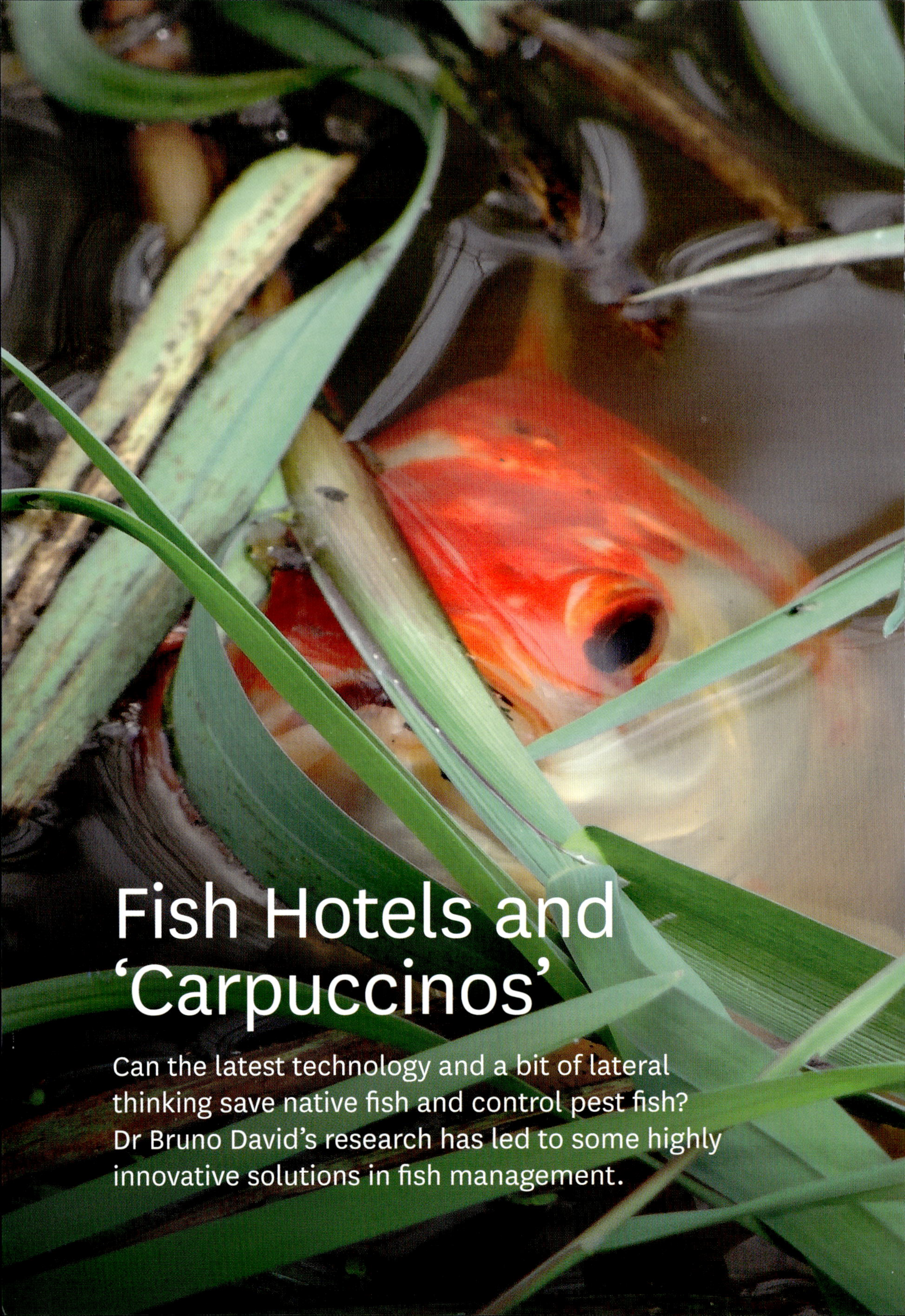

Fish Hotels and 'Carpuccinos'

Can the latest technology and a bit of lateral thinking save native fish and control pest fish? Dr Bruno David's research has led to some highly innovative solutions in fish management.

Dr Bruno David

Specialist field: Freshwater ecology, mudfish ecology, native and invasive fish biology

Currently works for: Waikato Regional Council

Bruno David, BSc (Hons), PhD, has almost 20 years' experience applying scientific methods, modern technology and practical solutions to the study and management of native and invasive fish. His work has involved research into behaviour and life history, otolith microchemistry, radiotelemetry, bioacoustics, fish passage and habitat remediation, invasive fish management, food web energy transfer, and the development of consistent fish-sampling methodologies and network design. During his PhD studies, he described fish communities throughout the Sinclair Wetlands and lower Taieri Plain in New Zealand's South Island, and researched the (at that time little-known) life history and behaviour of the giant kōkopu (*Galaxias argenteus*), the world's largest galaxiid and a species endemic to Aotearoa New Zealand. He has since worked as a freshwater ecologist for the Department of Conservation and Waikato Regional Council. Bruno has contributed to the development of standardised national fish monitoring protocols and has been a member of New Zealand's threatened fish species panel since 2005. He is currently testing the potential of environmental DNA in aquatic ecosystem assessment.

BRUNO DAVID is an avid cyclist. He's not the Lycra-wearing type of cyclist (although he does enjoy mountain-biking and spent many years cycling 40 kilometres to and from work), but a cyclist in the sense that he thinks in cycles. Life cycles. Food cycles. Nutrient cycles. Cradle-to-grave product cycles. Recycling. He applies the philosophy of the circular economy — reduce waste and pollution, recycle products and materials, and regenerate natural systems. Bruno's a big thinker who understands the full picture, and uses keen observation, lateral thinking, a broad worldview and the latest technology, along with low-tech, low-cost resources and waste products, to solve some pretty gnarly problems.

Migratory fish struggling to get past perched culverts? Build them a ladder. Pollution spills flowing downriver threatening to kill fish? Make them a safe lay-by. Too many invasive carp stirring up sediments? Just turn them into fertiliser.

Bruno's philosophy is that there is a solution for every problem if you examine it from every angle, apply common sense, think outside the box and look for new ways to use emerging technology. An example is his application of acoustic ecology — using underwater audio recorders to capture nature's sounds — to help thousands of tuna (native eels) complete their life cycle.

Snooping on fish

If an eel goes through a flood pump and no one is around to hear it, does it make a sound? Apparently, the answer is yes — a sickening clattering, grinding crunch, crunch, crunch, like a lawnmower going over a branch — as they get sliced and diced into sushi segments by the spinning blades of the pump. The culprits are older-style axial-flow pumps — basically just a pipe housing a propeller driven by an electric motor.

During heavy rainstorms, low-lying farms on former peatlands become floodplains. Overwhelmed streams and drains burst their banks and spill over into paddocks, temporarily reverting them to wetlands. It's rich feeding ground for eels, with easy pickings as drowned soil invertebrates rise to the surface. The ephemeral wetlands may be nirvana for eels, but they are a nightmare for farmers. This is when flood pumps are cranked up by councils and landowners to draw the water off flooded grazing land. Heavy autumn rains also coincide with the very time of year that huge eels are migrating downstream for their once-in-a-lifetime chance to breed far out at sea. Sadly, the chance of a full-sized mature female eel surviving passage through an axial-flow pump without being turned into mincemeat is virtually zero.

THREATENED SPECIES ON TOAST, ANYONE?

It's an oddity of New Zealand's wildlife protection laws that the only native freshwater fish with absolute protection has been extinct for almost 100 years. Of the 50 or so species still swimming in waterways across Aotearoa New Zealand, three-quarters are in such low numbers that they are classified as threatened with extinction or at risk of becoming threatened. One of the two native species of eel is as threatened as brown kiwi.

In New Zealand you could face jail time for catching a non-native trout and selling it to your neighbour, but you are legally entitled to kill a native whitebait fish (such as the īnanga, shown above) on the brink of extinction and spread it on your breakfast toast. Eels are treated with even more disdain, caught commercially and canned for pet food. Luckily, public appreciation of the heritage value of our native fish species is growing, and many efforts are being made by agencies, groups and individuals to protect their habitats and migration routes.

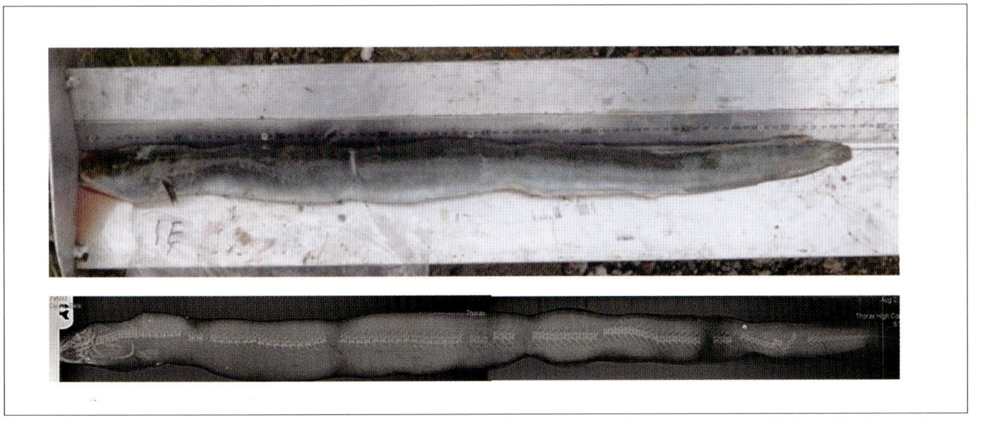

A photograph and x-ray of an eel killed by an axial-flow pump, showing multiple fractures caused by the spinning blades.

It sounds gruesome, and it is, but Bruno has been recording the sounds made when eels pass through a flood pump. He describes it as a 'visceral' way to understand what's happening, to document what we're doing to places and species. But it's also a clever way to reduce eel deaths:

> Now that we know what the sound is, we can start monitoring flood pumps for eel traffic. With a listening device attached to a mobile phone system, you can collect information in real time. You can use it in different ways. We can identify the most problematic pumps, so they become the first priority to replace with fish-friendly pumps. Or let's say you have a listening device in these pumps, and you get two or three eel crunches in a row. That indicates there's a migration happening, and the system could automatically lower the pump speed, or trigger a manual trap and transfer or something else to protect the eels.

With one of the nation's two native eel species significantly threatened, the Waikato Regional Council is testing eel-friendly Archimedes-type pumps — which turn at a lower speed — with the aim of replacing their 120 or so 'slicer-dicer' pumps. Trials are going well, according to Bruno: 'We had 30 large eels go through the new pump one week. The previous pump station would have resulted in 100 per cent mortality of those eels. With the new pump we had 30 go through and 29 came out alive.'

What other sounds are being made below the surface of New Zealand's freshwaters? No one knows. Bruno is dabbling in this area:

> This is really cool stuff. It's called bioacoustics, the science of natural sounds produced by living organisms. It's been going for quite a while in terrestrial systems, and for recording whale songs, but hardly been used in freshwater.

> Fish can't actually sing, but there's all sorts of different sounds that animals underwater can make . . . a kōura crayfish* munching on something . . . a mussel siphoning water . . . a koi carp sucking away at a streambank.

All of these sounds can be characterised, and because sound travels so far underwater, it's possible to document the aquatic soundscape over a wide area. Bruno gives an example of how we can use that technology. He recorded the sound made by freshwater mussels siphoning water in an aquarium. From the data, he reckons that with a bit of experimentation and testing you could potentially deploy a recorder in a lake and estimate the density of mussels based on the intensity of the siphoning sound. Bruno explains the potential:

> For me, some of the most interesting things are things we can't see or hear. There are so many other animal adaptations that are beyond our capabilities . . . think about the way bats or dolphins communicate using echolocation, or the incredible sensitivity of smell that dogs and bees have . . . there are so many sights and sounds 'hidden' from us, but they are happening all the time . . . and now we have technology to eavesdrop and learn!

And yes, some day there will be an app for that. Imagine strolling along a river and dipping a small waterproof microphone attached to your smartphone into the water. Shazam! Your phone starts identifying all the animals in the water and sends you links to pictures and audio files of the unseen aquatic critters below.

Social climbers

Some of Bruno's ideas are lower tech than his plans for bioacoustics, but they are no less ingenious. Take, for instance, his recycled mussel-rope fish ladder invention.

While your average goldfish might seem content swimming circles in a bowl, our native fish have places to go. New Zealand's wildlife migrations don't have the same visual drama as majestic herds of wildebeest sweeping across the African plains, but they are pretty remarkable in their own way. Of our 50 or so native freshwater fish species, about a third set off on their own Middle Earth-style epic journeys.** Some start their life in saltwater, spending a few months at sea before travelling inland to find the perfect wetland or waterway to mature in. They may stay put or move around their watery highways and byways during their adult life, eventually returning to the sea — sometimes in large groups — to breed and start the life cycle over. New Zealand is a mountainous country with abundant fast-flowing streams and waterfalls. Larger waterfalls are insurmountable for native fish, but many species have evolved to become adept climbers,

---

* Kōura (*Paranephrops planifrons*) is an endemic freshwater crayfish.

** These migrations are part of the life cycle of diadromous fish (those that spend part of their life in freshwater and part in seawater). Many native fish species have also adapted to spend their entire life cycle in freshwater, so they no longer have to undertake a migratory journey to breed.

capable of scaling small rocky waterfalls. This allows them to travel far inland, even up into alpine zones.

However, many don't get that far. In the lowlands, humans have installed thousands of culverts to funnel waterways under roads, farm crossings and urban areas. A short tunnel is not a major issue for most migrating fish, but for those heading upstream, a long, sloping culvert pipe that forces water to gush through at high speed, or a perched culvert — one that juts out into mid-air on the downstream side, pouring out water like a giant hose — is a showstopper.

Enter Bruno and his recycled mussel-rope fish ladders:

> [On a field trip with some colleagues] we noticed eels climbing up pōhutukawa tree roots to get over a waterfall and that got us thinking, Huh, they kinda look like ropes. Can we just use ropes to help fish climb perched culverts? And what about long, narrow culvert pipes? A road could be, what, 15–60 metres wide? That's a pretty long, hard swim through a pipe for a little fish. Could we feed rope through those pipes to help fish get through?

Banded kōkopu fry climb mussel spat ropes fitted to a simulated culvert during early laboratory trials.

Bruno and his colleagues set up some laboratory experiments. After initial trials in the council basement proved that discarded spat ropes could help capable climbers such as banded kōkopu (*Galaxias fasciatus*), travel through culvert pipes, they decided to test their idea on less athletic fish.[1] They chose introduced rainbow trout (*Oncorhynchus mykiss*; average swimmers), native īnanga (*Galaxias maculatus*; weak swimmers) and paratya shrimps (*Paratya curvirostris*, New Zealand's only freshwater shrimp species; described by Bruno as 'junk' when it comes to swimming through pipes).

They then created a fish obstacle course, enticing the fish to swim through pipes measuring 3 metres and 6 metres in length, with and without the ropes. They varied the rates of water flow (0.24 and 0.75 litres per second) and pipe angles (1.5 and 3 degrees). They also tested perched culverts to see if the fish could literally climb up the ropes through the falling water to the pipe opening.

As predicted, the shrimps proved to be the worst athletes. Even with the slope almost dead flat and the slowest water flow, they could not get through the bare pipes. Īnanga struggled even to get into the opening. But with a rope to assist them, both species managed to get through.[2]

So how does an old bit of mussel-spat rope help a fish swim through a pipe? The ropes absorb a bit of the flow and slow down the speed of the water, but Bruno noticed something else:

> You have all these little strands of fibre that stick out, so the rope's kind of fuzzy, a bit like a German shepherd's tail, and when the water flows through, these fibres make little standing waves against the edge of the culvert pipes. We watched the īnanga come into the pipe and scoot underneath the ropes, and realised there's a sort of dead zone the whole length of the culvert where the water flow is really slow. So they come into the pipe and get under the rope, but then we saw some of them suddenly pop out from under the rope and zip to the edge of the culvert, and in the blink of an eye, boost themselves from one standing wave to the next. Bop, bop, bop, boomp. There are these little negative pressure points in the wave, and the fish are surfing these and within 10 seconds they've made it right through the pipe.

Other īnanga had a different technique. They would burrow under the strand to have a rest, then swim a short burst, burrow again, and have another go. The paratya shrimps grabbed hold of the rope with their claws and just pulled themselves along.

The ropes Bruno used as ladders were made from UV-stabilised polypropylene — very strong, long-lasting, easy to install and able to be retrofitted into existing pipes.* Not only that, but they're also a waste product, discarded by industry after a couple of years' use. Instead of heading to landfill, they are being repurposed for a long-time service of keeping fish highways open.

---

* The ropes tend to develop biofilms and moss, which protects the surface from scour and likely the erosive forces of sediment, reducing the risk of any microplastic fibres being released into the waterway.

AGEING FISH

If fish could talk, they could save themselves the injustice of being euthanised in the name of science. As it is, the best way to extract information from a fish is to check inside its ears.

Otoliths — Greek for 'ear stone' — are tiny shell-like structures found in the inner ear of all vertebrates, humans included. They float around in a hair-lined, fluid-filled pouch. When an animal moves, the structures bump up against the hairs, telling the brain whether the body is moving, stationary, upright, lying down or upside down. If you've ever thrown up on a roller coaster, blame your otoliths.

Fish otoliths keep growing throughout the animal's life and, much like tree rings, can be used to determine its age — and not just to the year, but to the *day*. Otoliths can also tell us much more, including helping us identify the species. That might not be useful if you have just extracted the otolith from a whole fish, but as these little identification cards are indigestible and fossilise well, they call tell us about the diets of animals that eat fish, and the species composition of ancient ecosystems.

Last but by no means least, otoliths can also act like GPS tracking devices, storing information about where a fish has been. Fish otoliths need water to grow — the water the animal is swimming in. Waterbodies have unique chemical signatures, based on the geology of their catchment, and these chemicals get bound up in the alternating calcium carbonate and gelatinous rings of the otolith. This creates a permanent in-ear record of the waterbodies the fish has lived in. Bruno explains: 'With a laser you can beam into different parts of the otolith at different parts of its development and look at the chemistry in there. If you have a good "chemscape" of the wider natural landscape, you can match up the chemicals in the otoliths to the landscape and work out where the fish has been.'

Bruno used this information to solve a mystery about kōkopu, several species of native galaxiid fish (named for the starry galaxy-like pattern on their scale-less skin). 'We were finding young fish in Hamilton city streams a couple of months earlier than when they should have been arriving if they had travelled up from the ocean 50 kilometres downstream. We suspected they must have been breeding in our inland lake systems, instead of the estuaries where they would normally spawn.'

But which lake? The answer was in the otoliths: 'Lake Waahi's water chemistry proved a good match for the otoliths. We trawled the lake with a larval net and found abundant larvae of both giant and banded kōkopu.'

As a result of this work, Waikato Regional Council is now directing funding to this location. The council is also considering a number of proposals to improve fish habitat at Lake Waahi, including placing woody debris in the water to provide cover, bank contouring, and streambank planting along inflowing tributaries to enhance egg retention and development after spawning.

Fish hotels — helping the homeless

Having helped solve the fish traffic problem, Bruno turned to another issue. He dubs this 'The great Hamilton urban fish-housing crisis'.

City engineers like their streams to be efficient, to carry water away quickly and keep their paved walkways and streets dry. Anything that clogs the culverts where streams are funnelled under pavements, roads and footpaths is anathema to engineers. The fish have a different point of view. Submerged logs, branches, rotting leaves — these are all home essentials for a fish. As we build more urban areas, add more impervious surfaces that channel overheated and polluted water into nearby streams, funnel watercourses through concrete pipes, and armour riverbanks with rock rip rap or gabion baskets to stop them eroding, our waterways become more and more compromised for aquatic life.

Bruno ponders these sorts of tensions between human and fish needs, and again begins to think laterally to get a better outcome for both:

> The council[s] will take all of the fish habitat out of streams: anything that's going to jam culverts and create flooding. So we came up with this idea of building kōkopu condos, tuna townhouses, and kōura cribs. The councils won't accept flooding, so the only way to address this problem in urban streams that still support aquatic life is to go sideways — cut into the streambank and build underground burrows for fish. You can make a 3D environment under your feet — it's still connected to the stream, but it's out of the main current. It's like a portal to a secret buried lair! People don't even know it's there — it's all underground.

Bruno and his team came up with several designs for different situations. The simplest fish house, the 'kōura crib', is simply two cinder blocks placed side by side and connected by a snug-fitting section of plastic drainage coil, curved into a U-shape through the holes in the blocks. The cribs can be dug into the banks of smaller streams or placed directly into a stream. Eels can comfortably curl themselves inside the U-shaped tubes, and the double entrance gives smaller fish an escape route if a predator drops by.

For larger urban streams with streambank armouring works (gabion baskets or rock rip rap), the team designed multi-storey 'tuna townhouses' — simply using longer and larger drainage coils curved in the same double-entry U-shape and tucked in among the armouring material at different heights on the bank. Some of these structures are placed out of the water on purpose, so that if a big flood comes through and the fish need a refuge, they can hopefully find them.

The 'kōkopu condo' design is a longer, straight length of pipe leading to a concrete chamber filled with wood to create a multi-living situation for more social species. Finally, the team designed a 'pallet palace' using recycled material — untreated timber packing pallets, free from many big-box stores and haulage companies. Bruno describes their use: 'You can break out the bottom boards and pin them onto the riverbed, and you put hessian sacking on top, then plant them in wetland sedges — they are like a low-profile, undercut riverbank.'

The tiny homes proved so popular that some were occupied within weeks of installation. And there's an added bonus: as well as providing new fish residences, these piscine cul-de-sacs create safe zones. If there's a pollution spill in the river or a large flood pulse threatens to flush fish away, they can duck out of the main flow and take refuge in one of Bruno's lay-bys.

So Bruno has designed fish ladders from recycled mussel ropes, and fish homes from recycled timber pallets, but what about recycling fish themselves?

Going Carp-N Neutral

Someone in Bruno's Australian homeland once coined a memorable and fair dinkum phrase: 'Ornamental pets can become monumental pests.' Goldfish and koi carp may be pretty fancy in hotel lobbies, but they are pretty damaging to the environment when they get into natural waterways, where they prey on native fish, stir up sediment and pollute the water. Koi carp now make up 80 per cent of the fish biomass in the waterways and wetlands of the lower Waikato Te Awa system. They can reach 10 kilograms in size, their body mass built from the excess nutrients flowing off farmland and into waterways. When koi die, those nutrients remain in the aquatic environment, cycling into another part of the food web and continuing to pollute freshwater.

Bruno and his colleagues have come up with some clever ways to turn these problem fish into prized commodities. They recycle the 'little packets of swimming nutrients' into organic fertiliser for dune restoration projects, enticing lures for rat traps, and stinky repellents to deter rabbits from eating young plants.

The scheme, cleverly named the Carp-N Neutral Project, started with the aim of reducing the number of koi carp in the lower Waikato River system because of the environmental problems they cause. But for Bruno, simply catching and discarding the fish was not good enough. As he says, 'It's frustrating to see a lot of pest programmes kill animals but never reuse them.'

It's illegal to breed, sell or move live koi carp in New Zealand, and they're no good for eating, tasting of the mud and stagnant water they fossick in. So what do you do with thousands of tonnes of stinking dead koi? Bruno had a solution: 'We aimed to generate products from the koi, products that were themselves ethical, including from a carbon footprint point of view. The product ideas just kind of grew as we went along . . . and we tested their effectiveness through robust, peer-reviewed science, which we later published in a high-ranking international journal.'[3] The idea was to show how environmental problems and solutions are intertwined, to turn a negative environmental problem into multiple positive environmental solutions, and develop a cost-recoverable programme at the same time.

NOT SO KOI NOW

With technicolour coats like a scale-clad calico kitty — all orange, black, white and gold — and reaching the size of a miniature schnauzer, koi carp are anything but coy creatures. An ornamental variant of the European carp (*Cyprinus carpio*), koi now dominate the fish biomass in New Zealand's Waikato waterways. Despite being highly prized in some countries and cultures, they are considered by the International Union for Conservation of Nature to be one of the most invasive freshwater fish in the world.[4]

Koi carp are thought to have become established in New Zealand's waterways in the 1960s. In terms of how, theories vary. They may have been flushed into a stream from a garden pond during a flood, or deliberately released by 'Smithy', a notorious amateur fish breeder and covert 'liberator' — or perhaps a bit of both. Bruno describes koi as 'aquatic vacuum cleaners':

> They use their feelers and big 'botoxed' lips to fossick around in the sediment after small organisms, stirring up mud and ripping out aquatic plants. Their diet directly overlaps with eels — in some of our Waikato waterways the condition of eels is absolutely shocking; their fat content is so low. They're very skinny. If we could get rid of catfish and carp, there'd be more food for eels. At the lower end of the food chain, juvenile catfish and carp also feed heavily on the zooplankton — that's a food source that's been taken away from native fish like īnanga, as well as dabbling ducks and waterfowl.

Step 1 in the process of turning pest carp into useful products is to catch them — by the thousands — without harming any native fish. Bruno knew where the koi 'hotspots' were in the Waikato Region, and where he could corral them in massive numbers. Each spring, thousands of koi carp leave lowland Waikato streams and wetlands and head to lakes to spawn. Lake Waikare, which drains through a constructed canal (Pungarehu) into the internationally significant Whangamarino Ramsar wetland, was the perfect spot to catch them, as Bruno points out:

> The trapping part was pretty easy, as much of the technology for capturing invasive fish had already been developed in Australia to control carp over there. I just adjusted and adapted the automated traps to suit New Zealand conditions and species. We developed a cage designed to let native fish through, but selectively trap the smallest invasive fish that we could. Over five years we only caught around seven individual non-target fish in that trap, and the rest were all invasives — carp, catfish, goldfish, rudd, hybrids of those — so it worked really well.

Breeding carp heading upstream from the Whangamarino wetland into Lake Waikare encounter the metal trap, which spans the full width of a small channel constructed beside the Pungarehu Canal to provide fish passage when the canal outlet gates are closed. Along the base of the trap is a series of metal fingers, like the teeth of a giant flea comb. Koi like to push their nose into things, and as they push on the cage, individual teeth swing up like a one-way cat flap, immediately closing behind them by the force of the water flow. Koi can get in, but they can't get out. Over a four-year period, around 35 tonnes of invasive fish have been trapped and prevented from entering Lake Waikare.

Step 2 is to process the product. As Bruno says, 'So we had this trap and were catching all these fish. Then it was, "Well, what are we going to do with them all?"' Originally, the team loaded up the back of a council ute with koi and drove them an hour north to a company that processes waste food into organic fertiliser. However, a truck stinking of dead fish wasn't very popular with other council staff who shared the vehicle, and the method wasn't particularly carbon-friendly either. The solution was to develop New Zealand's first 'invasive fish recycling centre' right there at Lake Waikare. In 2012, with funding from the Waikato River Clean Up Trust, they did just that.*

A motorised arm periodically lifts the koi-filled trap out of the water and the fish are quickly and humanely euthanised. They are then tipped into a thermophilic biodigester, which may sound like something designed by Willy Wonka for his chocolate factory, but is actually little more than an old cement-mixer bowl. Protein-eating *Bacillus subtilis* bacteria are added — this is the thermophilic (heat-loving) part. The bacteria operate at temperatures of 60–70 degrees Celsius and generate their own heat — enough to destroy any pathogens that may be present. The biodigester drum rotates to ensure there is plenty of oxygen for the bacteria, and the aerobic digestion happens quickly. As Bruno tells it, 'In the space of 48–72 hours, 2–3 tonnes of fish turn into beautiful brown crumbly fertiliser mix — I call it "carpuccino" because it looks like coffee grounds!'

Bruno had the carpuccino grains analysed for heavy metals and other contaminants, and found that there was nothing in there that would render the mix unsuitable for use, even on organically certified orchards or farms. The team started trying different mixes, adding recycled untreated sawdust and locally sourced biochar (incinerated waste corn cobs) to bind the carpuccino grains into pellets and provide beneficial properties for soil bacteria.

> We were able to compress the grains — with no artificial binders — into carp pellets, which doubled the density and halved the volume, making transport easier and more carbon-efficient. We packaged the pellets in

* The Waikato River Clean Up Trust is administered by the Waikato River Authority, an independent statutory body set up as a co-governance and co-management framework between the Crown and river iwi. The Clean Up Trust fund was established to restore and rehabilitate the Waikato River and its associated tributaries, catchments and wetlands. See https://waikatoriver.org.nz.

> paper bags so there was no plastic waste. This was a whole cradle-to-grave project — taking an invasive fish and repurposing the nutrients into a part of the landscape where the nutrients were once quite high but currently aren't there anymore.

Again thinking laterally, Bruno thought of sand-dune systems, which were once high-nutrient systems enriched by seabird guano (essentially excreted fish protein) before invasive pests decimated seabird populations.[5] People involved in dune restoration were planting native dune species but using imported fertiliser tablets to get the plants growing. Bruno worked out that the carbon footprint of those imported fertiliser tablets would be higher than the amount of carbon each plant would sequester over its life — it wasn't carbon neutral. The team ran replicated trials in local dune restoration projects, which confirmed that recycled koi nutrients were just as effective at establishing dune plants as imported artificial fertiliser, only more carbon-friendly.

The Carp-N Neutral team also developed koi-based lures for animal pest control. They tested them at the Rotopiko Wetland Centre south of Hamilton, where a

New Zealand's first invasive fish recycling centre on the shores of Lake Waikare. A semi-automated invasive fish trap is on the left and the thermophilic biodigester that makes the 'carpuccino' is located within the yellow shipping container on the right.

community group was trapping introduced mammalian pests around the perimeter of a peat lake to protect rare native birds. The volunteers were game enough to deploy the stinky carp baits and sprays, putting koi lures in half the traps and their standard lure in the other half. Stoats seemed to prefer chicken eggs and dried rabbit, but more rats and hedgehogs were caught in the traps baited with koi lures. Bruno notes, 'We found out that dune plants love invasive fish meal, and so it seems do hedgehogs!'

Having developed a potential market for the waste product, perhaps even selling the end products to fund the processing operation, the next question is: what happens when you run out of fish? Carp produce so many eggs that it is unlikely they will ever be eliminated from the lower Waikato wetlands. But by targeting bottlenecks of breeding fish, it may be possible to get the biomass of invasive fish below the threshold known to have deleterious effects on the ecosystem (around 100–300 kilograms of fish per hectare). Bruno elaborates: 'The idea was not to kill fish willy-nilly but to be very targeted around different environment types, remove one of the stressors and self-fund so that you could essentially do that in perpetuity.'

Career highlight

The Carp-N Neutral project has been highly acclaimed. It was the recipient of the 2014 Green Ribbon Award in the 'Caring for our water' category and a Kudos innovation award for science excellence.* Bruno shares his secrets to success with budding scientists:

> Be prepared for people not sharing your same love, passion and understanding of the value and importance of ecosystem services and functions . . . Think laterally about how to convey your important science messages, but always underpinned by robust data. Use all the senses; videos and audios are a great way to get people to understand. The other thing I would say is get out in the environment as often as you can . . . careful observation and time in nature is the most important thing to improve your own understanding. Ecosystems are so incredibly complex, we really know very little about how things interact and function . . . but spend enough time, and some bits of the jigsaw suddenly start to fit together.

* The Kudos Science Trust aims to help scientists across the greater Waikato region and Bay of Plenty gain national recognition, and it awards prize money to help scientists pursue more research, attend international conferences and buy new equipment. The organisation also recognises and rewards top regional secondary school science students and aims to create opportunities that allow their scientific careers to prosper. See www.thekudos.org.nz.

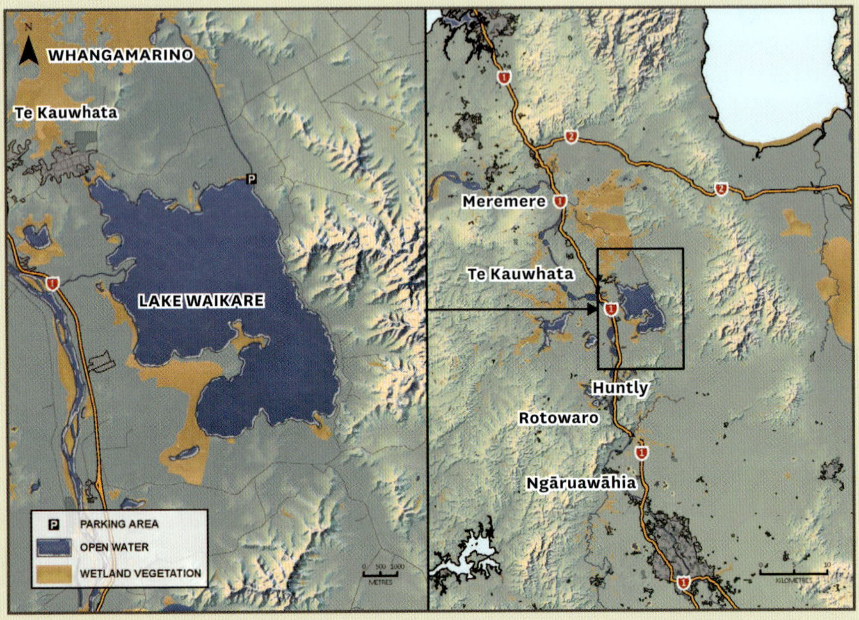

Lake Waikare

Riverine lake

Lake Waikare is the largest lake in the lower Waikato catchment but also one of the shallowest — a tall person could practically walk across it. Its condition has deteriorated in living memory. Local Māori recall collecting freshwater mussels and swimming in the lake, but it was partially drained in 1965 to provide capacity to store floodwaters during big rain events. This lowered its average water level by around 1 metre down to a maximum of 1.8 metres depth, and dried up most of the surrounding wetlands. Evidence of old swamp areas can be seen in the remaining stands of once seasonally flooded kahikatea forest, now high and dry on farmland over 2 metres above the channel. The lake is now so shallow that wind whips up the water, continually stirring up mud from the lakebed.

The lake also periodically turns an alarming rusty orange-red colour, signalling high levels of naturally occurring algae (*Monoraphidium* species). It's not really a picnic destination, and definitely not safe for a swim these days, but it's well worth a quick visit for a glimpse of the fish recycling centre.

The lakebed has been returned to the guardianship of tangata whenua Waikato-Tainui,[6] who are working with the Waikato Regional Council and other agencies to address the many issues facing the lake.

Highlights

You can't take a tour, but you can see the fish recycling facility through a mesh gate and read an information board about the lake's history and potential future. You can also see the Pungarehu Canal, dug to lower the lake level, and there is a good chance of spotting koi fish congregating near the outlet gates.

Access

Lake Waikare is east of State Highway 1, some 75 kilometres south of Tāmaki Makaurau Auckland. From State Highway 1, drive through Te Kauwhata and head east along Waerenga Road. Turn right into Waikare Road, and right again into Ruahine Road to view the water-control structure, where Pungarehu Canal exits the lake.

A public parking area is on the left just before the canal. You can walk across the canal to view the control gates; beyond is the constructed fish-passage channel, within which the 'carpuccino' fish trap lies.

https://lakes380.com/lakes/lake-waikare

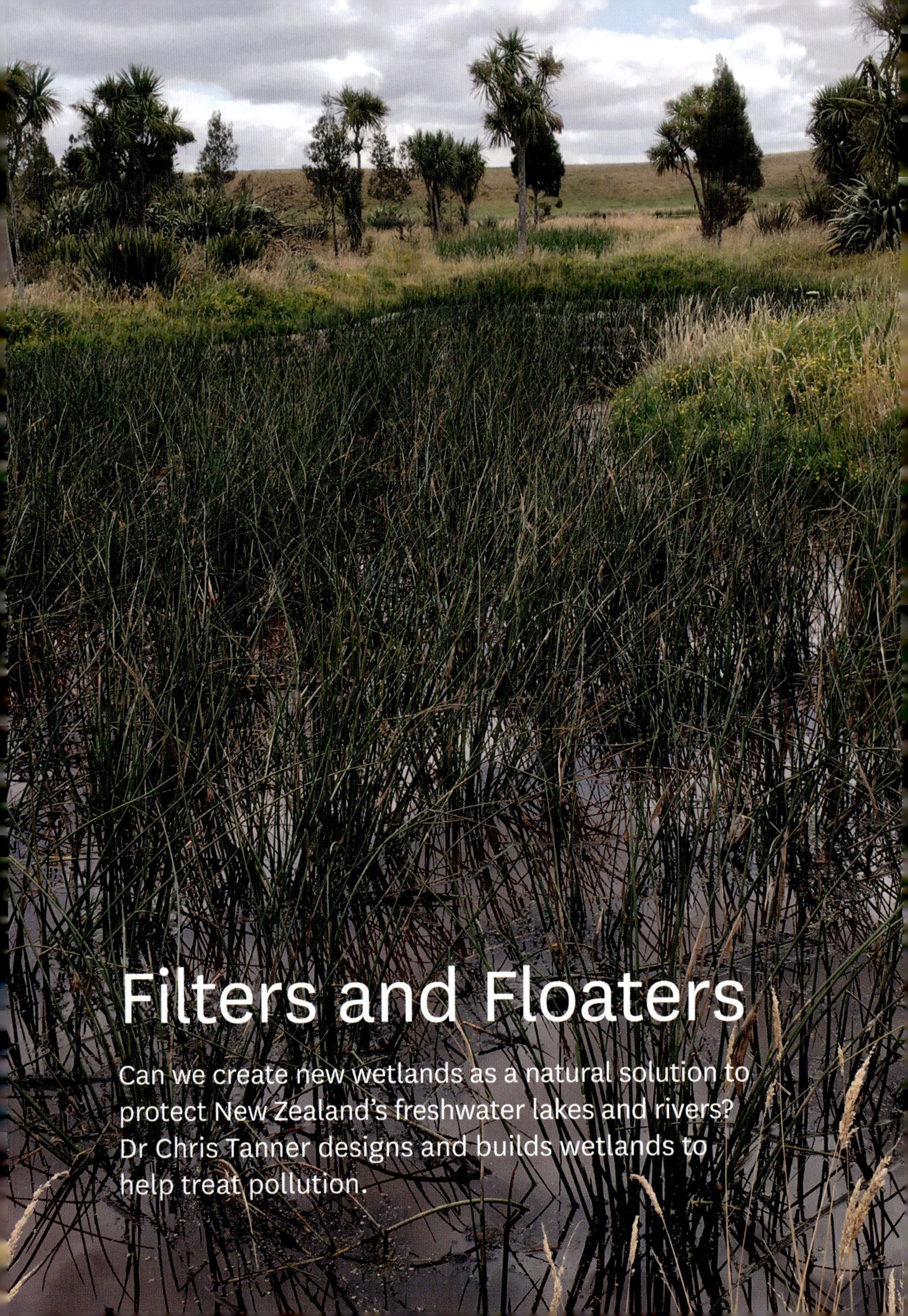

Filters and Floaters

Can we create new wetlands as a natural solution to protect New Zealand's freshwater lakes and rivers? Dr Chris Tanner designs and builds wetlands to help treat pollution.

Dr Chris Tanner

Specialist field: Ecology, design and management of constructed treatment wetlands

Currently works for: NIWA (National Institute of Water and Atmospheric Research Taihoro Nukurangi), Kirikiriroa Hamilton, as Principal Scientist — Aquatic Pollution

Chris Tanner has spent more than three decades researching the ecology and functioning of natural and constructed wetlands. He works widely with engineers, industry and government to design site-specific wetlands to treat urban, industrial and agricultural wastewaters and to diffuse pollution. He has over 90 research publications and book chapters to his name, and has led demonstration projects throughout Aotearoa New Zealand on farms, in small communities, towns, papa kāinga (villages) and marae, and in Fijian villages. His research has been translated into guidelines to promote best practice and enhance the practical uptake of eco-technologies.

WHEN JONI MITCHELL sang that we don't know what we've got till it's gone, she could easily have been referring to wetlands. It took a long time for Western science to understand what wetlands do for us, and by the time it did, 90 per cent of the wetlands in Aotearoa New Zealand had been destroyed in a quest to 'tidy up' the landscape or make way for development: cities, towns, farms, plantation forests, crops, marinas and more.

We've paid a huge price — wholesale loss of habitat for many unusual and unique native plants and animals (many driven towards extinction), increased flooding in some areas, and prolonged droughts in others, coastline erosion, and degraded freshwater in many lowland waterways. Much of what wetlands once did for us at no cost now requires expensive engineering solutions — stopbanks, sea walls and water treatment stations.

If the landscape were a body, then wetlands would be a vital organ. A healthy wetland ecosystem functions much like a kidney: it removes waste products (pollutants), so that the water leaving it is generally in much better condition than when it entered.[1] While wetlands are an important tool for dealing with water pollution, our dwindling number of remaining natural wetlands shouldn't be used as nature's garbage cans. The solution seems deceptively easy — make new wetlands. This is where Chris Tanner comes in. His research work involves designing and testing 'treatment' wetlands, specifically built for the job of purifying water.

As a young man, Chris was 'full of questions and hungry to learn'. A lucky break led to an 'amazing' series of summer jobs with the Ministry of Agriculture and Fisheries Aquatic Weeds section (later the Aquatic Plants section). There was plenty of hands-on, practical work: putting together experimental set-ups with tanks and pumps, and exploring lakes from north to south by boat and scuba to study lake weed invasions and control options. These summer jobs eventually led to more permanent work with the same research

POLLUTION CONTRIBUTIONS

Pollution is anything that negatively changes the way water looks, smells or tastes, or that makes it unsafe for creatures to live in or for us to swim in and drink or collect food from. Common pollutants are nutrients, pathogens, toxins and sediment.

In the right quantities, nutrients are essential for plant growth, but too much washed off the land can send freshwater systems into overdrive. The result is blooms of toxic algae, death of native plants and fish, and dramatic changes in water colour and clarity. Fingers are generally readily pointed at the dairy industry — it's an easy target, because the figures are widely available. Alarmingly, we know that nitrogen applied in the form of fertiliser increased more than sixfold between 1990 and 2015, and that the number of dairy cattle ballooned from 3.4 million to 6.3 million (an 82 per cent increase) between 1990 and 2019.[2]

But farmland isn't the only source of pollution. Poorly maintained household septic tank systems are an often forgotten source of pollution — although they're designed to capture the solid bits, overflows can send unwelcome nutrients, pathogens, and faecal contaminants (such as the bacterium *Escherichia coli*) into streams, rivers, lakes and the ocean.[3] Roads, too, can contribute to water degradation. Every downpour flushes particles of tyre and brake linings (containing lead, cadmium and zinc), and residues of engine oil, petrol and cooling liquids (containing sulphur, chlorine and cyanide), into waterways.[4]

team, but as Chris explains, 'There was always a topic on the end of our research agenda that kept falling off the to do list — it was about the potential to use aquatic plants to clean up water pollution.' This was to become his major career path.

Model wetlands

Putting wetlands back into the landscape to help cleanse freshwater is a logical solution, but to ensure they are cost-effective and built to work as well as their natural counterparts, it's important to understand how naturally occurring wetlands treat water. Wetlands are like sponges: they absorb water and release it slowly. This allows sediments and attached pollutants to settle out, improving water clarity and locking up toxins. Pathogens are killed off in open water areas by exposure to the natural ultraviolet light in sunlight. But the real engine room of a wetland, where nutrients are stripped out, is the alliance of plants, soils and beneficial microbes. Between them, these elements cleanse and purify pollution-laden water that enters a wetland. Understanding how they work together is central to Chris's work.

Over the past two decades he has focused on many treatment wetland designs for processing pollution and enhancing water quality. Different designs will suit different settings — some are constructed on land, while others are designed for open water. On land, polluted water can be directed through plants in a shallow pond (surface flow), through planted gravel beds just underneath the surface (horizontal sub-surface

A small establishing 'horseshoe' wetland constructed to intercept stormwater before it enters a waterway.

flow) or vertically drip-fed through soil (vertical sub-surface flow). Floating wetlands are installed in open water where polluted water circulating underneath plant-filled rafts is treated as it passes through the plant roots and their attached microbes.

The hard part of designing a fit-for-purpose treatment wetland, says Chris, is determining the best type and size to carry out the pollution-processing task required. He often carries out smaller-scale 'mesocosm' experiments in tanks or in the field to better understand and measure how well a design will perform. This allows different design approaches to be compared in a way that's also cost-effective. He uses the results from the mesocosm experiments to predict how well the scaled-up wetland will perform when it's constructed in a real-world setting. 'Wetlands are quite hard to tease apart and understand, because there are so many things going on within them at any one time,' says Chris. 'Mesocosm experiments help you look inside the "green box" of a wetland so that you can understand what's happening inside them.'

These medium-sized experiments fall somewhere between life-sized 'macrocosm' field trials and test-tube scale 'microcosm' investigations. Like a model railway, a mesocosm experiment radically scales down the size of a whole ecosystem while also simplifying it. In Chris's case, the experimental treatment wetland is based on just four key ingredients; plants; something to grow them in (such as soil, gravel or lake sediments); naturally occurring microbes and invertebrates; and polluted water.

At the NIWA research facility in Kirikiriroa Hamilton where he works, Chris may set up a series of experimental 'filter wetlands' made up of, for example, 200-litre tanks arranged in various combinations of plant species or amounts of pollutant added to the water. This helps him measure how well different species grow, how much pollution they process and whether one species is better than another. Or he may develop a 'treatment train' filter wetland, with polluted water flowing through different structures, such as deeper sediment traps, winding canals to reduce water flow, and shallow planted ponds, each designed to perform a specific role in removing pollutants and improving water quality. Working at the mesocosm scale makes it easier for Chris to assess outcomes and directly compare the different treatments.

The treatment combinations that show the most promising results can then be scaled up and tested in a real-world setting: on someone's farm to mop up excess nutrients; as part of an urban stormwater pond system to capture and process pollutants from a roading network; or integrated into a household wastewater treatment set-up to stop pollutants from getting into nearby water bodies. However, there are a few caveats when scaling up an experiment. Plants can grow more vigorously and densely in tanks than in large, open wetlands because they get more light coming from the sides of a tank and there are no pesky pūkeko (swamp hens) to eat them. Enhanced growth of young plants in a tub also means a greater uptake of pollutants — Chris cautions a trap for new players is to overestimate how effective the scaled-up wetland will be at pollution removal. He's well versed on the ins and outs of applying his research into everyday settings: 'A lot of what we're trying to do is work out practically what's the sustainable treatment performance, what's a cost-effective way we can build them and make them reliably do what we want them to do for a long time.' Size also matters, says Chris: 'A pocket handkerchief wetland can't perform magic. You need something large enough to hold the water long enough to remove the contaminants.'

Computer simulation models calibrated with real-world data are often run to establish the right size and design for a constructed wetland to carry out its water treatment works most effectively, but in the end, it's not a recipe. The wetland still has to align with planning rules (such as distance to property boundaries) while making the best use of the often limited space available. Tucked next to a motorway, for example, a wetland built to treat road run-off might end up being a long, skinny, plant-filled channel, while on a farm there may be more flexibility around size and shape. The design also needs to factor in soil type, slope and variable water flows — how will the system cope with a sudden heavy downpour, or when a big group of people arrive and generate profuse quantities of wastewater?

One of the most important functions of treatment wetlands is slowing down the flow of water. If polluted water is rushing through a wetland, there isn't enough time for heavier contaminants and sediments to settle out, or for the swards of plants and their microbial communities to work their nitrogen-stripping magic. Seen from the air, some of Chris's designs — particularly those on farms — are artful serpentine arrangements of deep open ponds (to settle out sediments) and shallow basins densely planted with native wetland species (to process nutrients). Plant-covered bunds made from compacted soil may jut into the ponds and planted areas to create a meandering pathway for the water. The designs for systems to treat household wastewater and septic tank effluent are more utilitarian — exiting the septic tank (the solids having settled as a sludge at the bottom of the tank), the polluted water percolates through a thick bed of gravel and tangle of wetland plant roots before eventually leaving the wetland as cleaner water that can safely soak into the soil.

The experiments don't stop at the mesocosm stage. Once a full-sized treatment wetland is built, Chris continues to check its performance and refine his design models based on real-world experience. So far, the research tells us that constructed wetlands can be enormously effective. In large catchments, if 2–5 per cent of the area is wetland in the right location, it can capture 30–50 per cent of leached nitrogen and phosphorus, and up to 90 per cent of sediment coming off the land.[5]

Wetlands adrift

Constructed filter wetlands work well when they're near the source of pollution, are big enough and are well designed. But what about removing pollution that's already in a waterbody? This is where constructed floating wetlands come into play.

Chris very modestly admits that his work, with Dr Tom Headley, on floating treatment wetlands helped spawn a major scientific effort around the world investigating their potential. The starting point is wetland plants that float naturally on water, drifting with

PLANT POWER

What superpowers do plants need to survive the demanding conditions in a treatment wetland? They need to be semi-aquatic, able to live with their roots in water and their tops in the air. They also need to be able to cope with fluctuating water depths — water levels in treatment wetlands can fluctuate wildly, evaporating under the hot sun and rising suddenly after a downpour. These plants also need to tolerate nutrient and pollutant levels much higher than those typically found in natural wetlands. On top of that, birds may decide to make their home in the wetland, so being able to tolerate some grazing and nest-building is also a must. Kūwāwā (lake clubrush, *Schoenoplectus tabernaemontani*, shown right) ticks all of these boxes and, crucially, it's a widespread native species, not a weed likely to invade other places.

The behaviour of this plant in different settings piqued Chris's interest: 'I found it really interesting . . . taking a plant out of its natural environment, growing it in really different things . . . like gravel, and supplying it with very high levels of nutrients and contaminants from wastewater to see how much that would affect its growth rates and nutrient content.'

During his PhD studies, Chris spent a lot of time measuring the growth of kūwāwā under different conditions to gain a better understanding of when and how the plant could survive and thrive. He says it's an ideal species for a treatment wetland — it's easy to propagate because the finger-like underground rhizomes can be broken up to make new plants, and it grows quickly. Structurally, it's quite basic too — tapering cylindrical shoots with a small tassel of seeds on the top. The vertical shoots sprout liberally from rhizomes, creating a pale blue-green thicket 1.5–2 metres tall. Under the nutrient-rich conditions found in wetlands created to

the wind. They can even look like a solid surface of vegetation — as many a surprised dog retrieving a wayward ball from a thick patch of floating red *Azolla* fern has discovered. Because their roots aren't anchored to the beds of streams, rivers or lakes, floating plants take up nutrients directly from the water.[7] When their roots do this, they're also cleansing the water. Scientists like Chris, working alongside engineers, have designed novel floating treatment wetland systems to boost what wetland plants already do.

To keep the wetlands afloat, a raft is made using buoyant materials (usually plastic, but more eco-friendly options are being researched) and densely planted with species that like having wet feet. Dense curtains of plant roots grow down through the raft and dangle in the water. The root mat is colonised by microbes, creating a slimy biofilm where most of the nutrients from the water are absorbed and broken down. It's also where pollution such as dissolved metals from road run-off can accumulate.[8]

treat agricultural wastewaters, kūwāwā can grow up to 1500 shoots per square metre and its shoots can lengthen by up to 30 millimetres each day. This adds up over the growing season, resulting in more than 2 kilometres of shoots per square metre of wetland! Cultural uses have also been documented: Māori wove the stems together to make floor mats and occasionally used them, along with the buoyant flowering stalks of harakeke (flax, *Phormium tenax*) to make rafts.[6]

To stop the rafts from floating away like a floristic Noah's ark, they're usually tethered to land, to rise and fall with changing water levels. Floating wetlands have been trialled in constructed urban ponds to remove metals from stormwater and treat urban and industrial wastewater,[9] and also deployed in natural lakes to enhance water quality. There are added benefits — they can inhibit the growth of nuisance algae in the water through shading, reduce wave erosion of the lake banks, and provide additional habitat for fish as well as nesting and resting areas for birds.

From 2009 to 2010, Chris was involved in New Zealand's first trials of floating treatment wetlands at Rotoehu, one of the 18 lakes of the Rotorua Te Arawa Lakes Programme. Rotoehu is Māori for 'murky water', so it's likely that the lake water has never been clear. But the shallow lake and its surrounds have always been valued: before European colonisation, many pā (fortified sites) were found nearby. Today, the shoreline is clustered with holiday homes and fly fishermen try their luck catching wily trout.

GREENING GREYWATER

Since 2014, Chris has led an initiative to help Māori communities strengthen their marae wastewater treatment. That's no mean feat. As important social and cultural venues, marae and community centres can be hives of activity. They host hui (meetings), unveilings, birthdays, weddings and reunions. Hundreds of people may arrive for tangihanga (funerals) and stay for several days. Some also double as kōhanga reo (early-childhood teaching centres), health clinics and, in crises, local civil defence centres. These events can create big pulses of blackwater (sewage) and greywater (hand basin and shower wastewater), which typically head into a septic tank. However, many older systems are too small for the job — they get blocked, overflow or just smell bad. Any new wastewater system has to reduce potential health risks, protect the environment and be culturally appropriate. For Māori, preventing human waste from entering waterways and the ocean is paramount. Culturally, it is more acceptable to allow Papatūānuku (the Earth) to purify polluted waters. Day to day, the ideal new system also needs to be budget-friendly, and simple to construct, run and maintain.

Working with hapū at the Kokiri Centre, a community educational centre run by Tainui Awhiro, near the mouth of Whāingaroa Harbour in the Waikato, it was clear that their septic system wasn't coping with the amount of wastewater during large events.[31] Connecting to the nearby town wastewater treatment plant was not culturally acceptable because it discharges effluent straight into the harbour — too close to Tainui Awhiro ancestral lands. Instead, Chris and his team designed a separate on-site filter wetland treatment system for the greywater (from showers and hand basins), leaving the existing septic system to manage the blackwater (from toilets and kitchen sinks). The greywater is now channelled into a filter treatment wetland system constructed downslope of the centre. Greywater first flows into a sump (to allow coarse particles in the water to settle out), before trickling through a gravel bed densely planted with native sedges once abundant in the local area: purei (*Carex secta*), pukio (*C. virgata*) and ūpoko-a-tangata (*Cyperus ustulatus*). The wetland plants, microbes and invertebrates help break down the waste to purify the water as it slowly percolates through the gravel bed. In the last stage of treatment, the now cleaner greywater is dispersed through an infiltration area planted up in harakeke (flax). An advantage of this system is that water travels below the surface throughout, which reduces the likelihood of anyone coming into contact with polluted water.

So, does it work? Yes — and extremely well! NIWA's monitoring over the first summer and winter of operation showed that almost 100 per cent of nitrogen and pathogens were removed. Added to this was a more than 90 per cent reduction of phosphorus and suspended sediments. The water leaving the wetland is almost pure.

In 1993, Rotoehu's water quality changed dramatically when nutrient and algae concentrations rose alarmingly after a long, warm summer and a drop in water levels.[10] Someone suggested installing a floating wetland, but this was still novel eco-technology — rigorous testing was needed first to measure actual rates of nutrient removal. So Chris and his team set up a year-long mesocosm experiment at Maero Stream, identified as a major source of nutrients to the lake. They needed a bigger set-up than for their filter wetland trails, so instead of tubs, they placed a modified shipping container next to the stream. A barrier welded along the middle of the container formed two chambers,[11] and a series of floating mats of pukio and ūpoko-a-tangata was installed in each side.

Stream water was pumped into one end of the shipping container at two different rates (high and low). The water then flowed slowly through the root mass under the floating wetlands towards the outlets and back into the stream. Nutrient levels measured at the inlets and outlets showed how effective the plants were at processing nutrients under the two different flow scenarios. Across the mesocosm's high-flow and low-flow scenarios, nutrient removal rates were between 45 per cent and 77 per cent for nitrogen, and 32 per cent and 35 per cent for phosphorus.[12] The results were promising enough to forge ahead with installing a 2800-square-metre floating wetland on the lake itself, near the stream mouth to intercept and treat the stream water.[13]

Using the knowledge gained from the Rotoehu floating wetland, the next step was to trial the technology on nearby, and much larger, Rotorua — the jewel in the crown of the Te Arawa lakes complex. For this government-sponsored project, the floating

A floating wetland planted in Carex *sedges, Lake Ōkareka, Rotorua.*

wetland spelled out 'ROTORUA', cleverly incorporating the region's branding features. The location was strategic: highly visible for passengers on scenic flights over the striking geothermal landscape or on aircraft landing at the nearby airport. Constructed in 2011, the design consisted of 20,000 native plants growing through a buoyant matrix made from recycled plastic bottles. The whole wetland was expected to remove 4 tonnes of nitrogen and 1 tonne of phosphorus from the lake each year over the nine-year lifetime of the project.[14] In 2012, the government celebrated the success of the Lake Rotorua floating wetland. Amy Adams (Environment Minister at the time) was reported saying, 'The floating wetland is helping to improve Rotorua's water quality, provide a habitat for wildlife and offer an original tourist attraction. It is great to see such an innovative solution to an environmental problem.'[15] However, in 2018 a huge storm battered the structure, resulting in dramatic news headlines: 'Lake Rotorua's floating wetland in "quite a state"!'[16] Two sections had broken away from the main structure and smaller pieces had drifted off — the ROTORUA sign resembled a barely readable spam-prevention CAPTCHA code. Contractors repaired the floating wetland, but in 2020 the ageing structure reached the end of its project life and was removed. Rather than replace it, the council chose to expand and enhance natural wetlands at the edge of the lake to treat water inflows.[17] With lakes being such dynamic places, subject to wind and wave action, Chris suggests that using floating treatment wetlands for stormwater ponds and drains is a safer and more effective option.

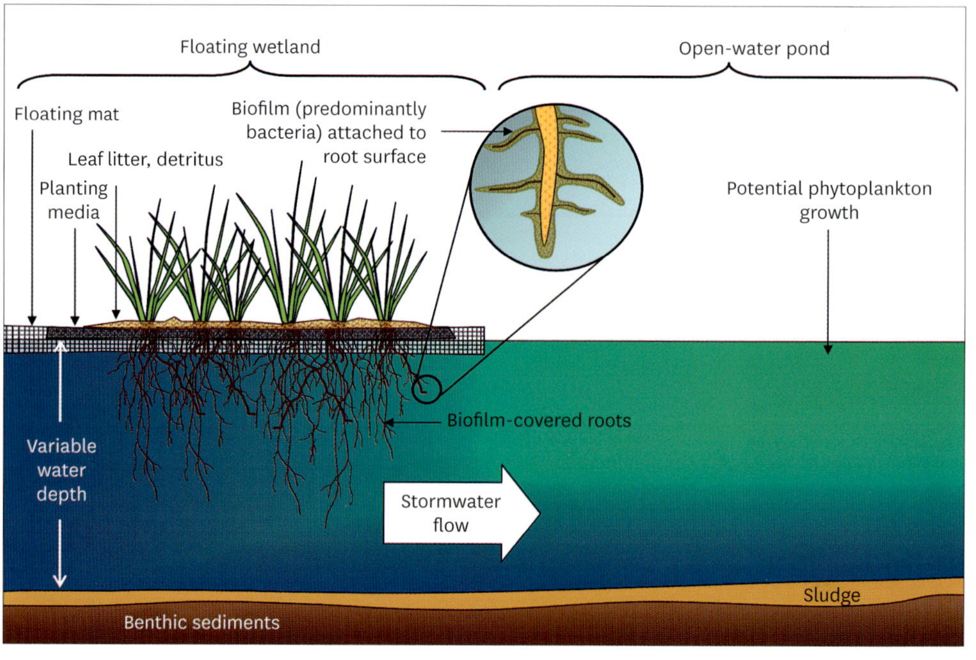

A diagram showing the main components of a floating treatment wetland.

Credit where credit's due

The various jobs that wetlands do so well, including absorbing and slowly releasing floodwaters and processing pollution, are known as ecosystem services. Generally, it's a term that engineers and purse-string holders can relate to because it helps quantify what wetlands do for us, for free . . . all day, every day, year in, year out. Placing an actual value on these services could encourage landowners to reinstate wetlands on their land (or construct new ones).

One possible incentive is to provide farmers with nutrient budget credits for the work their constructed wetlands do by way of treating nutrients. New Zealand planning rules are increasingly requiring farmers to account for the nutrients that leave their farm. Exceeding certain limits may trigger planning requirements that force famers to reduce their nutrient footprint — for instance, by decreasing the size of their herd or the amount of fertiliser they use.

To work out the amount of nutrient leaving their farm, farmers generate a nutrient budget using tools such as Overseer.* Chris and his colleagues have developed a series of guidelines to help farmers create effective treatment wetlands on their own properties.[18] He would like to see the value of these on-farm wetlands built into Overseer, to ensure that landowners get credit for the nutrient reductions they achieve, creating a win-win for the environment and for the farmer. Chris explains that if you calculate an expected range of nutrient reduction that relates to the size and design of the treatment wetland, then coming up with an actual value that could be translated into credits for the farmer should be feasible. With 90 per cent of wetlands obliterated nationwide over the past two centuries, providing farmers with nutrient credits could be a useful incentive for encouraging more wetlands to be either restored or constructed on the farm.

Chris's work is ongoing — there really is no end to places where new wetlands are needed to help knock back the worst effects of human-induced changes to the landscape. His work has helped shed much-needed light on the 'green box' of what happens inside wetlands. At the same time, he's also convinced landowners, managers and decision-makers that wetlands are worth investing in — if they're no longer there, then new wetlands can, and should be, constructed. Not only do they work, but they work well — and that's backed up by solid science. While Chris is a self-confessed advocate for constructed wetlands, he highlights the fact that there's still plenty to learn. He's realistic about when and where they are likely to function best, having seen them used to treat anything from livestock and human wastewater to stormwater and industrial wastewater. 'My thing is trying to make them a real tool that people can use to manage water quality. Constructed wetlands are a prime nature-based solution . . . as small-scale systems, we can live around them . . . they are part of creating green spaces in towns and cities where we can relax and commune with nature.'

---

* See www.overseer.org.nz.

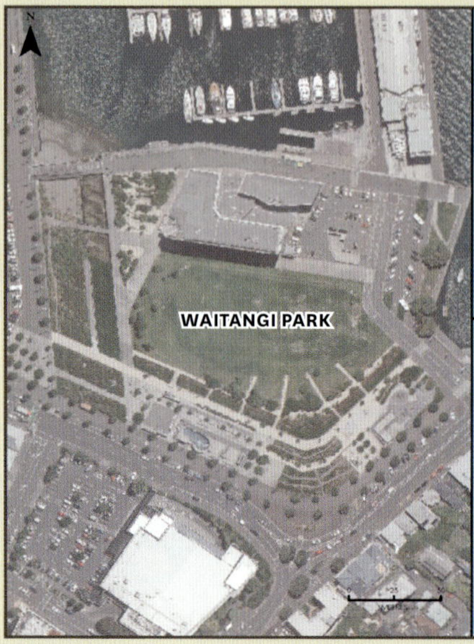

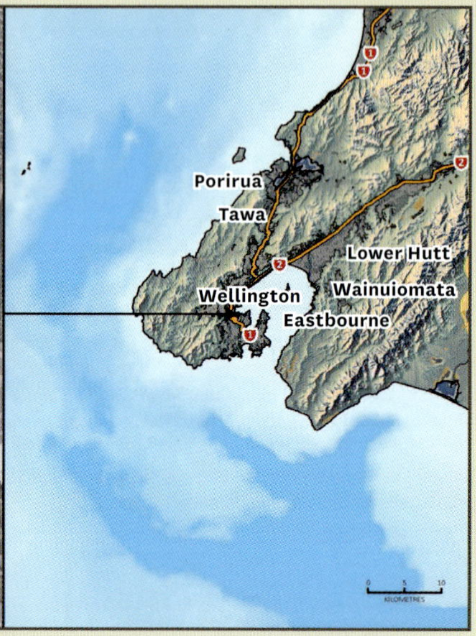

Waitangi Park

Urban treatment wetland

Urban treatment wetlands are incorporated into the award-winning Waitangi Park on Wellington's waterfront. Developed by landscape architects, it's a fascinating place that integrates snappy design with re-created natural areas using some of the eco-technologies that Chris has helped to mainstream.

Long before Te Whanganui-a-Tara Wellington became the bustling capital of New Zealand, Māori harvested abundant tuna (freshwater eel), fish and shellfish from a wetland along the waterfront, collected freshwater from the Waitangi Stream and used the site to launch their waka into the ocean. However, the area was dramatically altered in 1855 when a devastating earthquake lifted the land by 1.5 metres, forever changing the watercourse and the future of the wetlands. Pipes, concrete and tarmac eventually followed as this part of the harbour was 'reclaimed'. Waitangi Stream was buried in concrete, relegated to a wastewater drain. With the wetland and cultural history forgotten or ignored, the site first housed a morgue, then a bus park and giant incinerator.

Today, thanks to a radical rethink about how culture, ecology, history and sustainability can be combined, a major feature of the 6.5-hectare park is the Waitangi Stream.[19] Finally brought to the surface after decades underground, the stream first travels through a wide bed of densely planted native sedges and reeds,

where pollutants suspended in the water are settled out. The journey continues through several thickly planted, sloped terraces, to further 'polish' the water before it falls into a large pond. The pond combines shallow planted ledges and areas of deeper open water that, aside from its visual appeal, can also be used to irrigate the site over hot, dry summers. Once through the ponds, the Waitangi Stream empties into the harbour, much cleaner than when it entered the park.[45]

Highlights

- Large, multi-use recreational park on Wellington's waterfront
- Urban treatment wetland incorporating planted reedbeds, bio-retention 'tree pits' and a rain garden to strip pollution from road run-off and stormwater.

Access

The park's central location ensures easy access on foot and by public transport. Situated in downtown Wellington, close to the iconic Museum of New Zealand Te Papa Tongarewa, the park is bordered by Oriental Parade, Cable Street, Barnett Street and Herd Street. With multiple use in mind, it also contains a large grassed area, a skateboard park, areas for waka launching and pétanque, and a state-of-the-art playground, along with the re-created wetland. Dogs must be kept on a leash.

https://envirohistorynz.com/2010/12/12/waitangi-park-an-urban-wetland-recreated

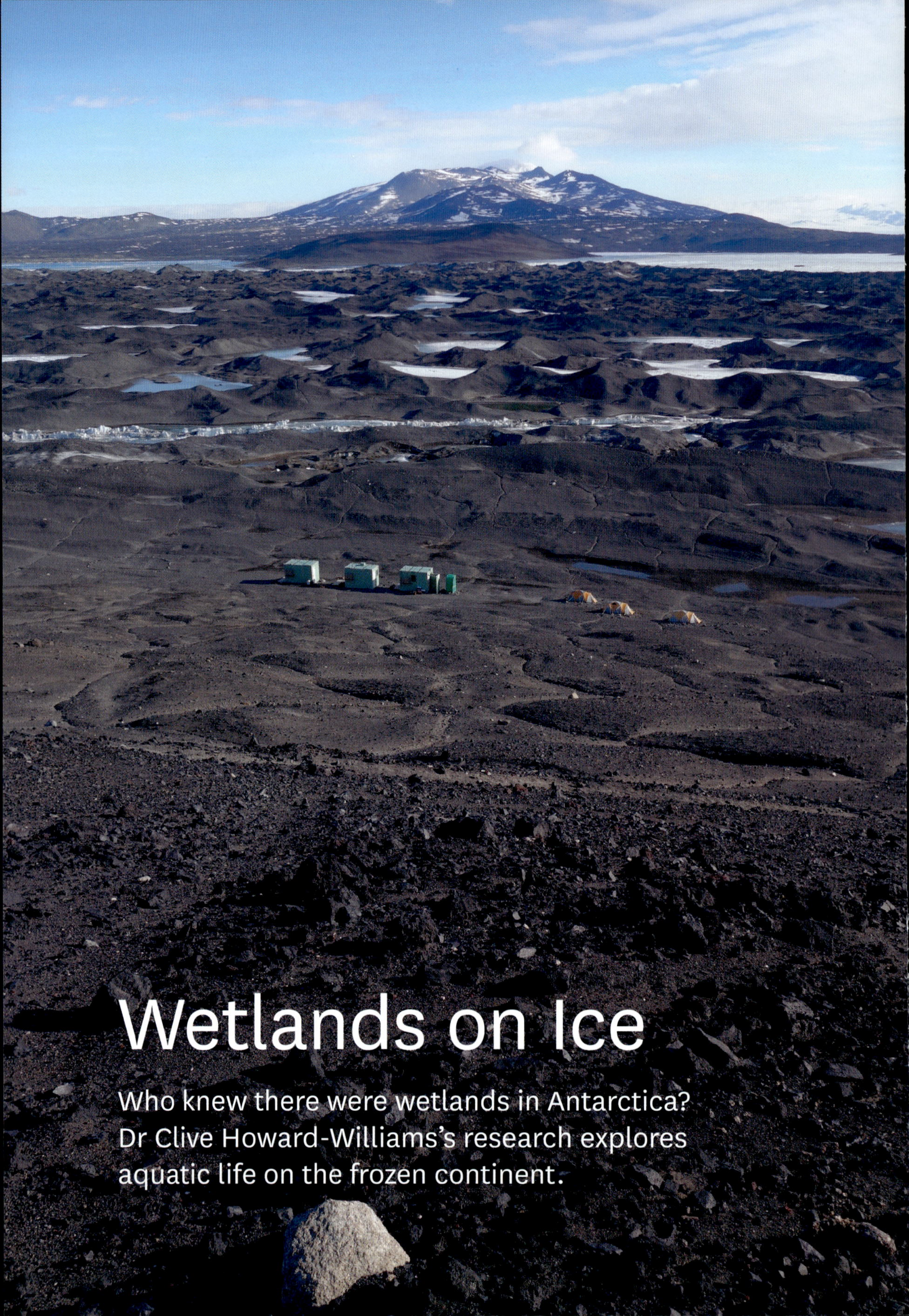

Wetlands on Ice

Who knew there were wetlands in Antarctica? Dr Clive Howard-Williams's research explores aquatic life on the frozen continent.

Dr Clive Howard-Williams

Specialist field: Wetland ecosystem processes, carbon and nutrient cycling

Currently works for: NIWA (National Institute of Water and Atmospheric Research Taihoro Nukurangi) as an emeritus scientist

Clive Howard-Williams, PhD, is an aquatic ecologist with more than 35 years' experience specialising in freshwater and estuarine water quality and aquatic ecosystem processes. His expertise results from research in a number of countries, from the tropics to the polar regions. With colleagues Dr Warwick Vincent (now at Laval University, Quebec Ciry, Canada) and Dr Ian Hawes (now at the University of Waikato), he initiated a long-term research project based at Bratina Island on the McMurdo Ice Shelf in 1988. This work has resulted in well over 50 peer-reviewed publications, a significant number of MSc and PhD theses, and a great many presentations. Clive's research into aquatic ecosystems has led to a Fellowship of the Royal Society of New Zealand and several awards, including the Antarctic Medal (Queen's New Year Honours 2007), and election as vice-president of the 32-nation Scientific Committee on Antarctic Research.

CLIVE HOWARD-WILLIAMS'S CAREER has taken him to the extremes. He began in the steamy tropics, completing a PhD on the swamps of East Africa's Lake Chilwa, teaching ecology at the University of Malawi and then researching wetlands in the central Amazon. Gradually, his work took him further and further from the equator, towards both polar regions, where he has studied freshwater ecosystems in some of the most remote places on Earth.

Not surprisingly, Clive's favourite organisms also love extremes; in fact, the collective term for them, extremophiles, means just that. They live in extreme environments like geothermal pools — and Antarctic wetlands.

Antarctica isn't generally famed for its wetlands, but the McMurdo Ice Shelf, south of New Zealand's Scott Base, supports around 1500 square kilometres of them.[1] This floating platform of frozen freshwater, about 50 metres thick, rides on top of the denser saltwater adjacent to the massive Ross Ice Shelf, where the Antarctic ice sheet inches its way down and across the Ross Sea.

ICE CONVEYOR BELT

The surface of McMurdo Ice Shelf is covered in rocky debris derived from an unusual source — the sea floor below the ice shelf. The way it gets on top of the ice is fascinating. Part of the ice lies above a relatively shallow area of ocean, so that the 50-metre-thick ice shelf touches the sea floor in many places. In these locations, seawater freezing onto the underside of the ice picks up seafloor sediments, and sometimes marine animals, and locks them into the ice. These sediments are then gradually transported upwards, towards the surface of the ice shelf, on a type of ice conveyor belt. As anyone who has put ice cubes in their drink will know, ice floats. As the warmer and exposed top of the ice shelf melts or erodes, the giant ice block floats up a little — lifting the trapped debris off the seabed — and a new layer of ice and sediments freezes onto the bottom, creating a sort of multi-layered rock and ice sandwich.

Clive and his colleague Ian Hawes explain the effect this debris has: 'Once at the surface, the trapped marine sediments are released during ice melt and add to the layer on the undulating surface of the ice shelf. The rock debris brought to the ice surface enhances the warming effect of the summer sun and increases the rate of melting. This strange process has produced a landscape of tens of thousands of ponds over the 1500 km^2 ice shelf.'[2]

The undulating surface of the McMurdo Ice Shelf is dusted with rock debris and peppered with thousands of shallow lakes and ponds, from a few metres across to hundreds of metres long. Some are freshwater, some brackish, some salty and some hypersaline (off-the-charts salty). They tend to be short-lived, appearing and disappearing over the space of years or decades. In winter, they are frozen, some of them completely so. But for a couple of months each year, during the perpetual sunlight of a polar summer, the smaller pools melt and warm up, some reaching 15 degrees Celsius — tolerable enough for a midsummer dip.

The wetland complex is a major centre of biodiversity in this part of Antarctica. As the pools thaw out, their frozen inhabitants come out of icy hibernation and spring to life. Rotifers, tardigrades, nematodes* — they might sound like mythical creatures, but these are the animals that live in Antarctic wetlands. And they all depend on Clive's favourite species — bottom-dwelling, mat-forming microbes. Yes, Clive's favourite species are what most people would rather uncharitably call 'pond scum'.

* These are groups of multicellular invertebrates. Tardigrades, rotifers and nematodes are commonly referred to as water bears, wheel animals and roundworms, respectively.

A (NOT SO LITTLE) SLICE OF NEW ZEALAND

No individual nation owns Antarctica, but seven nations have claimed a slice of it. The uninhabitable icy continent is governed internationally through the Antarctic Treaty, signed in 1959 by 12 countries who had scientists working in and around the continent at the time. The original signatories include seven countries who had, and still retain, territorial claims: Argentina, Australia, Chile, France, New Zealand, Norway and the United Kingdom. The treaty has since been agreed to by many other nations — 54 in total.

The key statements in the Antarctic Treaty are (in paraphrased form):

- Antarctica shall be used for peaceful purposes only
- Freedom of scientific investigation in Antarctica and cooperation towards that end . . . shall continue
- Scientific observations and results from Antarctica shall be exchanged and made freely available.[3]

Aotearoa New Zealand's slice of Antarctica, the Ross Dependency, is larger than New Zealand itself. It includes Ross Island, a volcanic landmass surrounded by almost perpetual sea ice that connects it to the continent. This is the location of Scott Base, the New Zealand Antarctic research facility. It was set up as a support base for field research and can accommodate 85 people over summer, with a skeleton staff of between 10 and 14 people remaining over the winter. A few kilometres away is the much larger United States Antarctic research station, McMurdo Station, capable of supporting up to 1250 residents.[4]

Antarctic wildlife

When visitors come to Antarctica, says Clive, they expect to see penguins, seals, whales and seabirds, but that's just the life found at the edge of this massive continent. As soon as you step away from the ocean, away from the clamorous penguin colonies, life is almost completely restricted to areas where there's liquid water, and it's essentially microbial.

You can't generally see microbes floating around in water — after all, the word means 'tiny life' — but some of them link up, forming visible colonies of floating chains or microbial mats, which are common in the Antarctic pools. To a human eye, microbial mats look like a layer of slime, but if you could shrink down to the size of a pinhead and enter one, it would be like visiting a biodome — a self-sustaining microbial city, busy with a population of individual cells, each with an important job to do. Multiple layers are formed by millions of micro-organisms of different species, interacting with one another and sharing energy and nutrients.[5] One group of species that form these mats are cyanobacteria, single-celled organisms capable of photosynthesising, in the way plants do.

Clive has studied cyanobacterial mats and discovered that distinct communities develop in different ponds, depending on salinity or other conditions that vary from pond to pond. A typical McMurdo pond mat (a few millimetres thick) has a dark black, brown or orange pigmented layer of cyanobacteria at the surface. The job of these cells is to screen out harmful ultraviolet (UV) radiation and protect a lower layer, which is usually intensely green. This is the engine room of the mat, containing the cells that photosynthesise to produce sugar to feed the colony. This layer becomes super-saturated with oxygen, and yet just a few millimetres below it oxygen is completely absent, having been consumed by bacteria breaking down dead cells. In this oxygen-deprived (anoxic) layer, other groups of bacteria grow — such as those using sulphate to produce sulphides, or those that convert carbon dioxide to methane. These bacteria recycle the nutrients needed by the overlying photosynthetic layer. According to Clive, 'Within the microbial mats at the base of the ponds, we find all of the microbial groups that occur as part of the biogeochemical cycles in more temperate regions.'[6]

OXYGEN MAKERS

Cyanobacteria have lived through it all. Dinosaurs. Ice ages. Snowball Earth. Mass extinctions. Human evolution. In fact, if it wasn't for cyanobacteria, it's possible we wouldn't have evolved at all. It's thought that every living thing on Earth is descended from a common ancestor, likely a bacterial form of life. American biochemist Douglas Theobald speculates that, 'to us, it would most likely look like some sort of froth, perhaps living at the edge of the ocean, or deep in the ocean on a geothermal vent. At the molecular level, I'm sure it would have looked as complex and beautiful as modern life.'[7]

Cyanobacteria are commonly referred to as blue-green algae, which is a convenient but not entirely accurate term. For one thing, they aren't algae; they are single-celled bacteria. For another, they're not always blue or green. Like algae, they contain chlorophyll, along with a blue pigment called phycocyanin, which means they can make their own food through photosynthesis and explains why they form a blue-green colour in some lakes. But in clear, shallow water in areas with high levels of sunlight, cyanobacteria can also come in a dazzling array of colours, which in some places has made them a major tourist attraction. Heard of Yellowstone's Grand Prismatic Spring? New Zealand's Rainbow Terrace? These famous geothermal features are visually stunning thanks to a single-celled organism often referred to as 'scum'. The colours are the result of nature's original sunscreen — the orange, pink, red and yellow pigments

A slice of life from the bottom of a pond on the McMurdo Ice Shelf. The orange-pigmented upper layers of cyanobacteria that make up the microbial mat protect against UV radiation. Below that are the darker layers of the mat, which are several millimetres thick.

make the bacteria highly resistant to enhanced UV radiation — which is very handy if you live in clear water.

This adaptation is an impressive trick for sure, but then cyanobacteria have had around 2.5 billion years to perfect it. It's also not the most remarkable thing they have done. They practically 'invented' oxygen-based photosynthesis. Their ancestors evolved mechanisms to take energy from sunlight and use it to turn water and carbon dioxide into sugar.[8] The waste product of this process happens to be vital for humans and all other breathing creatures — oxygen. Some researchers believe that cyanobacteria contributed to the so-called Great Oxidation Event — the build-up of oxygen in the Earth's atmosphere between 2.5 billion and 2.4 billion years ago.[9] It changed the entire planet and enabled the evolution of complex multicellular creatures, including humans.

Scientists hypothesise that during long-distant periods of climate extremes known as Snowball Earth, when almost the entire planet was frozen, cyanobacteria were the only surviving photosynthetic organisms, perhaps retreating to ice-melt refuges like those on the McMurdo Ice Shelf today, preserving the precious blueprints on how to convert water and carbon dioxide into oxygen.[10] Says Clive, 'It is tempting to believe that the ability of these mats to tolerate the extremes of Antarctica stems from the genetic lessons learnt at that time, perhaps as long ago as 2 billion years!'

It's probably time we stopped calling cyanobacteria 'scum' and started to appreciate what they have done and continue to do for us.

The ponds on the McMurdo Ice Shelf are a window into the past, offering a glimpse of how life on Earth may have looked 2.5 billion years ago, when microbial mats were pretty much the only life forms around. It's incredible that they're still around today.

Cyanobacteria occur naturally in wetlands, rivers and lakes across the world, from the tropics to the poles, and are the dominant life form in extreme environments where few other species can survive: steaming-hot geothermal areas or freezing-cold polar ponds; long periods of darkness in winter, or 24 hours of sunlight and high UV radiation levels in summer; hypersaline or chemically altered water. One explanation Clive offers for their tolerance is that, along with other mat-forming organisms, cyanobacteria have been around for a very long time, and they've lived through every possible type of climate and climate change scenario you can imagine. So they've learnt to adapt to extreme environments where other organisms struggle to survive or thrive. The Antarctic pools are almost entirely devoid of plants other than diatoms (micro algae), so the microbial mats form the basis of the food chain, supporting a community of tiny animals — tardigrades, rotifers and nematodes.

Clive's work on the McMurdo Ice Shelf has sought to add to our knowledge of Antarctica's biodiversity and natural ecosystems, but it also contributes to our understanding of both primeval and modern ecosystems, how microbes operate, and the important role they play in the life cycles that sustain all species on Earth.

The bed of this small pond on the McMurdo Ice Shelf (measuring approximately 3 metres across) is covered with a pigmented microbial mat. White salt deposits around the pond's edge (sodium chloride and sodium sulphate) point to the water being brackish or saline.

TENACIOUS TARDIGRADES AND RESOLUTE ROTIFERS

Rotifers (pictured below), tardigrades and nematodes are tough critters. These are the animals commonly found in the cyanobacterial mats of the McMurdo Ice Shelf and in moist areas on other ice-free Antarctic landscapes. They can withstand long periods of freezing and desiccation, significant changes in salinity, and even radiation.

Clive notes an interesting series of experiments that were carried out back in 1908 by the Scottish biologist James Murray on rotifers retrieved from cyanobacterial mats during explorer Ernest Shackleton's *Nimrod* Expedition. Murray carried out a series of experiments on the resistance of rotifers to freeze–thaw cycles, desiccation and increasing salt levels. He found that they survived almost all his treatments. In his 1909 book *The Heart of the Antarctic*, Shackleton wrote, 'We all became vastly interested in the rotifers during our stay and the work of the biologist in this respect was watched with keen attention . . . It became a contest between rotifers and scientist, and generally the rotifers seemed to triumph.'

Their findings still stand today — in 2021, scientists unearthed 24,000-year-old Siberian permafrost and found living (or at least revivable) rotifers there.[12]

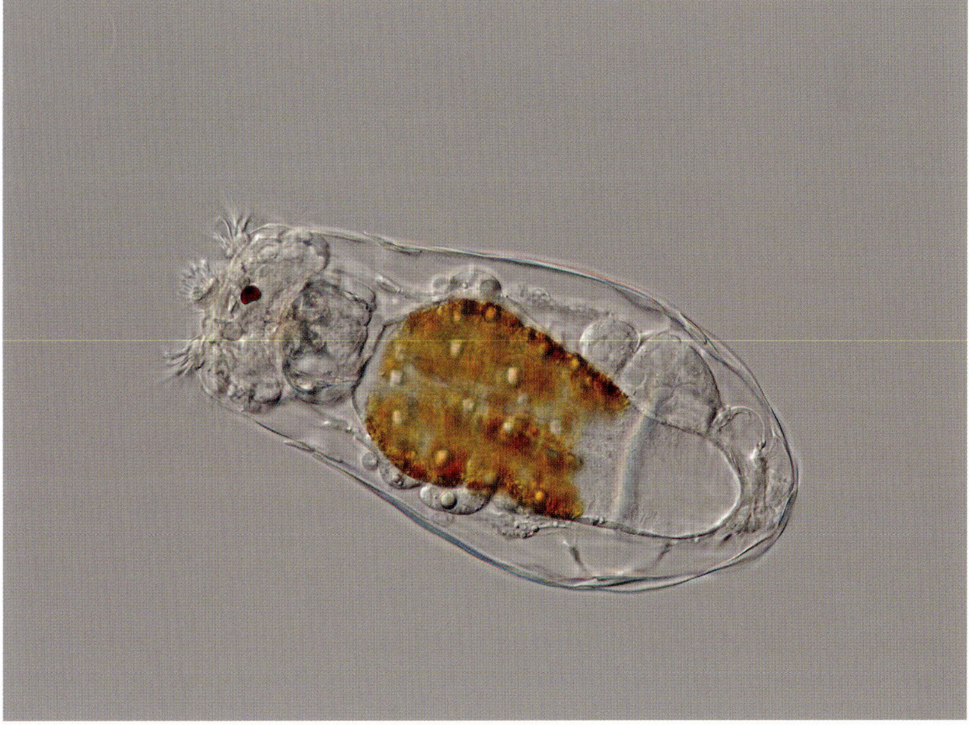

MYSTERY FOOD BOXES AND EXPLODING LENTILS

When you arrive as a new researcher at Scott Base, there's usually a day or so of orientation and field training, where you learn crucial survival skills to prepare you for the hazards of working and camping in such an unforgiving environment. Researchers are taught techniques of polar camping and safe use of Primus stoves in tents. They are also introduced to the New Zealand food box. Like the mystery box challenge in a television cooking show, you have to make do with what's in the box. That's all you have, so you have to be adventurous and inventive. Clive notes that times have moved on and the boxes include more fresh food and variety these days, but when he first went on an Antarctic field camp in the early 1980s it was pretty basic:

> We even had biscuits that were very similar to the ones of Captain Scott's day — they were fairly . . . solid. You couldn't have fresh bread, obviously, but plenty of flour, pasta, powdered egg, powdered milk and porridge. There were vitamin tablets to stop you getting scurvy, and dried herbs to add flavours. And there was an interesting series of recipes that came out in the *New Zealand Antarctic Research Programme Recipe Book* for how to deal with the contents of the food box!

Clive recalls one particularly memorable dining experience in Antarctica:

> While preparing dinner inside our shared tent, with sub-zero exterior temperatures, one of my field colleagues managed to over-pressurise a cooker. The entire contents were ejected into the tent at a high velocity. Both of us shot out of the tent like missiles with minor burns. The liquid fortunately extinguished the Primus cooker, but the whole inside of the tent was coated in lentil stew! It took several miserable hours in the wind and cold, pulling down the tent, turning it inside out, cleaning the walls and other items as best we could, and setting everything back up. Luckily, our sleeping bags had standard-issue canvas covers — but the tent smelt of lentil stew for the rest of the trip!

Nitrogen fixation

The processes that take place in ponds fascinate Clive, and have been a major focus of his research. Life on Earth relies on a number of essential processes and cycles to keep our planet habitable. These include food-cycle processes, such as photosynthesis, predation and decay; mineral-cycle processes, such as soil formation and erosion; and nutrient cycles, such as water, carbon, oxygen and nitrogen cycles. These are all essential for ecosystems to function successfully.

All living things need nitrogen to build amino acids and proteins. Nitrogen forms 78 per cent of the atmosphere, but plants and animals can't extract it from the air. Instead, they rely on bacteria to convert it to forms they can use. Liquid freshwater in Antarctica is typically low in nitrogen, so nitrogen fixation — turning nitrogen gas from the air into other forms such as ammonia and nitrates that can be taken up by plants and animals — is an important part of the whole nutrient cycle in these polar wetlands.

Cyanobacteria produce special cells called heterocysts. These capture nitrogen from the atmosphere and convert it to ammonia, which ends up as protein in the cyanobacteria. Rotifers and tardigrades eat the cyanobacteria, and nematodes eat the rotifers and tardigrades. So all animals that live in these habitats rely on cyanobacteria to provide them with nitrogen. Says Clive, 'As the global climate changes, cyanobacteria will continue to play a vital role in the functioning of many wetlands through processes such as photosynthesis and nitrogen fixation.'

Clive doesn't think that any of the predicted outcomes of the current climate forecasts will directly affect Antarctica's cyanobacteria — after all, they've already experienced extreme climate changes through their incredible life history. The real risks are loss of habitat and invasive species. In some places there will be more water, and life will probably move in from other areas:

> There might be more competition from other microbes coming in. Or if planktonic algae such as diatoms increase, they might cloud the water, resulting in less light in the bottom of the ponds, and the mats might disappear or survive only around the edges. But I don't think climate change is going to cause extinctions of the Antarctic's micro-organisms. A lot of them are widespread across the polar regions.

Working remotely

It's very expensive and logistically challenging to conduct fieldwork in Antarctica. You can't just 'come back Thursday' if you forget a bit of gear or some crucial piece of equipment breaks down. You've got one chance to get things right, or a major operation is wasted.

Field seasons are crammed into an intensive window of a month or two over summer — as a biologist, says Clive, 'you need to go there when there's liquid water'. Summer air temperatures are tolerable, hovering just above and below freezing point, and the sun never fully sets; it just rotates low on the horizon around the South Pole. The teams set up a tent camp near their field location and stay there for weeks at a time.

The sort of work Clive has conducted on the ice shelf to study microbial life is essentially the same as he has used in studying New Zealand's freshwaters — measuring photosynthesis, respiration, nitrogen cycling and nutrient uptake — and that requires complex experiments and sophisticated equipment. Microbial ecology and photosynthesis are fast-moving international fields, so for Clive and his team it was

Sampling a deep pond on the McMurdo Ice Shelf requires careful weight distribution and insulation when the ice is thin. Water samples are pumped up from a series of pond depths through the black tube inserted through the ice cover and stored for return and final analysis in New Zealand.

crucial to ensure their experiments and findings were up to date and publishable. The challenge lay in preparing the equipment for this type of work in the field, getting to the field camp and setting everything up knowing that there was no opportunity to get spares. As Clive points out, 'Innovation is key. It always amazed me what could be done with bamboo poles, duct tape, insulating tape, plastic bags, take-away dishes, and rolls of wire of varying thickness combined with electrical multi-meters. We seldom had complete failures and we always had a "Plan B".'

The researchers collected and prepared (filtered or preserved in ethanol, for instance) water and microbe samples in shipping-container laboratories, and then sent them back home for more-sophisticated analyses. They took samples from the same locations over many years, to measure changes, and conducted field experiments on site where possible. And they always carried out field tasks in pairs for safety and convenience — except for one 'hairy, never-to-be-repeated' occasion Clive recounts:

> Rather than wake one of my colleagues for the 2 a.m. sampling run (it was midsummer, 24 hours of daylight), I decided to head out alone. On the way back, an ice-bridge over a tide crack that we had crossed every day suddenly gave way, plunging me in full polar clothing into the freezing water. The sides of the tide crack were smooth ice. There was no point crying for

> help; I was too far from camp. I'm not exactly sure how I managed to get out, but when adrenaline surges you can do amazing things. I found a rough edge along the top of the tide crack that was enough to give my fingers purchase. That allowed my other hand to get to the top of the ice crack and I somehow pulled myself out. I knew that was my one chance, as at the time my polar clothing (including boots) was sodden and I still had a backpack on, weighing me down. It took me a while to recover from the shock and cold back at the camp, but never has a double-layered down sleeping bag felt so good! Safety lesson learnt — never go sampling in the field alone!

For Clive, science is always exciting — every new point on a graph and every new figure in a table constitutes progress, as long as it is written up and accessible, contributing to our global body of scientific knowledge. A number of New Zealand institutions have been involved in the research on the McMurdo Ice Shelf over the years, including Crown Research Institutes such as NIWA; Auckland, Waikato and Canterbury universities; the Cawthron Institute; and a significant international collaboration from the United Kingdom, United States, Germany, Canada and Spain.

Clive has hosted many distinguished visitors to Antarctic wetlands over the years, including mountaineer and explorer Sir Edmund Hillary, prime ministers and several ministers of the Crown. Explaining the significance of pond-bottom 'slimes' to such visitors was always a challenge, especially when it came to explaining the scientists' enthusiasm and why they would camp in a remote area for weeks just to study microbial life. As Clive puts it, 'It's always interesting when you pick up a bit of slime and put it in a visitor's hand, but the reality is, this is what life is away from the sea edge in Antarctica.'

The importance of wetland biodiversity in temperate and tropical landscapes has been known for a long time. However, loss of wetlands is continuing across the world, and the role of polar wetlands (including in the Arctic tundra) is still under study. In the journal *Antarctic Science*, Clive and two colleagues made a plea for better recognition and protection of Antarctica's wetland ecosystems, which Clive summarises:

> The international conservation literature is full of stories of the megafauna, the need for conserving penguin colonies, seals and birds. But conservation of inland water habitats has been neglected in Antarctica. Since the formation of the Antarctic Treaty, way back, protected areas have been an important component of Antarctic policy. Some are there for historical purposes, but most are to protect areas of wildlife breeding, terrestrial habitats, glacial or landscape features. We believe people need to think more about the importance of inland water ecosystems in considering Antarctica's biodiversity. However, there is currently limited protection offered to inland waters across the continent and we think that is something that needs to be concentrated on now, in terms of the designation of new protected areas.[11]

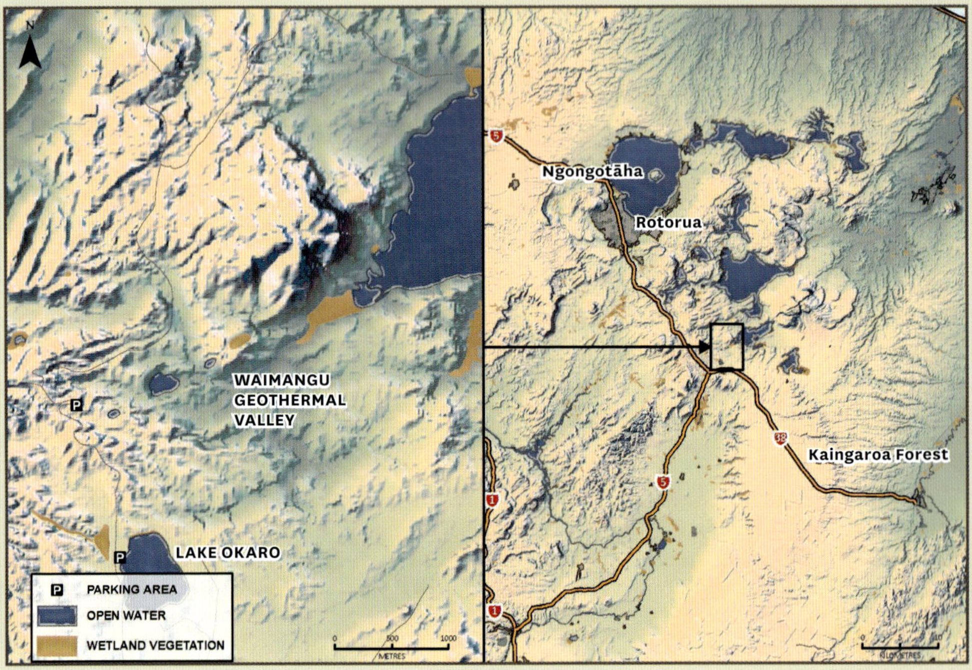

Waimangu Valley

Geothermal wetland

A visit to Antarctica is out of reach for most people, but you can experience extremophile cyanobacteria closer to home, in New Zealand's geothermal areas.

Says Clive, 'Waimangu Valley is one that I would really recommend people go to if they get a chance to have a look at mats of coloured cyanobacteria. These are in warm, shallow waters, at the other end of the temperature spectrum to the Antarctic mats, but they look pretty much the same.'

The valley's geothermal features were formed by the eruption of Mount Tarawera in 1886, a tragic event that claimed 150 lives. The massive eruption split the volcano in two, greatly expanded Lake Rotomahana, destroyed the famed Ōtūkapuarangi and Te Tarata Pink and White Terraces, and formed seven craters that now make up the Waimangu Volcanic Valley.

Highlights

- Brilliantly coloured cyanobacteria
- Inferno Crater, which rises and falls as it heats and cools
- Frying Pan Lake, one of the largest hot-water springs in the world

- Tiny springs and geysers, and delicate sinter terraces
- Rare and unusual plants growing in heated soil, including geothermal kānuka (*Kunzea tenuicaulis*) and thermal ladder fern (*Nephrolepis flexuosa*).

Access

There is an entry fee to access the geothermal features, which lie within Waimangu Scenic Reserve. Guided or self-guided walks through the valley are available, with an option to walk down the valley and bus back up to the car park. There is also an optional boat tour.

Entry to the reserve is at 587 Waimangu Road, a 20-minute drive south of Rotorua via State Highway 5.

www.hikespeak.com/trails/waimangu-volcanic-valley-rotorua-nz
www.waimangu.co.nz

Feathers versus Fur

Do wetlands keep invasive predators at bay? Dr Colin O'Donnell's research reveals just how vulnerable wetland bird species are to non-native mammalian predators.

Dr Colin O'Donnell

Specialist field: Ecology, conservation and management of our native birds and bats

Currently works for: Department of Conservation, Ōtautahi Christchurch, as a Principal Science Advisor for Ecosystems and Species

Colin O'Donnell, MSc, PhD, is a wildlife ecologist who over the past four decades has written 180 reports and scientific papers covering everything from threat classifications to bird and bat ecology, conservation and management, monitoring methods and the impacts of invasive species on indigenous fauna. Away from the comforts of his office, he spends as much time out in the field as possible — high up in the alpine zone, far away in remote forests and waist-deep in wetlands. Braided rivers, a special type of wetland, have been the focus of much of his work to protect rare birds.

WHEN COLIN O'DONNELL was young, his father announced: 'You'll never make a bloody living bloody birdwatching . . . get a real job!' Colin likes birds. A lot. So much so that a big part of his job involves studying the bloody things. It took some years for his father to get used to the idea, let alone talk to him about it again.

While he was still a teenager, a serendipitous meeting with Brian Bell (then working for the New Zealand Wildlife Service on threatened species management and conservation) convinced Colin that becoming a wildlife scientist was the way forward — a job he never even realised existed. Now that he is an established wildlife scientist tasked with ensuring the survival of vulnerable native birds, a primary question Colin grapples with is: 'How can we detect non-native predators and control them in wetlands?' The answer is complex. It starts with the mammals that were brought to our shores. They spread quickly throughout the country, spelling disaster for our naive native animals, which were unused to the teeth and claws of furry predators attacking from the ground.

The impacts of introduced predators on New Zealand's native species have been well studied in forest, riverine and coastal habitats. This is less true of wetlands, however, where research has been hampered by the perception that mammals don't like getting their feet wet.

UNWELCOME ARRIVALS

Before humans set foot in Aotearoa New Zealand, birds filled the forests, plains and wetlands with their songs, squabbles and squawks.[1] Free of land-based mammals save for three tiny pekapeka (species of insect- or nectar-feeding bats), the birds evolved in splendid isolation. They ranged in size from the terrifyingly huge Haast's eagle, which hunted flightless moa, down to tiny titipounamu (riflemen) weighing no more than a pencil. Without mammalian predators, many of our birds became ground-dwellers — their predators were other native birds. There was nothing to worry about on the ground until people turned up, along with their cargo of furry friends and fiends.

Extinctions began with the first Polynesian settlers around 700 years ago.[2] The kurī (Polynesian dog) and kiore (Pacific rat) they brought with them were responsible for wiping out some smaller species of birds, frogs and lizards, while larger species became food for humans. From the late eighteenth century, a stream of European ships brought cats, dogs, ferrets, weasels, stoats (shown above), possums and pigs. Once the ships docked, stowaway rats and mice simply flowed out into the landscape. The tally of destruction wrought by these mammalian predators is horrific: at least 51 species of birds, a bat, three frogs, three lizards, a freshwater fish, four species of plants and an unknown number of invertebrates.[3] The legacy of these unwelcome arrivals continues, with many more of our remaining indigenous fauna and flora still under threat today.

The 'wet feet' myth has been firmly dispelled by Colin and his colleagues, who have confirmed that furry predators such as rats, stoats, cats and possums roam freely in and around freshwater wetlands preying on native birds.[4] They first conducted a literature review to look for evidence, but not all records need to be formally published in a scientific journal to be useful. What scientists call 'grey literature' — old reports housed in creaking filing cabinets or piled onto shelves, unpublished datasets and information collected by citizen scientists — can contain valuable historical records. Bring it all together, and a compelling case can be made.

They started with predation records held by the Department of Conservation (DOC), then looked at wetland bird databases (around 15,000 records) and the decades of observations made by citizen scientists through the Ornithological Society of New Zealand recording schemes.* They leafed through books on animal behaviour, studied scientific journals and noted reports of dead wetland birds brought into houses by pets. They also found plenty of documented forensic evidence, including nests empty save for shattered shells, bones and feathers.

From their research, Colin and his colleagues confirmed the worst: the usual suspects are equally at home in wetlands as they are in other ecosystems. These include the voracious trilogy of mustelids — stoats, ferrets and weasels — along with rats, mice, possums and cats (both domestic and feral). Surprising to some people, the bumbling hedgehog is on the list too. This natural-born egg thief (partial also to chicks, insects and native lizards) is as comfortable in forests, wetlands and the high country as it is in your garden.

Sitting ducks

Colin's fascination with wetlands — particularly Canterbury's high-country braided rivers, lakes and wetlands — began with a summer job in the Fauna Survey Unit at the New Zealand Wildlife Service. The work centred on a major bird survey of the vast upper reaches of the Rakaia River. Resembling a military operation, the project saw a large team combing the shoreline while others scanned the landscape from a jetboat. They counted and identified every bird they saw — crucial information for establishing the area as a Site of Special Wildlife Interest.** However, giving an area a title and stating this in a report doesn't protect it from non-native predators.

Many New Zealand wetland birds nest on the ground. The species that inhabit braided rivers barely bother making a nest at all — just a slight hollow among the rounded stones and gravel will do. While being hidden away in a swamp might offer some protection, birds that nest in open-air colonies are sitting ducks for predators. Colin explains:

---

* www.birdsnz.org.nz/research/#projects.

** The Sites of Special Wildlife Interest (SSWI) is a system of ranking natural or semi-natural areas on the basis of the importance of their habitat for one or more species of wildlife. The system and standard ranking criteria were developed by the Fauna Survey Unit of the former New Zealand Wildlife Service and continue to be used today by DOC.

A lot of predation happens at night when the birds are probably asleep or incubating, and so birds can get chomped very easily, but even if they escape, their eggs get eaten. When a predator goes through a colony, most of the other birds abandon their nests because they realise it's bad. One instance really sticks in my mind — when we followed the nocturnal footprints of a cat through a colony of black-fronted terns on the Rangitata River. This single cat devastated the colony, either preying on adults and their eggs, or causing sitting adults to abandon their nests. The result was a loss of 76 per cent of nests during a single visit.

Colin and his team have used video surveillance cameras and, more recently, motion-sensor-activated trail cameras, to spy on predators in his treasured South Island braided rivers. Film footage and photographs reveal the main culprits: cats, stoats, ferrets, hedgehogs and the occasional rat. Identifying the main predators paves the way for using the right kind of traps, designing how they are placed in the landscape and choosing which types of bait and toxin to use.

They found that most of the predators — cats, and especially rats and mustelids — aren't put off by water at all. Swimming a kilometre isn't a big deal for a Norway rat,[5] and stoats can easily cover 1.5 kilometres (one athletic individual was spotted an impressive 5 kilometres offshore, making a beeline for the bird haven of Kapiti Island).[6] As for cats, most people still believe they don't like to get their feet wet, but Colin thinks personality also plays a part: 'One cat may get to a channel and simply back off. Another won't hesitate to leap in the water and swim decisively to an island across the channel.' The only predator stumped by water is hedgehogs — but that doesn't stop them from feasting on any riverside nests they encounter.

A game camera captures a rat raiding a dotterel nest at night.

Colin doesn't mince his words: 'If these predators aren't managed, they will lead to the extinction of some of our most unique species.' The first time he saw their impact was in the early 1980s, when kakī (black stilt) eggs were taken off their braided-river nests for safekeeping to protect them from predation. Dummy eggs were left in their place, with chicks returned once they had hatched. During one nest check, Colin spotted a dead weasel; unable to crunch the dummy egg, it had swallowed it whole!

While the fast-flowing channels of a braided river don't deter all predators from swimming to river islands, they do lower the chances of many getting across to nesting colonies. However, the establishment of weeds on the islands adds another dimension — literally. Once predators swim to an island, they may become resident there if the plants provide enough shelter for them to hide in. If there's no shelter, they soon move on.

There's also a rich tapestry of tenacious native plants that grow on the shifting islands and floodplains of braided rivers. These days, they're mostly found in the upper reaches where fewer changes have been wrought by humans in their quest to tame the land.* The plants still hanging on include vegetable sheep (*Raoulia* species), miniature sedges and herbs, grasses, annuals and endangered species that include a tiny forget-me-not. Colin, clearly more knowledgeable about birds than plants, summarises the flora as 'a real richness of obscure, small mat-like things that botanists like crawling around on their hands and knees to look at'.

Braided river basics

Braided rivers are just that: channels that meander and cross over one another, like a fraying rope that connects the mountains to the sea. Globally they're rare because a specific set of conditions needs to come together to create them: steep, crumbling mountains producing a never-ending supply of gravel, and enough water to carry the constant flow of rock and sediment across the land to the sea.[7] New Zealand has one of the best collections of braided rivers in the world, with more than 160 examples mapped.[8] Most of these are in the South Island, but there are a few in the North Island.

Never still, these restless rivers shift and change as gravel builds up, water spills over, and channels move from side to side looking for easy flow paths across the plains. The constant movement of gravels and shifting sediment creates a mobile landscape of islands, plains, river mouths and coastal beaches. Water flowing underground creates springs and wetlands that pop up and then disappear, and the whole riverscape constantly redefines itself, sculpting new islands, ponds and channels. As you fly over the Canterbury Plains, you can see ghost reaches of braided rivers creating a subtle topography under the farmland.

A unique flora and fauna has adapted to this changing environment. It's a risky place to nest, but surprisingly rich in food for native birds, fish and lizards. A huge number of bird species — around 80 — use braided rivers during some part of their life cycle.[9]

For Māori, braided rivers such as the Waimakariri north of Christchurch are historical

---

* Years of development have left a legacy of weeds in lowland sections of Canterbury's braided rivers.

pathways to Te Tai Poutini, the West Coast. The river itself is considered a tūpuna (ancestor), a taonga (treasure) and a mahinga kai (source of food). But to early European settlers, braided rivers were a 'dreaded enemy' because their floods threatened towns, farmland and livelihoods. This attitude sowed the seeds for how these rivers are managed today: constrained within stopbanks, and with commercial activities such as farming and quarrying covering the floodplains, moving ever closer to the main channels.

Birds on the move

Despite the seemingly inhospitable environment of braided rivers, they're home to some of our unique, highly threatened birds, some of which are only found in Aotearoa New Zealand. These include kakī (black stilt), tara pirohe (black-fronted tern), tarāpunga (black-billed gull), ngutu parore (wrybill) with its beak bent sideways (always to the right), globe-trotting tūturiwhatu (banded dotterel) and tōrea (South Island pied oystercatcher). Many of the birds that rely on braided rivers are there only in the breeding season. They overwinter in the more benign northern North Island harbours and coastal areas, such as the internationally significant Firth of Thames. Some tūturiwhatu, however, head offshore, travelling 1600 kilometres or more to Tasmania and southeastern mainland Australia.[10] The birds flock back to the braided rivers in early spring, when the snowmelt in the headwaters kick-starts seasonal production in the channels, leading to an abundance of aquatic insects and small fish.

Colin explains how these specialised birds are well adapted to breeding on braided rivers: 'They nest relatively early in the spring compared to most birds, with the earliest arriving in July to take advantage of the spring feeding conditions, and many would have finished by Christmas. In fact, the earliest migrants start moving off the rivers by November.'

To cope with frequent floods in spring, the braided-river birds can produce replacement broods incredibly quickly, even within days of losing a nest. To describe the chicks as precocious is an understatement; according to Colin, they can run and swim at birth and feed independently, managing to avoid some of the smaller floods. They can also move to new breeding sites if, by bad luck, the island they've made their home suddenly disappears in a flood. Added to this is their ability to find the unpredictable but rich seasonal bursts of food: 'When mayflies emerge en masse, black-fronted terns have the uncanny ability to find and zero in to feed on these patches. They do the same when the river rises following a hot day of snowmelt, when "drifts" of mayfly larvae are washed downriver after being dislodged from under stones.'*

But what happens when two problematic predators of braided-river birds are native species themselves? Both karoro (southern black-backed gull) and kāhu (swamp harrier) have caused conservation managers more than a few headaches. The problem with doing a good job of controlling introduced mammals (which mostly prey on nests early in the

---

* To read more about braided river wildlife, see www.doc.govt.nz/Documents/conservation/native-animals/birds/sea-and-shore/life-on-a-braided-river.pdf.

Black-fronted tern chicks, like this one on Rakahuri Ashley River in Canterbury, can run and swim from birth.

breeding season) is that karoro and kāhu can move in later, preying on the large chicks. The general view is that the natural predator–prey relationship is now out of balance, because the numbers of karoro and kāhu are now much higher than they would have been before the arrival of humans. Colin explains why: 'Nowadays, huge numbers of black-backed gulls are sustained by things like sewage-treatment plants and by some of the big, intensive feedlots on the Canterbury Plains. Pig farms attract zillions of them. We think the harrier numbers are high because there are roadkills everywhere . . . and rabbit carcasses and things like that. There wasn't originally that much food around for them.'

This means that both species (especially the gulls) need to be controlled in places where they threaten wildlife that has already been devastated by non-native predators. It also means pulling out all stops to prevent further declines in numbers of the special birds that Colin and many others around the country are dedicated to protecting. An important part of the equation is integrated predator control, where the full suite of potential predators is targeted rather than just focusing on one or two.

AVIAN ASSASSINS OF AOTEAROA

In an ecosystem once dominated by birds and not mammals, there are plenty of examples where one bird chooses another for a meal. The most awe-inspiring, if not completely terrifying, was Haast's eagle, which was nearly 1 metre tall and had a wingspan of up to 3 metres. These raptors haunted the skies and forests of Aotearoa in search of the flightless moa, locking them in their grasp with feet and claws as 'big as a modern-day tiger's'.[11] They had to be large and strong — one species of moa weighed up to 250 kilograms! In Māori oral tradition, the eagles are pouakai, huge birds that were large enough to attack human children. However, the splendid evolutionary isolation of Haast's eagle and moa ended when Māori settled in Aotearoa in the late thirteenth century. By the 1400s, all nine moa species and their eagle predator had become extinct.

The main avian predators today are an owl, a gull, a harrier, a falcon and, on occasion, a parrot. The diminutive ruru (morepork) is our only remaining owl after the whēkau (laughing owl) was extirpated less than a century ago through predation by the plague of stoats, ferrets and weasels. The ruru diet includes small mammals and birds, especially tauhou (silvereye), along with large insects. The kārearea (New Zealand falcon) can tackle birds larger than itself, including kawau (black shag), but usually sticks to smaller birds and the occasional rabbit or small hare.

The cheeky kea, the world's only alpine parrot, feeds mainly on invertebrates, fruit and nectar — and the rubber fittings on rental cars when holidaymakers stop to take photos of its stunning mountain habitats. But it has also been seen preying on Hutton's shearwater chicks, dragging them from their burrows.[12]

Karoro (southern black-backed gulls) and kāhu (swamp harriers) make a meal of braided-river birds. Karoro are the largest of all gulls in Aotearoa and the only native bird species that has no protection under the Wildlife Act 1953. They get in the way at airports, where they are responsible for most aviation bird-strikes, while on farms they sometimes attack newborn lambs. Their preferred native fare is tūturiwhatu (banded dotterel), tara iti (fairy tern) and tūturuatu (shore plover).

Kāhu are often seen picking at roadkill. They prey on mice, rats and rabbits, are also partial to poultry and gamebirds, and on occasion will kill newborn lambs. They are partially protected under the Wildlife Act, so it is legal to control them in special circumstances, such as when protecting other native birds.

Predator 'selfies' — native predators including kāhu (left) and karoro (right) have been captured on game cameras taking dotterel eggs or chicks.

Success versus failure

Predator control in wetlands (as elsewhere) is a process of continued trial and refinement. Which traps are effective? At what densities? What is the best timing of control? Which baits and lures are effective and in what seasons? There are logistical challenges too — such as when traps or bait stations are washed away or submerged in floods. Colin elaborates:

> We know that we don't get all predators with predator control. You might have a scenario where there's little predation from certain predators, but one or two are big-time predators, so you might amp up your control of those or target control for different things. Basically, you adapt your management as you go. Things like rats only turn up once in a blue moon on braided rivers, so you may need to introduce rat control just for that season. Usually the [trail] cameras give you that information, on when and what to target.

It was in the Eglinton Valley, north of Te Anau, that Colin and his team proved that predator trapping could make a real difference to the survival of native wetland birds.[13] Eglinton is one of the only road-accessible valleys in Fiordland National Park. It's a fabulously scenic area, along the route to world-renowned Piopiotahi Milford Sound and the setting for the much-photographed Mirror Lakes. Although the area is remote and the Eglinton looks like a pristine braided river, all of the usual introduced predators were common in the area before Colin and his team arrived in the early 1990s.

They began by setting up traps in just 50 hectares of the long, narrow valley. Today, a line of traps runs the whole length of the valley — about 50 kilometres. The results? A slow but steady increase in the numbers of the endangered tara pirohe (black-fronted tern, *Chlidonias albostriatus*), from just one or two pairs attempting to breed in 1992 to more than 100 pairs today.[14]

Further north, another braided river, the upper Ashburton, has been monitored for wetland birds since 1982. Colin says, 'It got quite depressing. Year after year we recorded slow but steady declines in numbers of dotterels, gulls and terns breeding on the river — and increasing weed infestation covering the nesting habitats.' Over the past two decades, however, a predator-trapping network set up in the river's upper reaches has led to a slow but steady increase in tūturiwhatu.[17] Turning back the tide of predators is a long game, as Colin reminds us: 'It's taken about 15 years for [banded] dotterel numbers to double, which on the one hand is a constant reminder that predator control needs to be intensive and ongoing — but also that projects like the Eglinton and upper Ashburton are proof that you can make a difference with these predator control programmes.'

While these are success stories for our native birds, there have also been failures. Over the past 30 years of putting programmes in place to control predators in and around

KAKĪ — A BIRD ON LIFE SUPPORT

Being the rarest wading bird in the world isn't really something to be proud of. When a species is down to its last 23 individuals, it's clear that serious intervention is needed. The kakī (black stilt, *Himantopus novaezelandiae*) has been intensively managed since 1981, when its population declined to that 'practically extinct' level.[15]

To protect the remaining birds from predators and boost breeding success, DOC set up a captive breeding centre near Twizel in Te Manahuna Mackenzie Basin. Kakī build precarious nests — a mere scrape among the gravel of a braided river — making them especially vulnerable to passing predators, the wheels of off-roaders and unseasonal flooding. The conservation work involves collecting and incubating eggs, hand-rearing the chicks and then releasing them back into the wild. At the last count, kakī numbers had jumped to 169 adults, with nearly 200 chicks released in 2019, another 100 in 2020 and 150 in 2021.[16] If these efforts are kept up, the kakī may get pushed off its perch as the world's rarest wader — and that will be worth celebrating.

braided rivers, many haven't been able to show clear gains for native birds. In some cases, the already tiny populations of endangered birds that conservation managers sought to protect have just got smaller.

In conservation, there's no single game-changer that flips a predator-control programme from failure to success. Decades of research by Colin and his colleagues have shown that we need to think big right from the start. The reason that cats, mustelids, hedgehogs and rats are so plentiful and widespread is because they are incredibly skilled at finding food — even if that food is a small bird population tucked into a remote corner of the landscape. Intensely controlling all predators at the landscape scale, and using all available humane methods, is the only way to go.

Then there's the science — the ongoing detailed monitoring of what, how and when predators go in for the kill. This is the vital information that underpins predator-control programmes, shaping and refining them. The pace of predator research has picked up markedly since Aotearoa made the bold pledge to go predator-free by 2050. Innovative

A pair of adult kakī wading in the Godley River, near Lake Tekapo. Coloured leg bands enable DOC staff to identify individual birds.

Colin O'Donnell releases a banded dotterel after placing identification bands on its legs.

ideas are being developed all the time, some of them designed to take advantage of the natural curiosity of predators. One effective example is a new lure that attracts stoats and weasels to traps. It's simply a scrap of fabric soaked in the urine of a bigger predator, but draws on that primal urge to know thine enemy.

In 2021, scientists from Manaaki Whenua — Landcare Research reported on another finding that might protect nesting wetland birds in areas where traps are not suitable: you can train wild predators to believe that the delicious aroma of a braided-river bird will end in nothing but disappointment.[18] Scientists made up a slurry from bird feathers — intoxicating to wild cats and ferrets, but they soon lost interest when they repeatedly discovered that there was no prey reward. By training the predators resident in an area before the start of the breeding season that a bird scent equals an empty stomach, they turn up their noses when the real birds arrive to nest and ignore them in favour of more reliable meals.

PREDATOR FREE 2050

Predator Free 2050 (PF2050) is a bold vision for the whole of Aotearoa New Zealand: to completely eradicate rats, stoats and possums, the country's three main introduced predators, by the year 2050. It's crazy, it's ambitious and it's been described as our 'moonshot'.

The programme, supported by government agencies and the science community, has captured the imagination of the public. Pest-trapping projects have sprung up nationwide, ranging from people laying traps in their backyard as part of a neighbourhood initiative, to citywide projects, to others spanning huge swathes of the countryside or even an entire inhabited island, such as Waiheke near Tāmaki Makaurau Auckland.

PF2050 builds on six decades of dedicated predator-control and eradication work, which has cleared more than 110 of our offshore islands, large and small, from predators.[19] Lessons have been applied to both fenced wildlife sanctuaries such as Zealandia in Te Whanganui a Tara Wellington, and unfenced sanctuaries such as the Eglinton Valley in Fiordland, where Colin has worked. Techniques for controlling and monitoring predators keep evolving, constantly refined to make them more efficient and to bring us closer to our goal of being a nation free of introduced predators.

Although Colin freely admits that parts of his long journey to protect native birds have been depressing, the successes and scientific advances along the way have been exhilarating.

Colin has come a long way from his self-directed studies as a geeky teenager mapping gull nests on the Boulder Bank in Whakatū Nelson, comparing the feeding behaviour of tūī and korimako (bellbirds), and recording the time of year ruru called near his home. Colin joined the Ornithological Society of New Zealand (Birds New Zealand Te Kāhui Mātai Manu o Aotearoa) at an early age and found their meetings and field trips inspiring, cementing his interest and desire to become a professional wildlife ecologist. He also worked as both a volunteer and on short-term contracts for the Wildlife Service and Forest Service (later amalgamated as DOC) while studying. This built valuable contacts and gave his CV greater depth beyond academic experience. He summarises his advice to budding scientists: 'Overall, it helps having something you are passionate about . . . so follow that passion if you have already discovered it . . . or experiment and experience things until you do.'

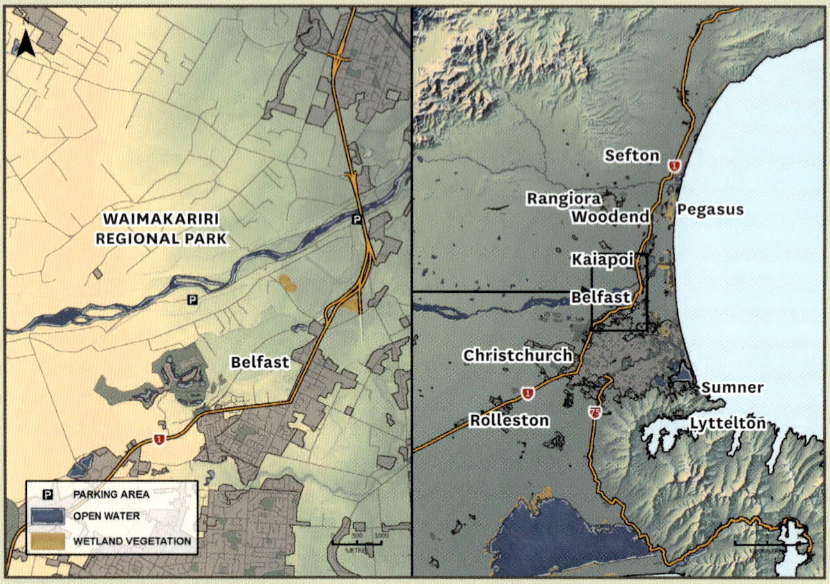

Lower Waimakariri River

Braided river and spring-fed wetland

The Waimakariri is one of the largest braided rivers flowing across the South Island's Canterbury Plains. The icy headwaters originating in Kā Tiritiri-o-te-Moana the Southern Alps gave the river its name: wai (water), makariri (cold). The river was a traditional mahinga kai (food-gathering area) for Ngāi Tūāhuriri and also formed a route for those travelling to Te Tai Poutini the West Coast or accessing the food-rich lakes of the Canterbury high country.

Much has changed since European colonisation. Before then, the vast river frequently spilled out over the surrounding land, including where the city of Ōtautahi Christchurch now sits. Although the lower reaches were straightened in the early twentieth century, creating faster, deeper channels, flooding still occurs, typically covering the lower shingle islands. The islands high enough to avoid flooding have become weedy and support colonies of karoro (southern black-backed gulls) — you might see them harassing other birds.

The establishment of the Waimakariri River Regional Park in 2005 recognised the importance of the river as bird breeding habitat but, being heavily modified and on the city fringe, it's a tough environment for native species. The Waimakariri Braids trail heads through the park. It also forms a section of the 130-kilometre Christchurch 360 Trail, developed by ecologist Dr Colin Meurk to showcase

the diversity of the city's surrounding landscapes. While on the track you will also pass through the Sanctuary wetlands, the largest wetland area on the lower Waimakariri River. The swamp is spring-fed and there are still remnant patches of the original vegetation under the introduced willows and poplars.

Highlights

- Along the Waimakariri Braids trail, look out for endangered tarāpuka (black-billed gull), ngutu parore (wrybill) and pārera (grey duck). Common species to see include pāpango (New Zealand scaup) and pūtangitangi (paradise shelduck)
- Through the Sanctuary wetlands, keep an eye out for koitareke (marsh crake) and matuku (Australasian bittern).

Access

The Waimakariri Braids section of the Christchurch 360 Trail is 18.8 kilometres long and will take around 7–8 hours to walk. It runs along the southern shores of the river and can be walked or cycled in either direction.

Start at Whites Crossing, next to the confluence of the Otukaikino Creek and the Waimakariri River, accessible from Coutts Island Road. This part of the Christchurch 360 Trail follows the existing Templars Island Track west through part of the Waimakariri River Regional Park.

www.kahurumanu.co.nz/atlas
http://christchurch360trail.org.nz/trail-segments/waimakariri-braids

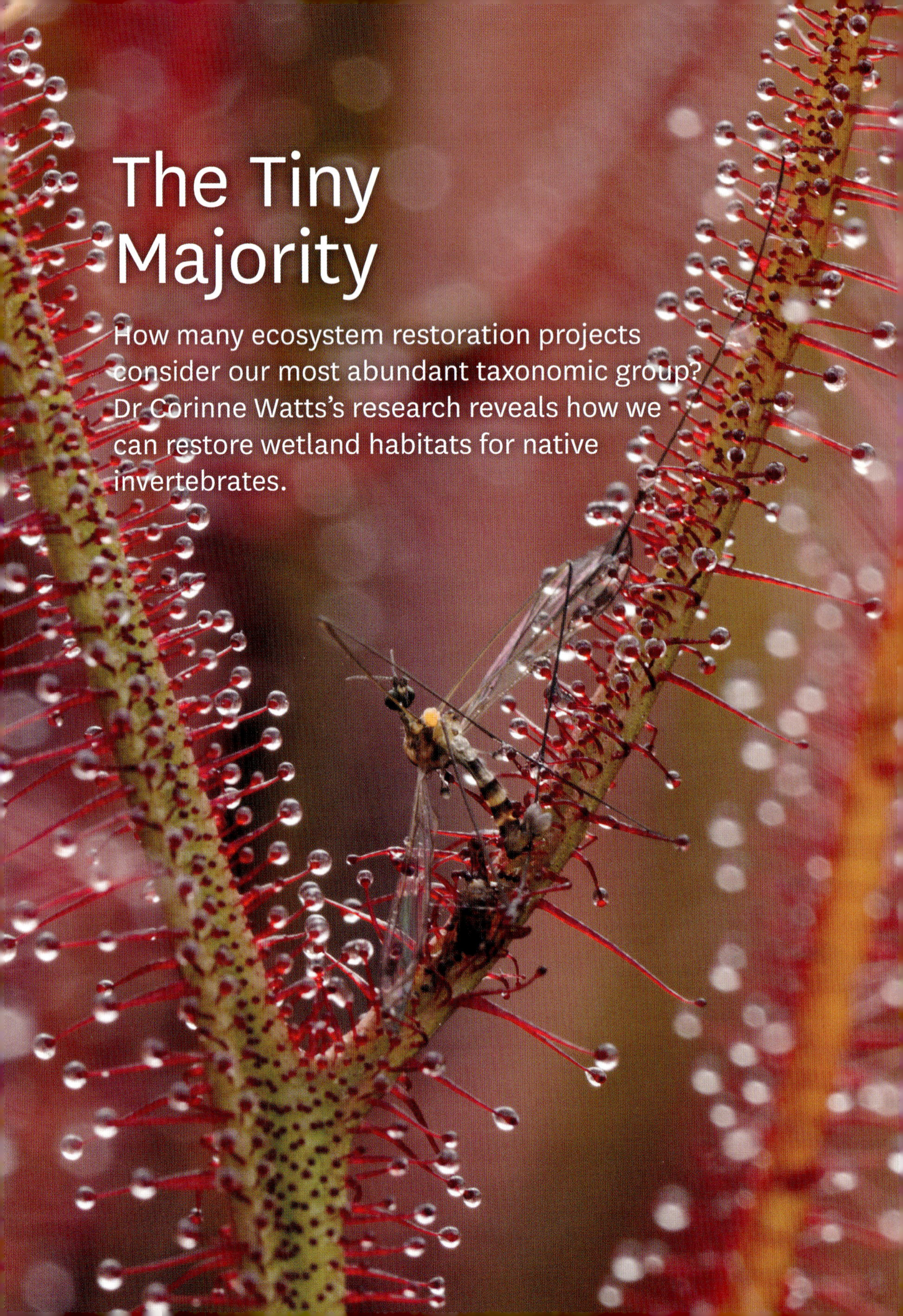

The Tiny Majority

How many ecosystem restoration projects consider our most abundant taxonomic group? Dr Corinne Watts's research reveals how we can restore wetland habitats for native invertebrates.

Dr Corinne Watts

Specialist field: Invertebrate ecologist specialising in wetland insects and their restoration

Currently works for: Manaaki Whenua — Landcare Research, Kirikiriroa Hamilton

Corinne Watts, MSc, PhD, is an applied ecologist focusing on invertebrate communities and their restoration. She has studied and published on invertebrate communities in wetlands, including the impacts on native invertebrates of biocontrol agents for the ground cover plant *Tradescantia* in Northland wetlands, invertebrate responses to willow removal and herbicide use, and invertebrate community restoration after peat mining. She discovered and described *Houdinia flexilissima* (aka Fred the Thread), the world's skinniest caterpillar.

MOST PEOPLE WHO consider themselves animal lovers don't extend their affection to the non-furry members of the animal kingdom. Corinne Watts isn't one of those people. She was one of those kids who collected monarch butterfly caterpillars and bugs for pets, but it was at university where, inspired by the enthusiasm of entomologist Dr George Gibbs, she became seriously fascinated with insects.

Professionally, says Corinne, while Gibbs is considered an entomologist — an expert on insects — 'actually, he pretty much knows all there is to know about the flora and fauna of New Zealand'. His ability to grasp the big picture and yet also specialise set Corinne on the path to becoming one of New Zealand's few invertebrate ecologists.

Corinne describes herself as someone with a broad general ecological knowledge, combined with a passion and deeper skill set centred around insects. 'A lot of entomologists I interact with have really got into the taxonomy of a certain group . . . that's never floated my boat.' Instead, she became an applied ecologist — someone who makes scientific findings and then applies them to an immediate real-world need. In her case, this has opened the door to a new area of science, namely restoring ecosystems for the most abundant and least appreciated animal group — insects, the tiny majority.

Up the creek without a backbone

Conservation in Aotearoa New Zealand has tended to focus on the 'K' species — kiwi, kākāpō, kōkako. If you've got feathers, you've got an audience. If you're a tiny, spineless critter, your chances of grabbing the limelight — and the conservation dollar — are pretty slim.

Wetlands have also struggled to capture the public's attention. Māori have valued wetlands for centuries, but it has taken a while for the rest of New Zealand to catch up. In the past decade, though, more people have become aware of the importance of freshwater wetlands and also their drastic reduction (down to the last 10 per cent), and there's been a groundswell of interest in habitat restoration. Wetlands are finally having their day in the sun, with hundreds of landowners, iwi and community groups now planting natives, dealing to pests and restoring water levels.[1]

It has often been assumed that once the land is re-wetted and the plants have become established, native animals will magically reappear. This assumption even has a name — the 'Field of Dreams' hypothesis, which comes from a classic line in the eponymous 1989 Kevin Costner movie: 'if you build it, he will come'.[2] However, our remaining wetlands are scattered far and wide. Long-ranging birds such as matuku (Australasian bittern) may well turn up at some stage, but less mobile or more isolated species like pāteke (brown teal) likely won't. To get them back into the landscape, some bird species have been deliberately bred or captured, airlifted across the country and then released into a restored wetland.

Recovery of invertebrates, though, is rarely considered in wetland restoration projects, despite them being essential workers in every ecosystem on Earth. In Aotearoa, an estimated 50,000 or so species of invertebrates toil daily, aerating soil, breaking down leaf litter, recycling nutrients, pollinating plants, spreading seeds, and feeding native birds, reptiles, frogs and fish.[3] It seems folly to rebuild a wetland without first ensuring that the ecosystem drivers can recover to play their part.

Corinne is one of the few scientists who has looked at how invertebrate communities recover in restored habitats. Her research has revealed that, even after 14 years, restored peat wetland sites are still markedly different from nearby undisturbed comparison sites.[4] Insects aren't generally long-haul travellers. Many native insects have poor dispersal abilities, and active intervention may be needed to reintroduce them to restoration sites — either by translocating the insects themselves, or by literally picking up their homes (features like dead wood or clumps of plants and soil) with the insects inside and moving these to the restoration site. Much of Corinne's wetland research has highlighted the fact that creating a fully fledged, restored wetland requires us to do more than 'just add water'.

Getting into wetlands

If it's hard enough to get wetland restorers into invertebrates, Corinne has found it equally hard to get invertebrate experts into wetlands. As she points out, plenty of entomological surveys have been carried out in forests, but not nearly as many in rarer habitats such as wetlands, gumlands and sand dunes:

> They're those marginal habitats that people just aren't interested in. People don't go into wetlands. They're wet. They're boggy. When I tried to get a taxonomist out into the field in a wetland, it took a *lot* of convincing:
> 'Am I going to need waders?'
> 'No, we're going to a peatland — you can wear your jandals if you want.'
> There's this perception that you're going to be up to your waist in water and it's going to be horrible!

When Corinne signed on as an ecological technician at Manaaki Whenua — Landcare Research's Kirikiriroa Hamilton office in 1999, one of her first tasks was to help sample invertebrates in the wetland research programme led by Dr Beverley Clarkson. Corinne recalls the excitement of those early days:

> We visited some amazing wetlands, all across New Zealand, including the Chatham Islands. Many had never been surveyed for their invertebrate fauna before. Because wetlands have been forgotten ecosystems, their insect fauna is poorly known, and they could be hiding plenty of unknown species. Every time we go into a wetland, we're finding new species and even new genera.

One of the sites they visited was Torehape peat mine near Ngatea in the Hauraki Plains. It would lead to a remarkable discovery — a previously unknown rare moth species, so distinct that it was given its own genus.

Rebuilding a bog

Torehape was once part of a massive bog complex that stretched for more than 30 kilometres across the Hauraki Plains of the Waikato region. The vegetation comprised mostly the wire rush (*Empodisma robustum*) and the giant cane rush (*Sporadanthus ferrugineus*), both peat-forming plants, along with associated bog species such as orchids, tangle fern, sphagnum moss, sedges and the like.

Drainage and land conversion since European settlement has fragmented the large wetland into the 10,000-hectare Kopuatai Ramsar site and several smaller remnants. One of the remnants is Torehape, a squarish site boxed in by dairy farms, comprising a 655-hectare wetland reserve and an adjacent 300-hectare mine site that harvests peat soil for garden potting mix. The mine operators scrape off the native bog vegetation and scoop up the top metre of the mildly acidic peat, helping gardeners across the country cultivate the perfect camellia blossom.

This may sound destructive, and it is, but it's not all bad news. Drainage of wetlands surrounding Torehape to create and maintain farmland has left the site high and dry, and vulnerable to weed invasion. Slicing off the top metre takes the peat surface down to the lowered groundwater level, creating a wetter site favoured by peatland plants. Plus, the resource consent issued by Waikato Regional Council requires the mine company to restore the native vegetation after it has harvested the peat, which will hopefully leave the wetland in a wetter, better state.

Beverley's wetland team was contracted to develop an effective way to kick-start the vegetation recovery on freshly mined surfaces. She and soil scientist Dr Louis Schipper began revegetation trials in 1998, experimenting with various treatments to see which combination worked best.[5] They tried four fertiliser treatments (adding nitrogen, phosphorus, nitrogen and phosphorus, or no fertiliser).* They also tried three seed treatments (mānuka seed, giant cane rush seed, or no seeds), and two water-table treatments (placing seeds directly on the wet mined surface or on drier raised islands of milled peat).**

The initial results showed that the best combination was to place mānuka seeds on raised peat islands with a plug of nitrogen and phosphorus fertiliser. After two years, however, the raised islands seeded with mānuka were at or close to 100 per cent cover with or without the addition of fertiliser. These results led to the standard practice now being used at a large scale across the mine site to stop further peat degradation and restore wetland habitat. The miners create islands of milled peat and cover them with freshly cut mānuka branches heavy with ripe seed. Mānuka seedlings grow quickly on the islands and form a protective canopy that allows peatland specialists such as the giant cane rush to grow. Once established on the peat islands, plants slowly spread via seeds or underground runners onto the lower, wetter mine surface. Within two years the islands are covered, and within six years the mine surface is revegetated with peatland plants.

While the scientists had cracked the revegetation solution, the effect of restoration on the invertebrate community remained unknown. Says Corinne, 'We discussed the importance of invertebrates in restoration projects, and how little is known about that, so I started sampling invertebrates at the Torehape trial site . . . and it turned into a PhD investigating the effects of peatland restoration on the invertebrates.'

Winners and losers

As part of her thesis, Corinne collected insects from the Torehape restoration sites (peat islands of various ages up to 13 years old) and compared these with the insects from the undisturbed areas of the Torehape bog (the reference site). Her studies revealed that during the first 2–6 years of peat mine restoration there is an initial increase in insect

---

* True bogs are naturally low in nutrients. By definition, bogs are wetlands quenched by rainwater only, so they don't receive nutrients from soil, fertiliser and livestock effluent flushed off the land. But when it comes to revegetating bare peat, the research team thought it worth testing whether a small boost of nutrient would give the plants a kick-start.

** Milled peat is loose and crumbly, created using spoon-type harrowers that drive across the surface, scooping up and turning the peat to dry it before harvest.

Tiger beetles were among the first insects to appear on the newly constructed peat islands.

species diversity, but even after 14 years the community composition on the restored peat mounds is still markedly different from those in undisturbed reference sites in the bog.[6]

Corinne found that some insects thrived on the recently mined peat surface, or on the milled peat islands, in the early stages of restoration. These tended to be species that favoured open habitats with sparse vegetation cover, such as the common tiger beetle (*Neocicindela tuberculata*), a leggy ground-dwelling carabid found only in Aotearoa. Tiger beetles should probably be called 'cheetah' beetles, as they are considered to be one of the world's fastest-running insects, with a recorded top speed of 9 kilometres per hour.[7] If that sounds unimpressive, consider that this represents 120 of their own body lengths per second. To match that feat, the world-record-holding sprinter Usain Bolt would have to run at 770 kilometres per hour. This superpower may be the secret to the tiger beetle's ability to recolonise the peat islands at Torehape.

Another beetle that flourished in the early stages of restoration at Torehape was the metallic green mānuka beetle (*Pyronota festiva*), which in Corinne's study turned up on the milled peat islands once mānuka, its preferred food source, had become established. Other species were less successful, particularly small, poorly dispersing native insects such as the rove beetle (*Physobryaxis inflata*) and the ground-dwelling carabid (*Aulacopodus calathoides*). These species may require specific habitats, environmental conditions or plant species that freshly created restoration islands lack in the early years of recovery.

THE PITFALLS OF SAMPLING WETLAND INSECTS

Corinne has mainly used two commonly applied methods to sample insects — pitfall traps for crawling insects and malaise traps (such as the one shown right) for flying ones — but each has its challenges when used in wetlands. 'I always do a recce, a reconnaissance, before we start the project to see how wet it is . . . yep, we can put pitfall traps in to do ground invertebrates or, nope, forget it they'll be floating and useless.'

As the name suggests, pitfall traps are holes dug into the ground that animals fall into and become trapped. To trap invertebrates, small cups are sunk into the ground with a liquid that quickly and humanely kills any captives. A small canopy — a piece of wood or the lid of an ice-cream container — keeps out rain and debris that might help victims escape. Traps are set for one month during summer, when there is a lower risk of captured insects being flushed out of the cup by floodwaters. Malaise traps are like open-sided mesh tents that trap flying insects. Corinne explains:

> Malaise traps are easy to set up — it's just like putting up a tent . . . but in a peatland you don't have trees to tie the guy ropes to. I completely redesigned the malaise trap — I shrunk it, put it on poles, and put it on a wooden platform to make it easier to carry and to withstand the weather and exposure in a wetland. I don't actually bother with standard malaise traps. The mini traps are so easy to flick up, you don't have to worry if there's a big storm coming because they'll survive that.

The insects collected in the traps are preserved in ethanol and taken back to the laboratory to be identified under a microscope. 'Many of the insects we found have never been seen before, so they don't even have a name.' According to Corinne, insect survey techniques are still pretty 'old school':

Both time and distance affect insect recovery at the Torehape mine. Detritivorous beetles took years to show up on the peat islands, probably because of the time needed for the layers of leaf litter and dead wood they feed on to build up under the establishing vegetation. To speed up the process, Corinne suggests that litter and dead wood could be transferred to the peat islands from the undisturbed peat bog, although the success of such interventions has had varied results and requires further research.

Corinne has also tested direct transfer of peatland habitat — lifting and shifting intact clumps of plants and peat from areas of the bog yet to be mined to the mined surface. This more intense restoration technique sped up the invertebrate community restoration, but at a cost of around $12,000 per hectare (compared with $3500 to create mānuka-seeded peat islands) it proved too expensive and unsustainable at a larger scale for the peat miners.

> You collect samples, sort them, stick them under a microscope, identify what you can, count them — it's expensive and time-consuming, and requires expertise in diagnostics. We're trying out new technology, using DNA. You collect a sample of insects, mush them all up and do a DNA test. It's pretty interesting because you get a much wider variety using DNA than what you do in the trap ... as you get the DNA of insects that the trapped insects have eaten!

This may be a breakthrough for community groups that want to see how invertebrates are faring in their restoration site. Emerging DNA metabarcoding techniques could potentially revolutionise invertebrate monitoring because entire communities can be identified from a single, easily collected sample of soil or water.

In another experiment, Corinne investigated whether the distance from areas of undisturbed peatland habitat affected the ability of insects to colonise restoration sites.[8] She placed potted giant cane rush plants at various distances up to 800 metres from an intact habitat (the source population), and for 18 weeks sampled the insects that appeared on them. Invertebrates soon turned up on the potted plants, but the plants furthest from the intact habitat had the lowest number of insects and fewest species. Of the total number of insect types captured on the potted plants, only 38 per cent were on plants at 800 metres from the intact peat bog, compared with 62 per cent on plants just 30 metres away from the bog. From this work, Corinne recommended a best-practice approach for the mine restoration: place peat islands at 30-metre intervals outwards from existing areas of intact habitat to create 'stepping stones' for insects.

The Torehape research has provided useful information for future restoration management practices, both in mined peatlands and in wetlands generally, but it was finding Fred that has been a career highlight for Corinne.

Finding Fred

At Torehape, thick areas of giant cane rush hid a tiny secret — the world's skinniest caterpillar. During her invertebrate sampling in 2003, Corinne noticed intricate brown squiggles on the giant cane rush stems and wondered what was causing them.

> I opened up some stems and found this skinny little orange caterpillar, less than a millimetre thick [and] about 3 centimetres long. Looking down the microscope, I thought, Oh, it looks like a thread. I've just found a thread. And I kept finding these things that looked like threads. I remember a work colleague saying to me, 'You've been poring over the microscope for weeks.' I said, 'Yeah, I've been looking at these caterpillars that look like threads! Kinda like Fred the Thread!'

Corinne shared the mysterious larvae with Dr Robert Hoare, a national moth expert at Manaaki Whenua — Landcare Research. Hoare was also stumped.

> I thought it was far too thin to be a moth caterpillar, so I gave it to our beetle specialist, and he said, 'It's not a beetle, but it might be a fly.' We couldn't find any fly experts to see whether it was a fly maggot or not. So eventually Corinne and I decided that we were going to go down and sort out the mystery ourselves once and for all. We went to the peat bog and collected a whole heap of *Sporodanthus* stems and put them in a closed plastic bag. We kept a close eye on them and waited for something to hatch out. And eventually a moth hatched out.
>
> It was very exciting to find out it was a moth because it meant it was a new species, and it meant that we got to name it. It was a very unusual discovery because the caterpillar was so thin, and in fact we believe it's the thinnest moth caterpillar in the whole world.[9]

Corinne wanted to find out more about this curious creature, including when the adult moths are active, how long the life cycle lasts, and how the adult moths get out of the *Sporadanthus* stem. To solve some of these mysteries, she and Hoare collected giant cane rush stems occupied by the moth larvae. These were easy to spot because of the squiggle-shaped scars caused by the larvae chewing cellulose from the inside wall of the stem.

They managed to rear both male and female moths in the lab, and discovered that adults hatch during spring, from October to December. The species seems to have a two-year life cycle, because they found caterpillars inside stems at the same time the adults were emerging. How the adult moths get out of the stems remains a mystery, since the scientists have not found any pupal remains in or on the stems.

NEW THREATENED SPECIES — THE LESSER-SPOTTED TAXONOMIST

There are potentially 50,000 species of invertebrate in Aotearoa. About 20,000 of them are insects, but only half have been formally named. Corinne explains:

> We don't know what proportion of New Zealand insects we think we have identified, but looking at DOC's threatened species list and looking at how many are data deficient — that would give you an indication of how much we just don't know . . . that's just the species that we know of but have little information about, especially for wetlands. There are at least two or three other genera in peat bogs that we've found that need describing that we haven't even got to yet.

These include another (yet-to-be-named) moth found at Torehape that Hoare calls BoB —short for 'Beast of *Baumea*', after the genus of rush it lives in. It's distantly related to Fred, in the same family but a different genus. A shortage of entomologists skilled in taxonomy means that many invertebrate species will remain undiscovered, or unnamed, and Corinne doesn't see this improving any time soon:

> At the moment there's a huge cohort of taxonomists that are retiring, or on the edge of retiring. I see there's a huge gap . . . a lack of university courses that specialise in insects in general . . . but for terrestrial invertebrates, I'm currently the only one in Manaaki Whenua — Landcare Research that people could go to for advice on how to design an insect-sampling project or restore their habitat.

A moment of frustration for Corinne led to another discovery — and explained why no one had seen the adult moths before:

> Robert needed more male and female moths to be able to formally describe the species, and he was really keen to find out what their behaviour was like in the field. But trying to find them in the field wasn't easy . . . we were poring over these giant cane rush plants and we just could not find the moths anywhere. I don't normally get angry, but I was just getting *really* frustrated because I knew that they were there somewhere. So I just stood up and gave the plant an almighty kick, and there were just clouds of them! They were sitting near the base of the plants on brown sheaths, just above the ground, so well camouflaged. It turned out to be a great technique, but then I thought, I can't really promote violence can I? I said to Robert,

NAMING CONVENTIONS

Getting to name a newly discovered species is almost every young biologist's dream. Getting to name a whole new genus is next level. Discovering and co-naming a new genus of moth has been a career highlight for Corinne.

Every known living organism on Earth has a scientific name, whether it is still around or long extinct. The system designed by Swedish botanist Carl Linnaeus in the eighteenth century and adopted worldwide requires use of the Latin language, italic font and a binomial, or two-part, name — a genus name (similar to the concept of a surname) and a specific name (similar to our use of a first name, but never starting with a capital letter). Modern humans belong to the genus *Homo* and within this genus to the species *Homo sapiens*.

Naming rights go to the first person who formally describes a newly discovered organism. Within the constraints of the international convention, rules and norms may vary between taxonomists. Botanists tend to be rather conservative, naming plants for mundane features such as leaf size (*grandiflora*, for example) or after their botanical heroes —about 80 plant species are called '*Something banksii*,' after Joseph Banks, the scientist who accompanied Captain James Cook on his first expedition in 1768–1771.

Zoologists, meanwhile, can be downright unimaginative — *Gorilla gorilla*, *Rattus rattus*. Entomologists seem to cut loose. Some of the most hilarious and clever scientific names have been applied to spiders or insects, including a spider named *Apopyllus now*, a beetle named *Agra phobia*, a fly named *Pieza rhea* and a wasp named *Verae peculya*.

Fred the Thread's scientific name is no less catchy. Corinne and Robert chose to call the genus *Houdinia*, after the great escape artist Harry Houdini. The name has a double meaning — first, because the caterpillar stage must 'escape' from the narrow cane rush stem in which it pupates into a moth, and second, because it escaped the attention of entomologists until Corinne first noticed its wiggly patterns on cane rush stems in 2003. The specific name they chose, *flexilissima*, refers to the caterpillar's flexible body.

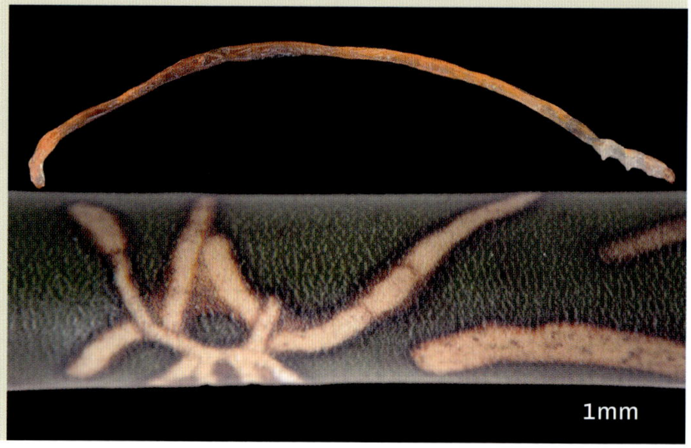

> ‘I don’t know about writing about that . . .’ but he wrote in his taxonomy paper, ‘adults are best collected in the field from *Sporadanthus ferrugineus* by giving them a really good kick’. I thought this is never going to get through the reviewers, but they obviously read it and just laughed.

Sharing the science

Corinne places great importance on skilfully communicating science to the general public, to raise their awareness of the importance of our invertebrates. She gets frustrated when people just lump them together as ‘creepy-crawlies’. Giving Fred a quirky common name was an effective way to engage people.

Technically, Fred the Thread belongs to a group of insects collectively known as stem borers. Boring insect. It’s not exactly marketable. But by personalising the insect and telling a story about it — like the wacky life of Fred the Thread with his flip-top head — people start to relate to it. It’s even better when the insect has a nickname, says Corinne. ‘The media love it!’ Fred’s name has made him something of a celebrity. He has featured on national radio and has a dedicated activity station at the National Wetland Centre, and Hoare has even written a comical poem about him.

Given her love of insects, it’s somewhat ironic that Corinne’s favourite plant is an insect-eating carnivore — the forked sundew (*Drosera binata*). Corinne describes it as a native plant that:

> sprouts tiny green tendrils that unfold into bright red symmetrical lobes which are sticky with sweet secretions. Insects, attracted by the sweet secretions, land on the plant and are trapped. The plants need these victims for nutrients as the soils they are growing on are nutrient-poor. I first saw these plants at Torehape peat mine and was amazed at their tenacity, as they are one of the first plants to colonise the bare mined peat surface. I was so impressed with these meat-eating plants, I dug one up to take home . . . and promptly killed it!

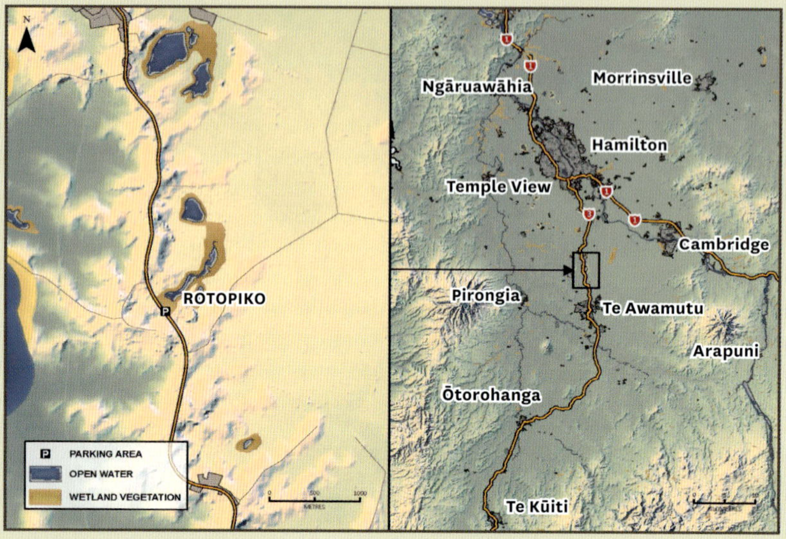

Rotopiko Lake Serpentine

Peat lake and bog complex

Much of Corinne's peatland research has been conducted at the Torehape peat mine. As a working mine, it's not accessible to the public, but it can be viewed from Torehape Road East and Central Road, southwest of Ngatea in the Hauraki Plains.

For a close-up view of the giant cane rush and the tell-tale stem scars caused by resident Fred the Thread caterpillars, head to Rotopiko, south of Hamilton Airport. Rotopiko is a peat lake with a swamp margin and adjacent mature kahikatea forest, all set on public reserve land.

In 2006, the NZ Landcare Trust conducted an experiment at Rotopiko to see if giant cane rush could be successfully transplanted into peat lake margins that had been previously cleared for farmland. At that time, the species was known to occur in only three places, all in the Waikato region — a precarious situation. To improve the chances of success and speed up the process, the trust transplanted clumps of peat with young cane rush plants, rather than grow plants from seed. This meant that Fred the Thread could secretly hitch a ride to the new sites — without knowing it, the trust had successfully translocated the rare insect to a new safe location. The trust sourced their plants from the Torehape peat mine and placed them on freshly exposed peat at the margins of two peat lakes retired from grazing. Rotopiko was one of the experimental sites, and the original giant cane rush plants are still present, complete with Fred inside.

Supplementary cane rush planting and species enrichment at Rotopiko

has been continued by the National Wetland Trust with the support of tangata whenua, Ngāti Apakura. The National Wetland Trust has also created walkways and installed interpretive signs and an interactive discovery trail for families and school groups to enjoy. One of the activity stations is centred on Fred the Thread, and visitors can see giant cane rush from the boardwalk.

Highlights

Special features at Rotopiko include:

- a 14,000-year-old peat lake with a floating pontoon
- the giant cane rush and resident Fred the Thread caterpillars
- a towering mature kahikatea forest
- habitat for rare black mudfish and pūweto (spotless crake)
- a pest-proof fence to exclude introduced predatory mammals.

Access

Turn into the reserve from the southbound lane of State Highway 3, approximately 3 kilometres south of the small town of Ōhaupō. Allow one hour to walk the easy tracks and enjoy some of the activity stations. The site accommodates groups by appointment and has the following facilities:

- all-weather gravel parking
- easy access for pushchairs
- floating pontoon with close-up views of the peat lake
- waterless toilet, but no onsite drinking water or rubbish collection.

www.wetlandtrust.org.nz/what-we-do/national-wetland-centre

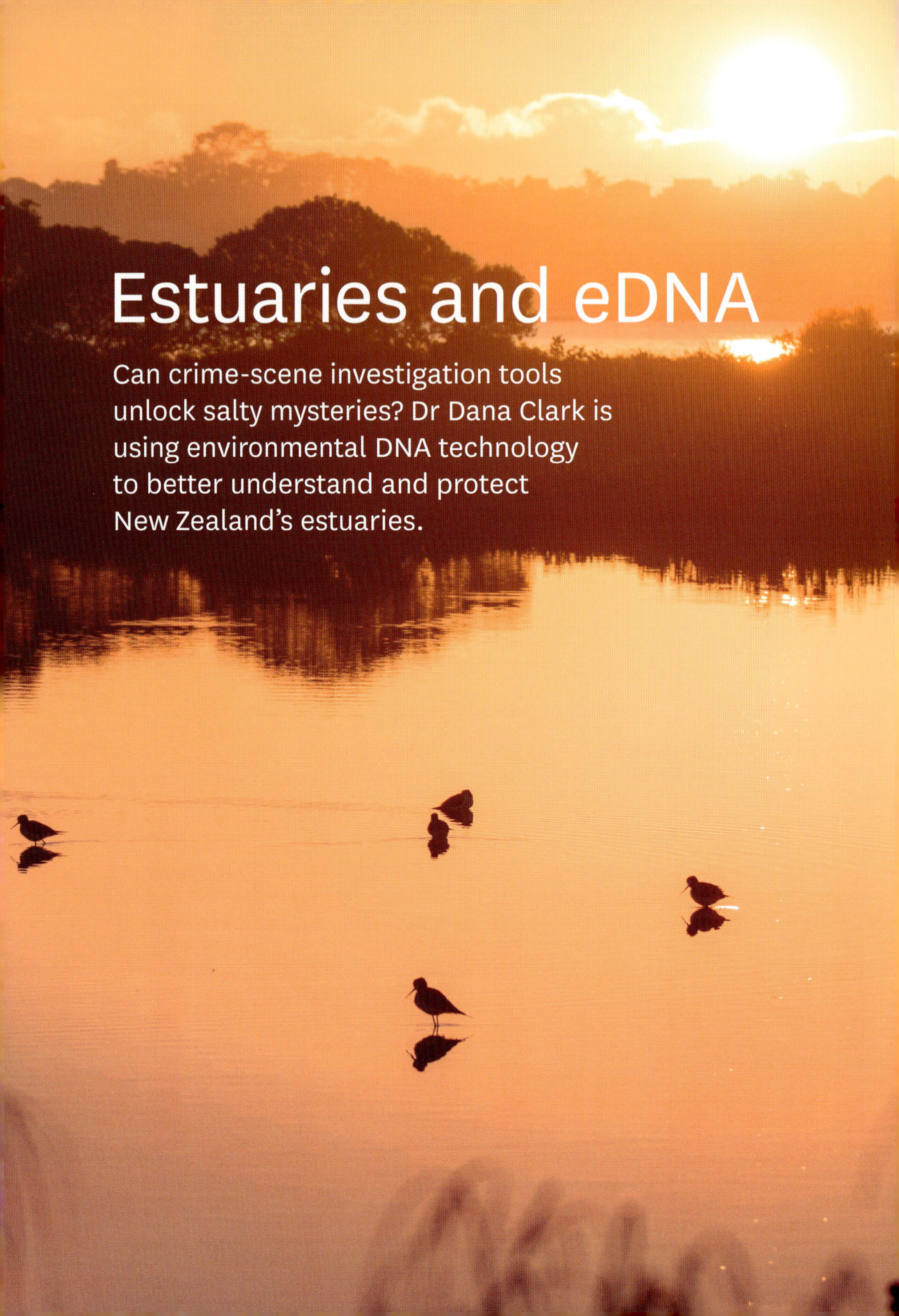

Estuaries and eDNA

Can crime-scene investigation tools unlock salty mysteries? Dr Dana Clark is using environmental DNA technology to better understand and protect New Zealand's estuaries.

Dr Dana Clark

Specialist field: Estuarine ecology, environmental DNA

Currently works for: Cawthron Institute, Whakatū Nelson

Dana Clark, MSc, PhD, is an ecologist specialising in coastal ecology and management. Her research aims to understand the drivers of change in coastal ecosystems to enable better management and thus protect the ecological, social and economic values of these ecosystems. She uses a range of statistical, molecular and mapping methods to develop applied tools for coastal management. As part of her recently completed PhD she explored how emerging molecular techniques could improve estuary health assessment.

MANY YOUNG PEOPLE dream of becoming a marine biologist, perhaps studying dolphins or coral reefs in balmy tropical waters. Dana Clark has done it for real, although when she began she didn't know it would involve using techniques first developed for criminal investigations.

Science was always Dana's best subject at school, so she knew early on that she would have a career in this field. Her dad was also a scientist, so you could say it ran in the family. However, it wasn't until she was in her teenage years that her attention was drawn to marine science. After spending summers as a lifeguard at Karioitahi Beach on Auckland's west coast, Dana had grown to love the ocean and wanted to find out more about the animals that lived there.

She completed her undergraduate degree and MSc at the University of Otago, and while researching her Master's thesis had the opportunity to travel to Antarctica to investigate how future climate change might affect Antarctic sea urchins, an important species in the ecosystem. Dana recalls visiting this vast, untouched landscape and says it inspired her love of field research: 'Visiting Scott's and Shackleton's huts, seeing the complicated world of penguin colonies and learning to drive a snowmobile — those were amazing experiences! I also remember contemplating the fortitude of the early Antarctic explorers — with only simple woollen clothing and wooden huts to keep them warm, while I was kitted out with a down jacket, enjoying the modern luxury of central heating.'

Following the completion of her MSc, Dana's taste for adventure led her to the Caribbean. There, she drew upon her diving and sailing skills to educate American college students and work with conservation organisations studying sea turtles and corals. After a couple of years spent travelling the world, she decided it was time to head home and get a 'real' job. So when she was offered the opportunity to join the Cawthron Institute in Whakatū Nelson as a marine ecologist, she jumped at the chance.

Although Dana's intention on joining Cawthron was to study rocky reefs thanks to her overseas diving experience, she soon found other sources of inspiration. She credits Dr Joanne Ellis, a senior scientist at Cawthron at the time, for igniting her passion for estuaries. Dana recalls how Joanne was genuinely surprised when her fellow scientists tried to explain to her why 'muddy old estuaries' aren't as 'sexy' as rocky reefs.

Through working with Jo, Dana says she came to realise the importance of saline wetlands in supporting coastal ecosystems, processing land-derived pollutants and protecting coasts. Now with a different viewpoint, she loves watching the rise and fall of the tides across estuary sandflats.

Ecosystems under threat

Worldwide, estuaries are degrading, and scientists are seeking the cause with ever greater urgency. These important but overlooked ecosystems require much more protection than they are currently afforded. Although they represent a tiny portion of the marine environment globally, they are critical to the health of some of the most productive ecosystems on Earth.[1]

Aotearoa New Zealand has more than 400 estuaries along its coastline, and they are enormously valuable to our economy, environment and society.[2] But not enough people see their value, and they are under threat from pollution, overharvesting of seafood and climate change.[3] It is important that we protect and restore our estuaries so that they can continue to provide us with the ecosystem services we rely on.

But how do you convince people of the value of an ecosystem when its beauty and ecological value are largely concealed beneath what some see as nothing more than swampy, stinky mudflats? How do you encourage investment in estuary research when there is little societal understanding of their importance and value? Dana met this challenge early on in her studies and has embraced it in her career so far. Her approach is to take schoolchildren into the estuary and show them how it works:

> Kids are curious about the environment and the animals and plants that live there. My hope is that by sparking their interest, they will start to understand that nature underpins much of the things we enjoy, even if that connection isn't immediately obvious. For example, shellfish are really important for keeping the estuary water nice and clear because they filter out suspended particles and stabilise the sediment. And kids love to share what they know, so by educating them you are also educating their parents, brothers and sisters, and friends.

Researchers add fertiliser to the sandflats in Waimea Inlet to mimic the effects of increased nutrients washing off the land into the estuary.

While working at Cawthron, Dana completed her PhD thesis, where she focused on learning more about the factors that might be changing the environment in estuaries and affecting the life within them. Dana says one of the greatest difficulties in estuary research is predicting how these ecosystems will respond to the increasing number of potential threats they are exposed to. She explains:

> Scientists have a pretty good idea of what stressors are impacting estuary ecosystems and how estuary health changes in response to these individual stressors. But these stressors rarely act in isolation. It's not as simple as summing the stressor effects, like 1 + 1 = 2. Interactions between stressors mean that sometimes their combined effect is better or worse than we would expect from their individual effects, so 1 + 1 might equal 3. And in a different estuary 1 + 1 might equal 10, or even 100! These cumulative effects can also result in unexpected shifts, or tipping points, where a relatively small change in a stressor results in large change in the ecosystem. It's understanding how these stressors are affecting and interacting with each other and how their

effects combine to produce sometimes unpredictable outcomes that drives much of my research. It's a puzzle we're trying to put together, and up until now we haven't really had all the pieces.

The lightbulb moment

As DNA technology has developed, scientists have found new ways of identifying species using a technique called eDNA metabarcoding. This reads fragments of DNA much the same way as a scanner reads the barcodes on your groceries. Very few researchers had attempted to apply this technique in an estuary environment until a thought occurred to Dana. What if we could use eDNA to determine how and which stressors are causing declines in estuary health? She explains: 'Because eDNA captures the response of thousands of different species (instead of just one, tens or hundreds as in conventional studies), there is the potential to develop indicators that can differentiate between sources of stress, because species have varying responses. For example, a certain type of bacteria might love mud but hate nutrients.'

Seeing how eDNA was transforming environmental monitoring in other fields of science, Dana and her colleagues at the Cawthron Institute and the University of Waikato decided to see if eDNA metabarcoding could also be used to detect ecosystem changes in estuaries in response to a stressor. They focused on one of the key stressors in New Zealand's saline wetlands, nutrient loading, which can be difficult to manage because it occurs over large areas and often gradually over a period of time. Incremental changes in nutrient loads have the potential to cause substantial changes to estuary health and functioning.

Dana says that eDNA metabarcoding could be used to pick up subtle changes in estuary health before the habitat becomes too degraded: 'Tiny organisms like bacteria play important roles in ecosystem functioning and are very responsive to changes in the environment. But up until now, we haven't been able to track changes in these communities. Using eDNA metabarcoding we can use these communities to provide early warning signals of stress before detrimental changes occur in the estuary ecosystem that affect larger animals like fish and birds.'

If eDNA data could be used to detect early signs of disturbance and give us more clarity on the most important stressors to focus on, it could have a big impact on how we monitor and manage estuaries. Dana can imagine the effect that eDNA tools could have on wetland management across Aotearoa, and how they could empower a wide range of people to protect estuaries and safeguard the ecosystem services they provide for future generations.

Putting eDNA to the test

The best way to determine whether eDNA metabarcoding could be used to detect ecological changes in estuaries was to conduct an experiment. Dana and her team decided to test the idea by adding standard garden fertiliser to estuary sediments to mimic the effect of increased nutrients washing off the land into the estuary.

STRESSED ESTUARIES

Stressors are environmental changes that harm ecosystems, many of them caused or exacerbated by humans. One example is sedimentation, the flushing of soil from areas of land disturbed or cleared for forestry, farming and construction into waterways, where they muddy the water and infill estuaries. Other stressors caused by humans include contamination from heavy metals or sewage, and excess nutrients from farm animals and fertiliser. Almost 30 per cent of New Zealand estuaries now have a high or very high risk of nutrient overloading.[4] Activities occurring in coastal areas can also act as stressors, including fishing, dredging, and building ports, marinas and other coastal structures.

Ecosystem tipping point in Chesapeake Bay

A good example of how the cumulative effects of environmental stressors can alter an estuary can be found in Chesapeake Bay,[5] which is the largest estuary in the United States. The bay has been altered dramatically over the past 200 years in response to urbanisation and agricultural land-use changes. Declines in water clarity initiated a shift from a healthy ecosystem with seagrass meadows to one dominated by phytoplankton blooms and low-oxygen 'dead zones'. These tipping points can be difficult to predict and are often hard to reverse. In the case of Chesapeake Bay, the seagrass beds previously trapped sediment, keeping the water clear and allowing the seagrass plants to photosynthesise. However, as the bay became more enriched with nutrients through land run-off, phytoplankton blooms made the water cloudier, meaning less light was available for the seagrass. In a downward spiral, this led to a decline in seagrass, which in turn made the water even more turbid as the sediment was no longer being stabilised, thereby preventing the re-establishment of seagrass. The most recent report on the state of the bay said that while the dead zone is shrinking, 'The Bay that 18 million of us call home is still a system dangerously out of balance. The road to restoration is steep, and the clock is ticking.'[6] While there has been a slow, steady recovery of seagrass since the 1980s, between 2018 and 2020 there was an abrupt decline, resulting in an environmental score of D minus.

Much of Dana's PhD research was carried out in two estuaries, Delaware and Waimea Inlets, both near Whakatū Nelson in the South Island. She says, 'These estuaries are shallow, with a huge tidal range, so they look different every time I visit. Sometimes they are covered with water, while other times they are almost completely drained.'

The first thing Dana does when planning a day out in the field is to check the tides. She is constantly trying to figure out when her sampling site will be uncovered and how much time she'll have to sample before the tide comes back in. A day in the field is usually long because Dana wants to collect as much information as possible, but she says it's a happy

eDNA — A NEXT-GENERATION ENVIRONMENTAL MONITORING TOOL

We've all seen television shows where forensic investigators don gloves and comb through crime scenes with black light looking for a stray hair, a smear of blood or any other biological traces containing DNA that could identify a suspect. These invisible signatures have solved many famous crimes since the 1980s. Today, biologists are using very similar techniques to make their research easier and more precise.

Imagine you are a scientist who needs to make a list of all the species that live in your field site, whether it's a lake, a forest, a stream, an estuary or whatever. For hundreds of years scientists have followed the same basic steps. First, go to the site and collect samples of plants, animals, insects, water, soil and the like. Second, painstakingly sort through your samples and try to name each species, usually with the help of experts and a good microscope. Maybe spend hours looking at almost invisible features, such as how many hairs it has on its back leg or whether it has a valvate median oesophageal bulb (yes, it's a real thing). This approach takes a lot of time, and it misses all the organisms that are too small to see with the naked eye, such as bacteria.

What if, instead of spending hours hunched over that microscope trying to figure out what species the tiny blob in front of you belongs to, you could use those DNA traces much beloved by trial lawyers and crime fiction writers? What if you could just put a droplet of water or a teaspoon of mud into a machine and press a button? It turns out we absolutely can.

We know that people leave traces of their DNA behind them in the environment when they move around — hair, skin cells, nails, saliva, waste — but so do other living organisms such as plants, insects and even bacteria. Since each species has a specific DNA code, these traces can act as identifiers for different species — much like a barcode identifies an item in a shop. DNA collected from the environment is called environmental DNA (eDNA). This DNA can be extracted from small water or sediment samples and used to tell us what species, and roughly how many, are living in that location.

trade-off because she gets to enjoy being outdoors, immersed in the environment she is working hard to protect.

In Dana's experiment, the band of researchers ventured out into each estuary decked out in gumboots, waders or dive booties to keep the mud out of their toes. They made their way around channels of deep water at low tide, towing sleds laden with sampling equipment behind them. They set up experimental plots (3 metre by 3 metre squares, marked out using short metal pegs) and then dug small holes, into which they poured known amounts of fertiliser. In some holes they added just a small amount of fertiliser (150 grams of nitrogen for each square metre), in some they added a high load (600 grams of nitrogen for each square metre) and in some they added no fertiliser. Plots with no fertiliser were control plots, to allow the researchers to compare the normal state with the simulated pollution events. The aim of this experiment was not to make the sediment toxic, but to test how well the ecosystem could process the additional nutrients.

Seven months later, the team returned and took samples of the sediment, along with the animals and plants that were living in it, to see what changes the addition of fertiliser had caused. Back in the lab, Dana extracted DNA from the sediment samples. She explains the process:

> We have to do this in a special room that is very clean, so that we don't contaminate our samples with other DNA. Once the DNA is extracted, we carry out a process called PCR, short for polymerase chain reaction, to multiply a specific part of the DNA many times. Since each species has a particular DNA code, these short sequences can act as identifiers for different species — much like a barcode identifies an item in a shop. These samples were sent away to Auckland University, where they have a DNA sequencing machine that reads these barcodes. The barcodes are matched to a reference library that tells us what organisms are present in the sample and roughly how many of them there are.

To everyone's delight, they found that small organisms such as bacteria, eukaryotes and diatoms are extremely sensitive to changes in estuary health, even at low levels of nutrient enrichment. There were slightly different responses in different estuaries — in one site, bacteria and diatoms were more sensitive, while in another, eukaryotes were. These findings suggested that it may be possible to develop indicators that can detect subtle changes in the health of our estuaries before they become too degraded. However, the different results at different sites suggest that Dana needs to expand the study so she can be sure they will hold true across a range of estuaries.

Another interesting finding was that certain bacteria were found in the high-nutrient treatment plots but not in the medium-nutrient or control plots. This suggests that these bacteria were favoured once nutrient loads reached a certain level. These bacteria appeared at both sites, meaning researchers could develop broadly applicable indicators for use in a range of estuaries.

Since that time, Dana and her colleagues have published research that shares these results, and they are continuing to work on the development of tools that can be easily used in the field by land managers, researchers, environmental monitoring organisations, kaitiaki (local site guardians) and others. The proposed monitoring tool is likely to include three steps: data collection, laboratory analysis and health scores. Data collection is easy, simply requiring someone to collect a small sample of mud or sand (less than 1 gram) while wearing gloves to prevent DNA contamination. Samples would then be frozen and sent to a laboratory, where the DNA would be extracted and sequenced. Based on the eDNA data, the tool will determine what stressors are affecting the estuary and provide a score of how

SUPER SEAGRASS

Dana's favourite wetland species is a marine plant called seagrass. Aotearoa has only one species of seagrass, *Zostera muelleri*, which primarily grows on sandflats in estuaries.

Seagrass meadows are one of the world's most valuable habitats, with 1 hectare estimated to be worth NZ$40,000 per year.[7] Their leafy canopy provides shelter for small animals such as crabs and juvenile fish, supporting coastal fisheries and creating habitat for many more species than would be found on bare sand. As seagrasses grow, they absorb carbon and store it in their tissues, helping to offset the effects of climate change — Dana estimates that every year 1 hectare of seagrass can capture and store the same amount of carbon that would be emitted by a car travelling the length of New Zealand ten times.[8] They also absorb nutrients and trap sediments, helping to improve the clarity and quality of our coastal waters.

healthy it is with respect to each of those stressors. The intention is to have the health scores (a bit like a car warrant of fitness) available on a website or smartphone app.

Estuary health means different things to different people because it is based on what an individual values about the environment. Dana considers an estuary to be healthy if it can carry out the functions that underpin the benefits people derive from it, and is resilient enough to cope if environmental conditions change — for example, warming temperatures or the introduction of a disease. This resilience relies on the estuary supporting lots of different species, so that if one is wiped out by a disease then another can take over its role in the ecosystem.

But this is just one view of estuary health, and Dana says it's important to recognise that other people may view the ecosystem in a different way. She wants to find out whether eDNA can be used to track shifts in estuary health from a te ao Māori perspective. This new technology could, for instance, be used by iwi to monitor changes in taonga species and mahinga kai (customary food-gathering) areas, helping them manage and make decisions about the environment.

The future

Dana says there is still much to learn about how environmental stressors are affecting our estuaries and the long-term effects. Estuaries are complicated environments where various stressors can combine to create unexpected environmental changes that are more than just the sum of their parts. However, with the discovery that eDNA technology can provide some of the missing pieces in the puzzle, Dana is confident our knowledge will increase, and we'll be better informed about how to protect these vitally important ecosystems.

She says she thinks the best way for people to learn about estuaries and oceans is to get out and enjoy being in them — that when people take the time to look, listen and note the incredible things happening around them, they'll see it's all connected. As Dana points out, 'So much of what makes our oceans special is hidden beneath the water and the seafloor sediments. Our environment is under threat, and our oceans particularly so, because people aren't motivated to protect something they cannot see or do not understand.' Honing an appreciation for nature is a key part of motivating people to make the changes required to protect it. Dana says, 'When we take the time to truly appreciate all the things nature does for us, it becomes clear why we need to do all we can to protect it.'

Although the task can seem overwhelmingly big, there are daily reminders for Dana about why her work matters. Once, while she was diving in Marlborough's Te Hoiere Pelorus Sound to assess the impact of salmon farms on a nearby reef, two fur seals swam over to see what she was up to. Dana recounts how the curious seals watched her as she worked and accompanied her as she swam back up to the surface. It was a casual encounter, common in a day's work for Dana, but easy to take for granted. Dana says experiences like this are a reminder of the beauty and mystery of ocean ecosystems and how fundamentally important they are to all life on Earth.

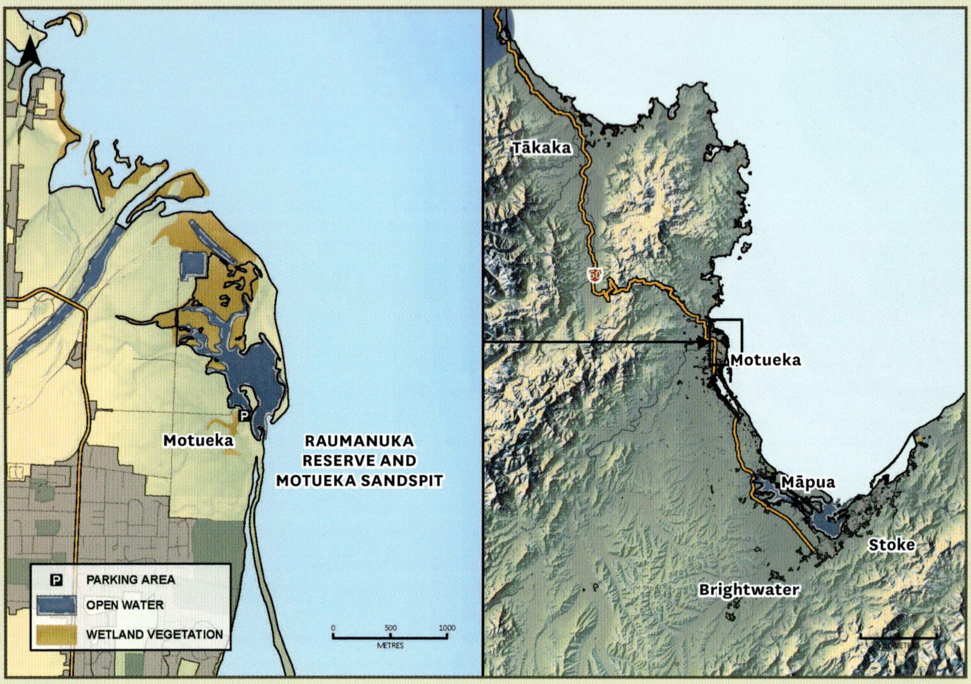

Motueka Estuary and Sandspit

Estuary

A great place to visit in Dana's neck of the woods is the Motueka Estuary and Sandspit, a significant site for local and migrant shorebirds. According to the Department of Conservation's website, it has extensive areas of rushland and saltmarsh where native īnanga (whitebait) spawn. It is also rich in shellfish, which makes it a major feeding ground for wading birds. In summer, up to 10,000 birds have been recorded feeding in the estuary or roosting on the sandspit.

Highlights

- Extensive views of Te Tai-o-Aorere Tasman Bay, Rangitoto ki te Tonga D'Urville Island and the inland Richmond Range
- Major feeding ground for up to 10,000 wading birds in summer, including bar-tailed godwits, eastern curlews, variable oystercatchers and South Island pied oystercatchers
- Keep an eye out for banded dotterels, royal spoonbills, black-billed gulls and black-fronted terns

- The rusting hull of the shipwrecked *Janie Seddon*, the last surviving military ship to have served in both world wars, lies on the foreshore near Motueka quay.

Access

You can enjoy views of the Motueka Estuary from a walk along the foreshore, starting at the old Motueka wharf. To access the Motueka Sandspit, follow the signpost marking the start of the walk at the end of Staples Street, just north of Motueka.

www.doc.govt.nz/parks-and-recreation/places-to-go/nelson-tasman/places/motueka-area/things-to-do/tracks/motueka-sandspit

Salty and Sweet

How can a fish live in two realms? Dr Don Jellyman's research unveils the mysterious lives of fish that live in two worlds — the salty brine of the ocean and the sweet freshwater of inland waterways.

Dr Don Jellyman

Specialist field: Biology of freshwater eels and lampreys; migrations between freshwater and saltwater that are undertaken by many native fish species

Currently works for: NIWA (National Institute of Water and Atmospheric Research Taihoro Nukurangi) as an emeritus scientist

Don Jellyman, PhD, is a freshwater fisheries scientist who has led NIWA's research on freshwater fish and fisheries. His specialist area is the biology of freshwater eels, which has led to considerable contract research for government agencies and hydropower companies. He has also been involved in studies of other native fish species, especially lampreys, and has carried out research on headwater trout fisheries throughout Aotearoa New Zealand. His research on eels has led to strong engagement with Māori, and he is heavily involved within New Zealand and internationally on issues related to eel management.

DON JELLYMAN HAS SPENT a lifetime studying New Zealand's freshwater eels (tuna). Over many thousands of hours he has pulled on his chest waders to watch, catch, count, measure, track and occasionally dissect eels, unravelling their secret, complicated lives.

More than 600 scientific papers and reports bear Don's name as either first or supporting author. You'd think he would know almost everything there is to know about freshwater eels, but the truth is, despite them being one of the most researched fish species on the planet and of huge cultural and economic significance in many countries, there's plenty about eels that no one knows. There are still many secrets we are yet to discover about these fascinating creatures, which have successfully occupied habitats across the world for millions of years.

Of the 200,000 or so published scientific papers that are listed on Google Scholar about the world's 19 eel species, nearly 4000 include the word 'mystery' or 'mysterious' somewhere in the text. There's also a lot of 'probably' and 'thought to' and 'likely to' statements. A large part of this uncertainty is down to the complex life history of freshwater eels. Despite their name, they all start their life in saltwater, spend most of it in freshwater, and then head back to their marine birthplace to breed and die — part of their life salty, part sweet. It makes them very difficult to study. These species are so complicated that Māori have more than 100 names for their different life stages,[1] and early European scientists thought the larvae and adults were different species!

Incredible journeys

All freshwater eels hatch in the ocean. For many species no one yet knows exactly where, but it's somewhere in the tropics. They then float on ocean currents for thousands of kilometres to the estuaries that their parents migrated through a year or so earlier. Following inborn instincts — perhaps scent memories embedded in their DNA — they swim upstream to freshwater habitats, where they live for anything up to 100 years. They then migrate back to their birthplace for a once-in-a-lifetime spawning event.

Throughout their life cycle freshwater eels transform dramatically from an egg, to a marine leptocephalus larvae (which looks like a transparent leaf), to a see-through glass eel, to a shoestring black elver, to an adult yellow eel, and finally to a torpedo-shaped

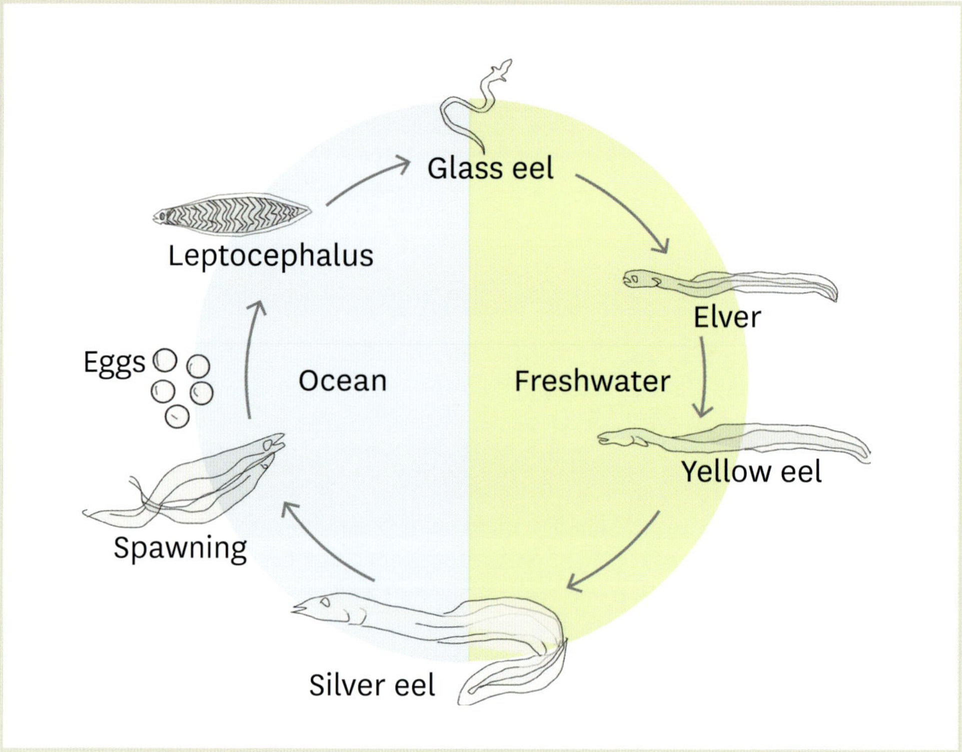

The complex life cycle of eels involves many changes in body shape as they move between fresh and salty water.

breeding silver eel, which stops eating and migrates to that mystery marine breeding ground thousands of kilometres away.

Tough tuna

New Zealand has two native species of eel, the shortfin eel (*Anguilla australis*) and longfin eel (*A. dieffenbachii*). Māori call both species tuna.

Adult shortfins live in a range of lowland wetland habitats, including swamps, rivers and lakes, in Aotearoa New Zealand, Australia and several Pacific Islands. They are shorter than their endemic cousin, not only in fin but also in lifespan. Male shortfin eels live for an average of 14 years and females for 25 years, while male longfin eels live on average 25 years and females 40 years.

Longfin eels are considered an 'at-risk' species, meaning their numbers are dropping.[2] This species spends its freshwater phase only in New Zealand, and tends to head far inland to stream headwaters and upland lakes. Because longfins have longer lifespans and travel greater distances yet still breed only once, there is a high chance that the longer-lived females won't survive long enough to make it back to their breeding ground — a contributing factor to the parlous state of this species.

Don has many adjectives to describe freshwater eels: resilient, adaptable, energetically conservative, and fecund (a single female eel can produce a staggering 30 million eggs).[3] By 'energetically conservative', he means they don't waste their energy. Part of the reason this ancient group of fish has been so successful for millions of years is because they are chilled out and extremely clever. They don't do anything they don't have to.

Yellow eels stay motionless for hours on end, feed sporadically, endure long periods without food, aren't too fussy about what they eat, locate their prey via an acute sense of smell (which saves having to search for them) and hunt by ambush (which means they mostly hang around quietly in nooks and crannies, waiting for food to come to them). 'They don't try and outdistance prey, they kind of sneak up on them,' says Don.

The marine larvae can swim, and do so with great elegance, but they mostly just float passively along in ocean currents — why exert yourself on a long journey if you don't have to? Elvers are energetic, and masters at scaling waterfalls, but silver eels migrating downstream learnt long ago to take advantage of flood flows and sensibly hitch a ride in the centre of river currents. On their migrations, silver eels travel enormous distances, but as Don says, 'they're built for a marathon, not a sprint'. One experiment showed that even after swimming non-stop without eating for five months, silver eels still had 40 per cent of their energy reserves.[4] This ensures that when they reach their mystical breeding ground, they have enough energy left in the tank to find other spawning eels and create the next generation.

They are also incredibly tough — 'resilient,' says Don. They can survive for years without food. They can survive in water temperatures approaching a stifling 40 degrees Celsius and as low as near-freezing, at which point they just enter a state of torpor and ride it out, waiting for things to warm up. They can survive drought by burrowing

MARINE SNOW

Adding to the intrigue of these amazing creatures is that once hatched from eggs somewhere at sea eels start life looking like a transparent leaf or a sort of see-through shoe insert with a tiny head. This is the larval life stage, known in technical terms as the leptocephalus (meaning 'thin head'). It looks like someone has taken a rolling pin to a glass eel and flattened it out, but the larvae swim vertically in the water, undulating in the most beautiful sinuous patterns.*

What eel larvae eat at sea remained a mystery for many years, until in 2011 it was confirmed that they feed on marine snow, a kind of biological dandruff (bits and pieces of zooplankton skins, excreta and bodies) that drifts slowly down through the water column like falling snow.[5] Someone has to eat it, right?

* For a video of leptocephali swimming, head to https://thekidshouldseethis.com/post/leptocephalus-transparent-eel-larva.

into mud, and they can breathe three ways: through their gills, through their skin or, if oxygen levels drop in their waterway, by gulping air. Eels are so tough that they are not useful as early warnings of declining water quality. 'If eels start dying, everything else is already dead,' says Don.

Freshwater eels are also adaptable. If a habitat isn't to their liking, they will up sticks and move somewhere else. If breeding eels find their pathway out to sea blocked, they can transform from a silver eel back into a yellow eel and resume eating, and just try again next year. They also wait 10 or 25 years before deciding whether to become a male or a female.[6] It's not really a choice as such — if eel population numbers are high, most individuals will develop into males, but if the population drops, more will develop into precious females, needed to produce eggs. In practice, it's usually difficult to identify the sex of most eels with any confidence until they are at least 35 centimetres long in the case of shortfins, and about 40 centimetres for longfins.[7]

Risky lifestyle

One of the mysteries Don has been considering is why, despite being adaptable, tolerant and apparently successful — found across the world in a variety of habitats — freshwater eels seem to be declining worldwide.[8] He puts it down to their risky lifestyle, a lifestyle that worked well until human pressures piled on. Fecundity has been the name of the game for freshwater eels. Live long, breed once and make millions of babies. Job done. Fingers crossed that of the millions of larvae produced enough individuals will survive the perils of an ocean drift across thousands of kilometres and a long life — during which any moment could be their last — to produce the next generation. It's called the surplus production hypothesis.[9]

Another key freshwater eel strategy is predator satiation. It evolved back in the day when predators were pretty much the only risk they faced. Basically, if enough larvae are produced and they bunch together in a form of synchronised swimming, creating an all-you-can-eat buffet of glass eels or elvers, predators will get full before the eels run out. As Don describes it, the young face a 'barrage of aquatic and avian predators', being attacked by birds from above and larger fish in the water from below. There is no escape other than through safety in numbers.

This strategy and lifestyle worked just fine for millions of years, but putting all their eggs in one salty basket has always been a risky approach, and the rules of the game have now changed. The reproductive tactic of the freshwater eel may no longer be fit for purpose.

New pressures have come along. Overfishing. Pollution. Massive destruction of wetlands. River straightening. Construction of dams, culverts, flood pumps and weirs. Combined, these have resulted in the loss of considerable habitat for eels and perilous migration journeys. The pressures on eels are now so great that many are not making it back to their breeding grounds, which in turn means that fewer young ones are arriving at our shores. For Don, this highlights the importance of the conservative harvesting of eels at all life stages, the protection and restoration of eel habitats and safe passage, and the need to transfer young eels far upstream into areas of low eel density, in the hope that this will increase the chance that they turn into females.

Restoring Te Waihora

Improving the odds of freshwater eels making it back to their breeding ground has become an important part of Don's research, including investigating ways to improve habitat and eel passage upstream and downstream.

One of Don's study sites is Te Waihora Lake Ellesmere, near Ōtautahi Christchurch. Te Waihora (meaning 'water spread out') is the largest coastal lagoon in New Zealand, separated from the sea by a natural gravel bar that is periodically breached by storm events. The lake can have a surface area of up to 30,000 hectares, depending on its water level. Historically, Te Waihora flooded vast areas as far as Christchurch.

Te Waihora has long been recognised as a valuable wildlife resource and traditional Māori and commercial fishery for shortfin eels.[10] Access to and from the sea is essential for both incoming juvenile elvers and outgoing migrating silver eels. However, the lake is now manually opened by the Canterbury Regional Council each year (usually in spring) to prevent flooding of adjacent farmland. Don has studied the crucial migration periods for eels, and recommended changing lake openings to meet the needs of migratory fish.[11]

Within living memory, the lake margins had clear water and wide stands of native aquatic plants, but during a large storm in 1968 almost all the plants were lost. They have not returned to any great extent, and the entire lake is now brown and murky.[12] It's still unclear how these changes have affected the ecology and fisheries of Te Waihora, but from Don's studies we know that juvenile shortfin eels seem to grow very slowly there, and in some years very few young eels join the population. Studies of shortfin eels in Te Waihora indicated that water snails were once an important food item, but by 2004 the diet of small eels was largely comprised of midges. Since the vegetation collapsed, snails are now rare in the lake. Don thinks the slow growth rates of small eels are a result of competition with common bullies for midge larvae.[13] Restoring lake vegetation may help restore snails for eels to eat.

NIWA scientists are now working with Ngāi Tahu at Te Waihora, experimenting with ways to restore the vegetation, including by installing tethered logs to create wave barriers and transplanting aquatic plants.[14] The trials are part of a larger programme in partnership with the iwi council, Te Rūnanga o Ngāi Tahu, along with Environment Canterbury and the Ministry for the Environment Manatū Mo Te Taiao, with the vision to restore the mauri (life force and ecological health and vitality) of Te Waihora and its catchment.

Māori knowledge

Much of Don's work on the timing and patterns of eel migrations is consistent with what Māori already knew about tuna through traditional knowledge. According to Don:

> It's been quite neat to come alongside and compare our findings with Māori understanding of the ecology of eels, and especially the migrations. Māori knew a lot about seaward migration, or the heke as they called it, and they knew a lot about the conditions that would lead to the onset of migrations. They knew the sequence of eel migrations. You get shortfin males first, then shortfin females, then longfin males. Finally, the big longfin females come downstream to the coast. Some of the technologies Māori had for capturing the downstream eels were quite amazing, in terms of weirs and traps and other methods.

Each of the freshwater stages in an eel's life have long been understood by Māori. They were able to predict good fishing times and knew when to harvest the right type of eel — from delicate glass eels in spring, to the fat, roe-filled breeding adults in autumn. Māori practised various forms of management, including imposing rāhui (fishing bans) on a

Transparent glass eels were a traditional Māori delicacy, harvested in spring as they entered river mouths on their upstream migration.

waterway for ceremonial or conservation reasons, and seeding areas with small eels.[15] They still follow these practices today.

Europeans had a very different view of eels. To early European settlers, eels were a pest species to be avoided, and Māori eel weirs were just a navigation hazard getting in the way of trade. As a result, many weirs were destroyed, obliterating not only a crucial source of sustenance for Māori but also a significant part of the cultural history of Aotearoa.[16]

Tuppence a tail

It's ironic that Don, who is a great advocate for putting limits on longfin eel catch because of declining populations, used to make pocket money as a boy by killing them. New Zealand's native freshwater fish tend to be small and active only at night — not ideal as sport fish. So, from the 1860s avid anglers brought in foreign fish.[17] Trout were particularly popular, and a lot of time and money were spent by the New Zealand Acclimatisation Society (now Fish & Game New Zealand) to rear and release trout into many rivers around New Zealand. Initially the eels were fine with it, viewing the trout as a new variety of food to add to their diet, but when the anglers caught on that eels were eating their sport fish, they were not well pleased.

As Don notes, 'Anglers were encouraged to kill as many eels as possible, and in the 1930s, eel destruction campaigns were started.' Even during the 1960s, after the intensive campaigns had finished, eels were still considered a nuisance and many regions kept a bounty on them, encouraging young boys, in particular, to kill them. 'As a kid growing up in Blenheim in the 1950s and '60s, I used to get paid tuppence an eel tail,' recalls Don.

There was no real understanding at the time of the ecological importance of freshwater eels, but an experiment in the 1970s changed everything. It involved removing eels from a trout stream for five years, and showed that eels actually improve a trout fishery. Big longfin eels are the apex predator, keeping the trout population in check. Without eels, you certainly get more trout, but they are smaller because they're competing with one another for food. With eels feeding on young trout, the streams are less crowded and individual trout can grow bigger, much to the surprise and delight of anglers.[18] This knowledge effectively stopped the bounty system on eels.

However, a report released by Parliamentary Commissioner for the Environment Dr Jan Wright in 2013 warned that the native longfin eel was on a 'slow path to extinction'.[19] In 2016, the government initiated controls on the number of longfin eels that could be caught in certain areas of Aotearoa. Headlines such as 'Longfin eel turned into dog tucker' have highlighted the growing unease among many New Zealanders that a native species as rare as kiwi can be legally caught and exported for private profit. Some are calling for a rāhui on commercial harvesting. People have tried to farm eels, says Don, but it's never really taken off, for various technical, economic and social reasons.

At the same time, attitudes are changing, especially as the wider community has come to recognise how important eels are to Māori and also more aware of the rarity of longfin eels.[20] In 2015, a man in Canterbury publicly apologised for catching a 1.4-metre, 10-kilogram female longfin eel, believed to be about 90 years old.[21] When he learnt from a Fish & Game officer that longfin eels are at risk of extinction, he said, 'If I'd known . . . I'd have said [to my son], just tickle it or take pictures.' More New Zealanders are now viewing eels as part of our heritage — just as iconic as kiwi and kererū.

Tracking tuna

Don says that discovering oceanic spawning areas is the holy grail of eel research. 'It has little economic or ecological importance, but has exercised many great minds over many years. It's very much a black box that we're still investigating.'

All 19 of the world's species of freshwater eel are 'hard-wired' for tropical spawning, but locating their spawning areas in the open ocean is very challenging. Spawning sites have been found for around half of the species but remain unknown for the New Zealand ones. For years, scientists have tried to pinpoint them by heading out to sea with fine-mesh nets

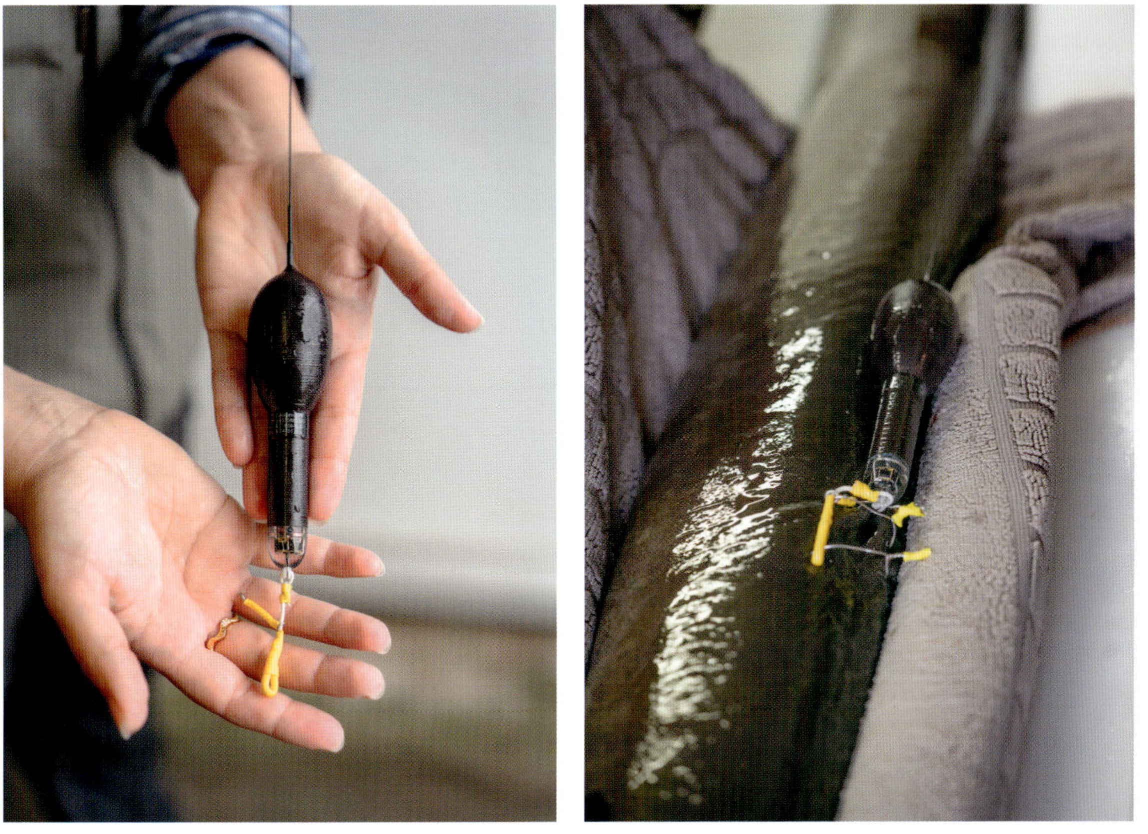

A pop-up satellite tracking tag is attached to a silver migrating eel. These tags record data on temperature, depth and location, and eventually transmit that information to a satellite.

to locate progressively smaller larvae. Don knows they will be somewhere in the tropics, and the general area has been narrowed down to somewhere near Fiji or New Caledonia:

> Even eels like ours, that we regard as temperate species, will all migrate back to the tropics. That's where they originated. So finding where those areas are has been something which has intrigued fishery scientists for perhaps over 100 years. These days we rely more on satellite-tracking tags. So, you can tag eels and just virtually sit at your desk and get the result sent back via satellites.

At this point, data are coming in from satellite-tracked longfins, and here Don can hardly contain his excitement: 'the likelihood is that we have discovered their spawning area!' It's not often that a scientist gets to discover the holy grail in their field of research, but in Don's case the decades of dedication and use of new technology have paid off.

SLIPPERY SUCKERS

Eels aren't the only species Don has studied in New Zealand. With their long, slippery bodies, lampreys — also called kanakana or piharau — may look a bit like eels, but these unusual and ancient fish are very different in many ways. They are boneless, having cartilage like a shark. They are also jawless. Instead of a mouth, they have a permanently open suction disc, decked with concentric rings of vicious-looking teeth, evolved to latch onto bigger fish and marine mammals and ghoulishly suck their blood. As they have been around for over 360 million years, it's tempting to think they may once have latched onto the hide of a plesiosaur.

More a marine species than a wetland one, lampreys still make use of both saltwater and freshwater, but their lifestyle is the opposite of that chosen by freshwater eels. Lampreys spend their adult life at sea and move into freshwater streams to spawn. The young spend 3–4 years filter-feeding from the sediments in the bed of their rocky stream, then migrate to the ocean. There they become piscine vampires or, as the scientists prefer to say, parasites. Adults spend another 3–4 years at sea. They return to freshwater between late autumn and early spring, and spend 14–16 months maturing before they spawn and die.[22]

Aotearoa New Zealand has one species of lamprey, *Geotria australis*, which is also found in Australia. It is the only species in the family Geotriidae,[23] meaning it has no very close relatives. Lampreys are widely distributed across New Zealand.

Their New Zealand breeding grounds remained a mystery until 2013, when Dr Cindy Baker, Don and several other NIWA scientists, along with colleagues from the USA, set out to spy on their movements. They inserted passive integrated transponder tags (a bit like a transmitting microchip) into the body cavities of adult lampreys migrating into freshwater, and tracked them to find their spawning sites. During two years of tracing their movements, the team found six lamprey nests, each containing eggs and an adult pair. They were all located under large boulders, with the eggs clustered together in a creamy gelatinous blob stuck to the underside of the boulder. Like salmon and eels, lampreys breed only once, but the team discovered that both the males and the females survived more than

three months after spawning, with the males guarding the eggs. The research team report that their observations offered 'the first insights to the long-standing mystery of Southern Hemisphere lamprey reproduction.'[24]

Māori consider lampreys a taonga, a valued species, and developed sophisticated methods to catch and prepare the delicacy. But this is a species in decline, and in 2011 NIWA began leading a six-year research project seeking to increase our understanding of piharau using mātauranga Māori, social science and biophysical science approaches. Their research included interviews with Māori piharau fishers to gain a better insight into the past and present abundance and distribution of this taonga. One part of the study is trialling what the researchers call lamprey love lofts — artificial nests to increase safe spawning habitat for the species. Baker explains: 'They're basically an upside-down kitchen sink made with fibreglass that mimics the underside of a boulder. There is room for a camera, so the lamprey can be easily observed. We've put them in areas we know they go to so we know they will come in contact with them — but whether they use them or not is still to be discovered.'[25]

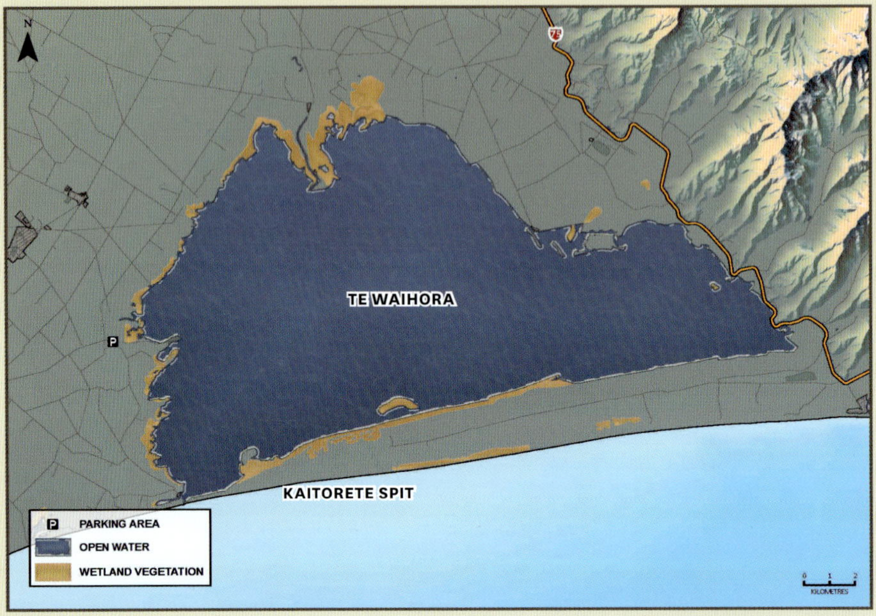

Te Waihora

Coastal lake

An hour's drive south of Christchurch, Te Waihora Lake Ellesmere is Don's favourite wetland. It's not a pretty lake, as the water is discoloured, but it is vast and very productive for fish, waterfowl and the threatened matuku (Australasian bittern). The lagoon is shallow (just 1.4 metres deep on average), semi-salty and usually closed to the sea by the 26-kilometre-long Kaitorete Spit.

Highlights

- More than 160 species of bird and 37 species of native fish
- High numbers of migratory shorebirds — up to 98,000 at any one time
- On Kaitorete Spit, the nation's largest remaining area of pīngao (golden sand sedge), a native sand binder prized by Māori for weaving
- Habitat for species that are unique to this location, including a flightless moth and a furry-leaved herb that is yet to be formally identified and is currently simply called *Craspedia* 'Kaitorete'
- You may also spot native katipō spiders and lizards.

Access

The lake is best explored from its margins, as boating can be tricky if it gets windy and swimming is not advised as the lagoon may experience toxic algal blooms. The Waihora Ellesmere Trust has produced a

brochure describing the walks and boat access points, some of which are described below.[26]

- Harts Creek Track on the western side of the lake is an easy 2.7-kilometre return walk to a bird hide, where you can view abundant birdlife. It's a great walk for children and families. The track starts at Harts Road, off Timber Yard Road (which leads to the Lakeside Domain and campsite, on a promontory on the lake). Allow an hour and a half
- You can also cycle along an old railway embankment adjacent to the eastern side of the lake and to nearby Te Roto o Wairewa Lake Forsyth. The cycleway starts at the restored Motukarara Railway Station
- For a longer excursion, you can access Kaitorete Spit via several unformed paper roads leading off Bayleys Road.

www.wetlandtrust.org.nz/get-involved/visit-wetlands/wetlands-you-can-visit-in-the-canterbury-region

Eel spotting

For an easy close-up view of longfin eels, head to the concrete terrace steps on the Ōtakaro Avon River in central Ōtautahi Christchurch. The eels living under the steps here have become used to people and may come out to greet you if you tap the water gently. Please do not feed them, as unnatural food like bread is not healthy for wildlife.

Mysterious Matuku

How do you study a bird that is a master of disguise? Dr Emma Williams has developed techniques to track cryptic wetland birds, including the mysterious matuku.

Dr Emma Williams

Specialist field: Ecology, conservation and management of native birds

Currently works for: Department of Conservation, Ōtautahi Christchurch, as a Science Advisor for Mobile Threatened Species

Emma Williams, PhD, is a wildlife ecologist who specialises in cryptic birds — well-camouflaged wetland species, including the rare pūweto (spotless crake) and Nationally Critical matuku (Australasian bittern). She spends substantial time in the field studying their populations and distributions, so that the birds and their habitats can be better understood and better managed.

WHEN MALE MATUKU (Australasian bitterns, *Botaurus poiciloptilus*) call, it's described as booming. Like the boom of a distant canon? Not at all. Emma Williams does her best to imitate how bitterns really sound. Well-practised birds let forth a sequence of long, deep 'wwwwrrrooomp' calls. It's a sound that emanates from deep down inside. However, the younger birds, like teenagers with their voices breaking, make a pained 'uuuuuuhn' sound. It's hard not to laugh at Emma's impersonation, but she explains that their whole body has to change — their oesophagus needs to thicken up to cope with the amount of air rushing out. She adds that the largely silent females become interested only when the young ones have progressed from 'uuuuuuhn' to a resonant 'wwwwrrrooomp'.

Matuku are among the biggest birds in Aotearoa New Zealand, with healthy males weighing up to 2 kilograms. They belong to the heron family, and so have an elegantly elongated body and long legs. Like their endemic cousin, the kōtuku (white heron), matuku are classed as Nationally Critical — in other words, the species faces an immediate high risk of extinction. By draining wetlands all over the country to make way for cities and human food production, matuku habitat has shrunk radically along with the places they forage for their food, which includes fish, tuna (freshwater eels), molluscs, kōura (freshwater crayfish), spiders, insects, worms, frogs and lizards. The last 40 years in particular have been really tough, with rising predator numbers also to blame. The birds are far from safe deep in their swampy habitats — cats, stoats and rats don't mind getting their feet wet if it means a feast

A clutch of bittern chicks in their platform nest built of dead raupō leaves.

of matuku eggs and young chicks. Today, fewer than a thousand birds remain nationwide — a far cry from their status in the nineteenth century, when they were documented as being 'remarkably numerous' in many parts of the country.[1]

How to find a cryptic bird

To the untrained eye, many of New Zealand's wetland birds are virtually invisible. In scientific terms they're called cryptic species, blending seamlessly into the reed- and rush-filled parts of swamps thanks to their excellent camouflage.

For such large birds, matuku are surprisingly hard to spot even when they venture out to forage on stopbanks, flooded paddocks or roadsides, or in farm drains. Their hiding tactic is to freeze when threatened, standing perfectly still, and point their bill skyward so that their streaky feathers align perfectly with the vertical patterning of their reedland habitat. But they have another unexpected way of hiding, as Emma recounts:

> The first of many times they flummoxed me was at Whangamarino wetland in the North Island. We'd spotted one foraging along the edge of a paddock . . . As I lifted the binoculars to my eyes, I could clearly see it there with the naked eye — standing in freeze pose . . . but in the split second it took to look through the binoculars, it had disappeared. At first, I thought I must

> have been looking in the wrong place, so I took the binoculars away again, but still no bittern . . . there wasn't time for it to have flown or walked anywhere. Thinking I must have gone mad, I decided to walk down there and take a closer look . . . lo and behold, there was the bittern in the exact same spot where I'd seen it before but now lying flat on its belly like a pancake in the grass!

So how does a scientist study a bird that has a habit of vanishing? Emma set out to develop methods to detect matuku using four of their own characteristics: the birds' body heat, the sounds they make, their movements and their distinctive scent. Her aim was to find which method(s) would be most useful to find and track matuku, and under which conditions.

A bird's body heat can be picked up by thermal cameras. Thanks to bleeping radio transmitters previously placed on one particular matuku, Emma and her team knew that Tama Tomoana — named after a well-known champion of Māori performing arts — was present somewhere in the raupō at Lake Whatumā in Hawke's Bay. However, he was almost impossible to see. They borrowed a thermal camera from New Zealand Land Search and Rescue to pinpoint his exact location, but even with that, Tama was hard to find in the thick vegetation. To up the game with thermal cameras, they attached one to a drone and went looking for another tagged bird, Teddy (named by the local iwi), a resident of Little Waihi wetland in the Bay of Plenty. The team trialled the camera at different times of the day and night. Flying it over the wetland, they successfully picked up other bird species, a whole suite of predators (cats, stoats, rats) and even cows straying where they definitely weren't supposed to be. But the wily Teddy wasn't spotted, according to Emma, 'possibly because he could hear the drone and then just hunkered down into the vegetation'. Despite the thermal no-show, the camera did detect an untagged matuku along a farm drain, and from that, the team were able to determine the best time to detect birds — between sunrise and 10 a.m., when the temperature difference between the warm-blooded matuku and their surroundings is greatest.

Matuku calls — those distinctive low-frequency wwwwrrrooomp sounds — are the easiest way to find a bird, not least because they are loud and can be heard up to 4 kilometres away. Emma's work has shown that male matuku are most likely to be heard (and therefore found) an hour before sunrise or just after the sun has set. Counting calls is a great way to tell how many male matuku are in a wetland and whether they are trying to breed. However, listening for calls doesn't say much about the presence of females, chicks or the males that are too young to breed, as they're all mostly silent, so other detection methods are needed for them.

Matuku can also be betrayed by their movements. To detect birds slinking through Lake Whatumā and at Harts Creek in Canterbury, Emma set up game cameras along the routes radio-tagged birds were known to frequent and in areas where a female was suspected to be nesting. The cameras are triggered by movement of the passer-by — ideally a matuku, but also any predators prowling around for a meal. This is a great way

FROM KIWI KILLERS TO CONSERVATION CANINES

The first Māori to arrive in Aotearoa around the thirteenth century brought kurī with them in their waka. These stocky, long-haired dogs provided meat for eating, skin for cloaks, and bones to make fishhooks and ornaments. Kurī were also hunting dogs — trained to catch kiwi, pūkeko and other ground-based birds. They are extinct now after many years of interbreeding with the dogs of later settlers.[2]

But dogs still worry our wildlife — every year, more of our native birds fall prey to pet and feral dogs. Part of the solution is kiwi aversion training. The Department of Conservation has set up a programme to teach dogs that live near kiwi habitat, and pig-hunting dogs taken into native forests, to avoid the vulnerable birds.[3] The next step up is conservation dog training. Emma's dog Kimi is a graduate of the class of 2015.

It can take up to 18 months to fully train a conservation dog and its handler, but it's worth it. The dogs power through jobs that would take mere humans forever, sniffing out protected species (mainly birds) for surveying, monitoring or transfer to a safer environment. They're also adept at sniffing out unwanted species, including animal pests such as rats, mice, stoats, ferrets, weasels and Argentine ants, and can even use their sensitive snouts to detect plant pests.

Aotearoa New Zealand has a long history of training dogs to benefit conservation. In the late 1800s, pioneer conservationist Richard Henry began using dogs to find flightless kākāpō and kiwi on the mainland, so that he could transfer them to an offshore island free from introduced predators.[4] While Richard Henry made history with his world-leading work, the names of the dogs that made it all possible have largely been forgotten.

to detect females and chicks, especially if you put cameras on tracks or in places where there are lots of patches of powder down from the moulting chicks.*

As for detecting matuku by smell, this is where the superior snouts of dogs outshine any human endeavour. Emma can only describe the scent of matuku as 'fishy' on account of the birds' diet, but she suggests asking Kimi, her specially trained black Labrador: 'She looks at me very despairingly sometimes when I can't find the bird and we're within 10 metres of it. She'll be trying to show me where it is, because it's so well hidden, and she just looks at me like, "What is wrong with you? It's just there and it's really stinky!"'

---

* Powder down feathers are patches of special feathers that produce a dust, thought to be used by the bird to clean and waterproof other feathers and keep parasites at bay.

Emma and her sniffer dog Kimi, outfitted with life jacket and camera, track a bittern using a transmitter.

INVISIBLE INHABITANTS

New Zealand isn't known for its flashy avifauna. Africa has flamboyant flamingos. Asia has magnificent cartoon-coloured mandarin ducks. North America has tuxedo-clad loons. South American wetlands are brightened by the stunning red ibis. Here in Aotearoa we have the fernbird, which, with its straggly tail, looks like a sparrow that's been dragged through a hedge backwards!

Most of our native birds are on the medium to small side, and not very brightly coloured. But they're still taonga — we even identify ourselves as Kiwis. Of the 245 species of bird breeding in New Zealand before humans arrived, 71 per cent were endemic — in other words, not found anywhere else in the world.[5] Most New Zealanders would easily recognise the birds that pass through our backyards (kererū, tūī, pīwakawaka), along with our national icon, the flightless kiwi. Ask about our wetland birds, however, and many people will draw a blank. They may mention the pūkeko, memorable for its strident squawks, big red bill and fluffy white bottom. But because our other wetland birds prefer their privacy, they remain little known.

The late Waikato farmer and conservationist Gordon Stephenson, founder of the National Wetland Trust, once described wetlands as our 'shy places', and this is certainly true of our cryptic birds. The koitareke (marsh crake), pūweto (spotless crake), mioweka (banded rail, below left), mātātā (fernbird, below right) and matuku (Australasian bittern) all quietly blend into their waterlogged surroundings, each sporting a palette of wetland-coloured feathers. In the absence of mammalian predators, their colouration, patterning and inclination to freeze evolved as protection from sharp-eyed avian predators flying above — the kārearea (New Zealand falcon) and kāhu (swamp harrier). While it may be a clever disguise from aerial attack, it's useless against mammalian predators, which follow their nose to find a feathered feast behind a flax bush.

Emma found that all four methods to locate matuku at closer range worked, but none covered all the bases. Kimi's olfactory expertise was best for detecting birds along wetland edges because she could sit in a kayak and be positioned downwind. She was also good at finding females, chicks and nests — they're all pretty stinky too. Automatic sound-recording devices that can be left out in the field for weeks at a time were a very cost-effective and convenient way to detect wwwwrrrooomp-ing males during the breeding season. Trail cameras (triggered by movement and heat) proved useful for detecting less vocal birds (females and non-breeding males), but it's risky leaving expensive cameras in a wetland when rainy weather raises water levels and can drown equipment. Thermal-imagery drone cameras showed potential, but detectability was dependent on the birds' behaviour — they're not easy to pick up when they flatten themselves into a pancake in thick vegetation.

How to catch a cryptic bird

Just as each method of detecting a bird has its ins and outs, so catching matuku has been a trial-and-error process too. Fieldwork flops do occur, but there is always something useful to be learnt in the process. Catching a bird is where detailed studies can begin — taking measurements, documenting observations, attaching tracking devices and identification aids such as leg bands, and taking samples — before gently releasing the creature back to its habitat.

In the UK, where Emma grew up, large, almost invisible nets called mist nets are used to catch great bitterns, so it made sense to trial the same approach here. Using a net relies on the bird being impulsive — flying into the mesh to attack an imagined rival, only to be captured (untangling the birds from the filaments is another story). Emma explains: 'You put a speaker on either side of the net — once you've got the bird there at one speaker you switch speakers to the other side . . . they get kind of mad that the other "bird" has moved and they just shoot across, and that's when they go into the net. Well, that's the theory.'

The reality was rather different: 'We'd get the bird to come close to the net, and that was great! . . . But then it would just sit there and boom at the speaker for 45 minutes . . . not really making those impulsive movements into the net. They had a tendency to walk up and down the net and not go across. In short, it just didn't work for us.'

There are logistical challenges to netting as well. It's a night-time job and huge teams need to be hidden in the wetland, so it's cold, wet and awkward. On top of that, dampness can seep into the speaker electronics, rendering the recorded calls unappealingly feeble.

But there is a silver lining. The team learnt from mist netting that matuku bide their time on hearing another bird. They just seem to hang out for several hours, booming at each other before attempting to approach, meaning that traps shouldn't be checked too frequently for risk of scaring the birds away. The sensitivity of the species matters too, as Emma explains: 'With any shy species, you don't want to be in the area when you're trying to catch them. It's much better to just have a passive method and then come back at a set time or have a camera or something to make sure you know whether they've been caught.'

Big traps placed back to back with a speaker in between proved successful, the clincher being a mirror at the back of the trap opposite the entry door to trick the bittern into thinking its own reflection was a rival bird. This method is straightforward, says Emma: 'The bird hears the call and eventually flies closer. It lands on the track we've made to the trap, which is positioned to line the bird up perfectly with the mirrors. Once the matuku has spotted the interloper it believes to be calling (which is actually its own reflection), it enters the cage to investigate, triggering a simple treadle plate that releases the door.'

How to handle a large bird with a sharp beak

Extracting a disgruntled matuku from a cage is a two-person job — the birds are around the size of a small goose, with a wickedly strong, sharp beak. One brave soul needs to gently slide their hand through the bars of the trap to hold the head, while the other person opens the big door and slides a bag over the bird, and then slowly brings the feathery bundle out of the cage. Despite being hoodwinked, trapped matuku remain remarkably calm, as Emma explains: 'Once you've got their head, and then their wings, there isn't really much they can do, so they tend to relax, and then we can just get on with what we need to be doing.'

Field research in a swamp takes a lot of planning. Before setting the traps, the field team establish a dry area to process the bird and lay tracks to allow them to move around the wetland more easily. There are also basic safety considerations for the handlers: protective eyewear is 'definitely, absolutely needed!' An important part of the birds' diet is fish and, being a heron, they excel at snatching prey in a nanosecond. Whether it's a shiny fish or a shiny eye makes no difference to a matuku.

If a bird is being tagged, the first task is to assess its body condition — only healthy birds get to carry the 20-gram tracking tag and harness (20 grams is a bit less than the weight of an AA battery). The team attach a metal bracelet-like band with a unique number to one of its legs. If there's still time and the bird remains relaxed, they measure the length of the tarsus (the avian equivalent of a shin bone) and weigh the bird. The team then measures the wings and assesses the moult.* They also look at the plumage to tease out the age and gender of each bird, and feathers are sometimes taken for DNA testing to double-check the sex of the bird. The combination of different measures and observations provides a lot of information about the matuku population, as Emma explains:

> Like, have we got an ageing population in some areas? This would be a concern for us, because if they're getting too old to breed then we've got a problem, as it becomes harder for the population to recover. It's the same if

* The moult is the annual shedding of old and damaged feathers. Each of the 10 main feathers on the wing is given a score from 0 (old, ratty and worn) to 5 (fully grown and fresh).

Emma catches matuku by placing a mirror at the back of the trap to trick the bittern into thinking its own reflection is a rival bird.

> there's a sex bias in the population. If there aren't many females, then fewer offspring are produced each year. Because not that much is known about bittern . . . it takes a bit of time to work that out.

Matuku on the move

Better methods to detect and catch birds, coupled with improved technology, have revealed a few surprises. No one knew that matuku were so mobile! One bird clocked an impressive 330 kilometres in a single flight, while others regularly cover 100 kilometres flying back and forth between the 7000-hectare Whangamarino Ramsar wetland in the heart of the Waikato and wetlands in the neighbouring Bay of Plenty region. Why? Historically, matuku probably didn't need to move around as much or fly as far. But the wetlands they live in now are a far cry from what they used to be — they're smaller and there are fewer of them, so they're much further apart. They're also degraded: Whangamarino, for instance, is now managed as part of the flood-control scheme, so its water levels change much more now than they used to.

> We're only just starting to see how bird movements are changing across different years . . . For example, at Whangamarino wetland we notice a definite effect from changes in the water level. Birds move to different parts of the wetland based on water-level changes and leave the area completely if water levels in their favourite spots become too deep or too shallow. I imagine this is to do with food. Their favourite foods live in water, and they stalk and stab their prey, so they need to be able to see and catch it without diving. This means that they rely on clear and shallow water or prey being close to the surface for them to survive. As water systems are all connected, and wetlands are flat, water-level changes affect large areas (or catchments) at once. So if a site becomes unsuitable for food, then the birds must move long distances to find somewhere else that is suitable — another catchment or a site that has not been affected in the same way. Once you realise this, the fact that the birds are dependent on a network of sites, instead of just one or two, is not really that much of a surprise.

Despite decades of drainage, an influx of animal pests and weeds, and changes to natural water flows through weirs and floodgates, Whangamarino is still an important habitat for matuku. At the last count in 2015, the wetland had 15 booming males — not many, considering that 145 male bittern were recorded there in 1981. That could be a 90 per cent decline over 35 years, the caveat being that the monitoring methods differed slightly between the two counts.

Emma's catch-and-release studies show that matuku from Waikato and Hawke's Bay are generally heavier and bigger than those found in the Bay of Plenty and Canterbury. She's not totally sure why, but she does know that healthy birds make hearty, resonant

HIGH-TECH TRAVELLERS

The first tracking devices used for matuku were VHF radio transmitters. To collect data, a willing researcher had to trudge about with a hand-held receiver, waving an aerial around to pick up the signal. Aside from the effort needed, logistical constraints meant that birds could be tracked for only short distances — around 3 kilometres — until they disappeared out of range. A better device was needed, preferably GPS-based.

Avian tracking technology today is developing fast. The tags Emma now uses are superlight (less than 3 per cent of the bird's body weight), have generous battery life (2–4 years) and a solar cell, and are not too pricey. The technology allows the birds' journeys to be tracked in real time, as their travel data are downloaded every time they pass a cellphone tower.

The tracking tags are attached to the bird using a special harness with a built-in weak link made from linen. Being a natural fibre, the link degrades over time so that the matuku isn't stuck wearing a transmitter for life. It also means the harness can be shed if it gets caught on something. With so few birds, the welfare of every remaining matuku is vital.

'wwwwrrrooomp' calls. Those with poor body condition typically manage only a deflated 'uuuuuuhn' sound — just like the youngsters with their undeveloped oesophagi. Perhaps there is more or better food in the Waikato and Hawke's Bay compared with the Bay of Plenty and Canterbury?

Of all members of the heron family, matuku have one of the most specific feeding methods. Others, Emma explains, 'do all sorts of cool stuff, including diving, stalking and there's even some clever heron species that shade the water by putting their wings over their heads (like an umbrella) so they can see their prey better. But bitterns . . . well, they basically see their prey and then they stab it, and that's about it.'

Becoming a wildlife scientist

Carrying out novel research in wetlands following the unsociable schedules of our native species doesn't make for a great job advertisement. Imagine it: 'Applicant must enjoy being wet, uncomfortable and work long, random hours. Experience dodging large, sharp beaks a plus, but not essential.' It's not for everyone, but Emma is passionate about her work.

Like many researchers, her pathway towards a science career followed a familiar pattern — first a childhood spent getting muddy in pursuit of knowledge. Then, as a teen, filling notebooks with diagrams, ideas and observations, and joining a society. After school, stepping into university study with mentors and teachers to guide and

BIRD OF THE YEAR

The matuku was one of the 'Famous Five' birds nominated for Bird of the Year 2020 by the Wetland Alliance (of which the National Wetland Trust was a proud member).[6] The online competition, run by Forest & Bird since 2005, has proved an enormously popular way of showcasing our quirky, charismatic native species. Despite the overwhelming success of the competition (55,583 voters in 2020), 'fowl play' has been documented. International media was aflutter in 2020 with reports of fake news and dodgy voting tactics, drawing analogies to presidential election upsets that were occurring elsewhere around the globe at the time.[7]

A good number of the winners since 2005 have been either household names (tūī, kiwi), endearingly poorly behaved (kea), plain loud (pūkeko) or just plain cute (pīwakawaka), so being a cryptic introvert might not always be helpful. The Nationally Critical matuku scored a paltry 500 votes in 2017, while in the same year the ubiquitous kererū received 4500 votes.[8] There was excitement among the Famous Five alliance that matuku might finally have won when the 2020 headlines rang out 'OK, Boomer!', but it was a rival booming bird, the kākāpō, that took top billing.

Matuku clearly need a strong marketing team, because as a nation we're obviously not familiar enough with our wetland birds.

shape learning. And then, working in a job that brings a research area into focus. Unlike others, however, it took Emma a little longer to realise she was actually already on the way to becoming a wildlife scientist — no thanks to the limited view of her school careers advisor.

> I always loved science and wildlife related topics, but 'wildlife scientist' wasn't exactly a recognised career for someone growing up in Stoke-on-Trent, England. When I told the careers advisor at school that I wanted to do something with wildlife, she said, 'Well, I've got a card about becoming a veterinarian or one about animal husbandry, which would you like?' I was definitely one of those kids that left school feeling a bit lost and unsure what I was going to do with my life.

Emma did plenty of odd jobs while she was studying: sorting mail in the post office, working as a waitress, doing data entry for researchers. At least the income meant she could volunteer on the types of conservation projects that really struck a chord with her. She sums up her journey: 'In the end, I've discovered this career path by just following what interested me and felt right . . . Over time, my passion for conservation work and ecology developed, and I built up enough experience to start getting paid for the work I loved and believed in.'

Studying wetland birds takes determination and stamina, but Emma remains unfazed by spending a lot of time in swamps. The pursuit of learning drives her — it's exciting and unpredictable, and it helps flesh out the many unknowns about this intriguing bird. It's also helped her develop a unique skill set — not every wildlife ecologist can accurately mimic a mature bittern's 'wwwwrrrooomp'!

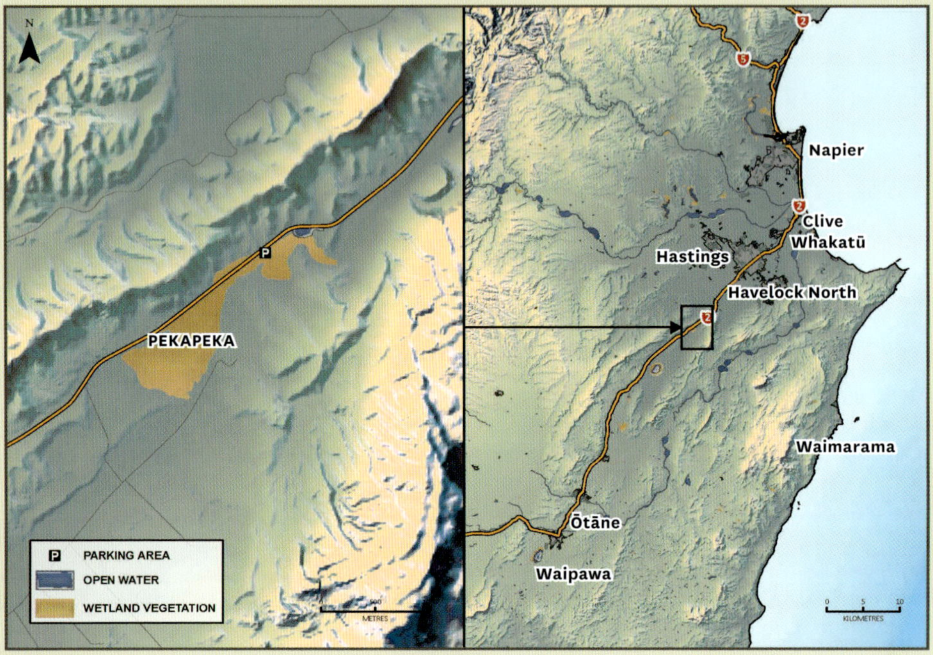

Pekapeka Swamp

Lowland swamp

One place you might spot a matuku is Pekapeka Swamp, south of Hastings in Hawke's Bay. This was once part of an extensive wetland complex, although buried tree roots suggest that the area was forested more than 10,000 years ago. Enormously significant to Māori, the site has wāhi tapu (sacred site) status under the Historic Places Act 1993. The name of the swamp, Pekapeka, means 'bat', after the native bats that roosted in caves nearby.

However, in recent times the site wasn't viewed as important at all — just a convenient place to dump rubbish, including debris from the demolition of two hotels! Strolling along boardwalks that snake their way across this picturesque wetland today, it's hard to comprehend the remarkable make-over that has occurred here since the 1990s. Tonnes of rubbish were removed, although some of the rubble and reinforcing rods have been left as reminders of how undervalued wetlands were in the past. Thickets of invasive willows are also gone, opening up broad vistas. As one of the few remaining wetlands of this type left in Hawke's Bay, Pekapeka is now a biodiversity hotspot.

Highlights

- If you are quiet, you may hear or spot matuku or one of the other wetland birds, including weweia (New Zealand dabchick), matuku moana (white-faced heron), tētē moroiti (grey teal), koitareke (marsh crake), pūweto (spotless crake), poaka (pied stilt), kōtare (sacred kingfisher) and several species of kawau (cormorant)
- Easy boardwalk
- Interpretation panels.

Access

Family-friendly Pekapeka Swamp is in Pekapeka Regional Park, a 12-kilometre drive south of Hastings on State Highway 2. There are wetland paths and boardwalks, picnic tables, bird hides, and information panels that give the wetland a voice, telling of its history, ecology and cultural value to Māori.

Note that there are no toilets in the park and no dogs are allowed, and Pekapeka is closed to the public during the first weekend of May each year for gamebird shooting.

www.wetlandtrust.org.nz/wetlands-to-visit-in-the-hawkes-bay-and-gisborne-regions

Tipping Points

How can we tell when a wetland is about to reach a point of no return? Dr Hugh Robertson's research is exploring environmental limits to help improve how we protect wetland health.

Dr Hugh Robertson

Specialist field: Wetland ecology, conservation, restoration and ecological limit setting

Currently works for: Department of Conservation, Whakatū Nelson

Hugh Robertson, BSc (Hons), PhD, is a Principal Freshwater Scientist for the Department of Conservation (DOC) and operates at the interface of science, policy and management. His role involves collaborating with specialists on research projects to improve our understanding of wetland ecological and hydrological processes, for application towards biodiversity protection, restoration strategies, defining limits of acceptable change for aquatic ecosystems, protected area establishment and monitoring, and assessing the impacts of agricultural and other development pressures on wetlands. He is the scientific lead of the DOC Arawai Kākāriki national wetland restoration programme, has been a national representative for the Ramsar Convention as New Zealand's Science and Technical Review Panel (STRP) National Focal Point since 2008, and was a member of the International STRP for the Ramsar Convention in 2019–2022, representing the Oceania region. Hugh has published more than 30 peer-reviewed scientific papers and reports on wetland ecology and freshwater conservation.

DEADLY TIGER SNAKES and mobs of kangaroos crashing through the reedlands caused Hugh Robertson more than one moment of existential threat while he worked on his PhD in Australia's Murray–Darling Basin wetlands. But back home in Aotearoa New Zealand, his work has been less about threats to his own existence and more about threats to wetlands' existence.

The politics and catchment-management challenges of wetland conservation in Australia sparked Hugh's interest in how science can influence wetland policy and management for the better. In his role as Principal Freshwater Scientist for the Department of Conservation (DOC), he has a broad mandate, looking out for wetlands from the top to the bottom of New Zealand, and holding international positions under the Ramsar Convention. In addition to leading research and monitoring projects in wetlands on public conservation land, he provides expert opinion on tricky issues and is often called on to evaluate whether proposed development projects may harm a wetland.

As an applied scientist, Hugh uses his broad range of wetland knowledge to help DOC respond to such proposals: 'We need to know where the limits are beyond which an activity will have a serious or irreversible negative effect on a wetland. So, when a large-scale resource consent is up for review, or there's a big regional plan change, we can go, well, what is the target; what is the environmental standard for this place?' Hugh needs to determine if the proposed activity or development will adversely affect a significant wetland, and if so, how much pressure that wetland can cope with.

Pushing wetlands to their limits

We rely on wetlands to do a wide range of things for us. They absorb floodwaters. They trap sediment. They take up excess nutrients. But are we asking too much and overworking them? How much is too much? Too many nutrients? Too much sediment? Too much drainage? Even too much water? If you push a waterbody too far, its ecological state can 'flip' — it becomes so overwhelmed that it simply can't recover.

This happened to Lake Waikare in the Waikato region when its water levels were lowered in the 1970s as part of a flood protection scheme. The shallow water was constantly stirred up by wind and became turbid and dark. The native submerged plants that oxygenated the lake were unable to grow in the gloom, and died. As they rotted, adding to the slurry of nutrients coming from the surrounding catchment, the lake became a haven for toxic cyanobacteria and algae — including a species that periodically turns the lake a disconcerting orange colour —making it almost impossible for light-demanding native plants to re-establish.

Should we try to resuscitate Waikare and other wetlands like it? Or should we try to recognise the early signs of stress and stop them from getting sick in the first place? While it's unlikely that many of New Zealand's degraded wetlands will ever return to an unmodified, pristine state, we need to consider what is an acceptable level of degradation. Hugh makes the best use of scientific data to help guide decisions that will hopefully prevent wetlands from degrading well before they reach the end of their tether. Five highly significant wetlands — Waituna Lagoon, Whangamarino wetland, Ō Tū Wharekai (Ashburton Basin wetlands), Moawhitu wetland and Kaimaumau-Motutangi peatland — have been the subject of Hugh's efforts to work out their environmental limits, so that he can encourage councils and local communities to take action before it's too late.

A lagoon on the edge — too many nutrients (and not enough water)

If you visit Waituna Lagoon on a warm, sunny day, count yourself lucky. This is the sort of place that can take your breath away, not just because it's in a stunning setting but also because the site — at the bottom of the South Island, about as close as you can get to Antarctica on the New Zealand mainland — is frequently blustery and bitingly cold. Hugh would know; over the years, he's spent many hours in the lagoon on a quest to understand how to preserve its natural values.

The lagoon is the site of a prolonged ancient battle between the forces of water. Strong

Awarua peatland and Waituna Lagoon, Southland.

RAMSAR ROLES

Hugh has long been involved in international wetland circles through his roles under the Ramsar Convention on Wetlands of International Importance Especially as Waterfowl Habitat (usually simply referred to as the Ramsar Convention). Each of the 171 nations that have signed up to the convention has an appointed Science and Technical Review Panel (STRP) Focal Point, a scientific expert who represents their government in matters relating to technical aspects of the convention. These might include identifying wetlands of international importance to add to the Ramsar list, developing ecological monitoring protocols and leading research at those sites, or considering the technical implications of proposals discussed at three-yearly Conference of the Contracting Parties meetings.

Hugh has been the New Zealand STRP Focal Point since 2008 and was appointed to the 18-member international STRP panel as the Oceania regional technical expert for the 2019–2022 term, representing eight Pacific nations. According to Hugh, 'It is a privilege to be on the Ramsar panel. I've been leading a project reviewing the global status of agriculture and wetlands. It is sobering stuff, collating information from various global assessments on how agricultural development is continuing to put immense pressure on the world's wetlands. We need transformational change to become more sustainable, and to protect wetlands.'

ocean currents sweep quartz gravel and sand up the coastline, building a barrier beach that blocks off the lagoon from the sea. The influx of freshwater from the Awarua Plains fills the lagoon and it becomes a coastal lake. When the pressure of the freshwater becomes too great, it overtops the gravel bar and bursts through to find its way out to sea. Seawater flushes back in, turning the lake into a brackish (semi-salty) lagoon. This process has happened over and over, and the plants and animals that occupy Waituna Lagoon have become accustomed to the occasional flips and flops. Hugh observes that the natural state has generally been more in favour of freshwater for prolonged periods, and that the lake would break through the gravel bar when it got to somewhere around 3–4 metres above sea level.

But now, if the lake rises 3 metres above sea level it backs up onto adjacent farmland. So when it begins to fill up, rather than waiting for nature to take its sweet time, excavators are brought in to open up a channel and let the water out. Generally, this measure swings into action when the lake gets to about 2.2 metres above sea level.

The transition between closed (freshwater) and open (brackish/estuarine) conditions has provided a habitat for a wide range of native plants and animals. This made Waituna Lagoon highly valued and well used by local Ngāi Tahu people, and contributed to its selection as one of New Zealand's first Ramsar sites. Every spring, around 40,000 wading birds haul themselves across the world from their northern hemisphere tundra breeding grounds to gorge on the abundant seafood on offer in the lagoon's shallows and nearby Awarua Bay. Even though Waituna now lies at the base of an intensively farmed (and drained) catchment, and is opened to the sea more often than nature intended, it still retains important ecological values, including a submerged meadow of native *Ruppia* plants.

Ruppia has some rather delightful common names — widgeon weed, ditch grass, widgeon grass — none of which are commonly, or in fact ever, used in New Zealand, mainly because most Kiwis have never heard of a widgeon, a North American duck with rather fetching plumage. Here, we prefer the name horse's mane. To be honest, the plant can look like a bit of a tangled mess, so it's not a bad moniker. Whacky names aside, *Ruppia* is a sort of watery canary in a coal mine — its sudden demise can indicate that something is seriously amiss.

In 2009, it became apparent that Waituna Lagoon was suffering. According to Hugh, 'We saw an increase in nuisance algae, which was partly attributed to the increasing amounts of nitrogen and phosphorus flushing in from the farmland, and also a drop in the amount of *Ruppia*. There was a risk that opening the lagoon too often, switching from high water levels and freshwater conditions to low water levels and tidal conditions, was disrupting the life cycle of submerged plants like *Ruppia*.'

Hugh and his colleagues set up a research programme to answer some basic questions: How does the artificial opening of Waituna Lagoon alter the plant community? What is the ideal water management regime for the lagoon? He explains: 'The challenge was to design a sampling scheme that could tell us how the amount of submerged aquatic plants changed each year, and if they were happier in some places than others. So we established a series of vegetation sampling transects across the lagoon, with 48 sampling sites.'

How do you sample bottom-dwelling plants in cold water that's up to 2 metres deep? The solution involves a common garden implement. You lean over the side of a boat with a super-sized long-handled garden hoe and just scoop up the submerged plants. This inspired method allowed the team to pull up intact *Ruppia* plants, so they could measure their size (length) and biomass (abundance) before putting them back where they found them. The results, says Hugh, were telling:

> When we did our surveys, it coincided with three lagoon-opening events, in 2008, 2009 and 2010. The lagoon was open to the sea for 43 days the first year, 67 days the second year and 181 days in the last year. While it was open, the *Ruppia* plants were subject to tidal conditions, and they declined from 69 per cent cover in 2009 to just 23 per cent by 2011.

Ruppia plants are indispensable, forming the basis of the lagoon's food chain. While they are widespread in brackish wetlands, for reproduction they seem to like the water to be more on the fresh side than salty. Opening Waituna not only increases salinity, but it also makes the lagoon tidal, leaving some *Ruppia* plants high and dry at low tide. The solution seemed clear: just buy up the adjacent farmland, let it flood, and stop opening up the beach barrier. If only things were that simple. The world has moved on since the *Ruppia* moved in. Thousands of hectares of the lagoon's catchment are now intensively farmed, and that means a lot of nutrients (mostly nitrogen and phosphorus) are being flushed into Waituna. If the *Ruppia* aren't keen on being subjected to tidal forces, they are even less keen on nutrient-rich water, leading to something of a stand-off. Do we shut the lagoon to maintain the freshwater to protect the *Ruppia*? Or do we open the lagoon to let the nutrient-rich water out to protect the *Ruppia*? Hugh adds:

> While lagoon-opening events cause extreme changes in water depth and salinity that can limit macrophyte* growth, they also provide a mechanism to reduce the effects of high nutrient levels. We need to understand these trade-offs to make sensible management decisions about opening the lagoon to the sea. It turns out that returning to a completely natural hydrological regime may not be in the best interests of *Ruppia*, in the short term at least.

Despite the conundrum, Hugh's collaborative science has nevertheless been useful already — and has led to action. In 2017, a new resource consent authorising regular lagoon openings to protect farmland was issued, but this time it had specific conditions to reduce the number of openings during spring and summer, which is the key growing season for *Ruppia*. In 2021 the consent came up for review, and new scientific data was applied to find a way to seek further improvements in how the Waituna Lagoon's water levels are managed.

A topsy-turvy wetland — too much sediment

Whangamarino wetland in the North Island's Waikato Region is another internationally significant Ramsar wetland that has experienced a massive change in how water moves through it. The direction from which it receives floodwater has flipped from north to south — its hydrology has been turned upside down.

Whangamarino lies east of Waikato Te Awa Waikato River, from which it is separated by a stretch of low hills and a four-lane state highway. The wetland is fed by the Whangamarino and Maramarua Rivers, which flow in from hills to the northeast and southeast, merging at the northern end of the wetland and discharging into the Waikato River. In its natural state, water also seeped into the wetland through a wide area of swamp from Lake Waikare to the south. However, during high floods, the Waikato River backed up into the wetland, sometimes all the way into Lake Waikare. As floodwaters subsided and the Waikato settled back into its channel, the waters eventually drained northwards again, back into the river.

---

* Macrophytes are aquatic plants large enough to be seen by the naked eye. They include species of duckweed, pondweed, reeds, water milfoils and stoneworts. They don't include microscopic plants such as diatoms or periphyton.

A SITE OF CULTURAL SIGNIFICANCE

Waituna is a very special place for the tangata whenua, Ngāi Tahu, and as such has been formally recognised in law under the Ngāi Tahu Claims Settlement Act 1998. This Act acknowledges that, as a result of the long history of use and occupation of the Waituna area, there are wāhi tapu and wāhi taonga (sacred sites and treasured sites) all along its shores. These places hold the memories, traditions, victories and defeats of Ngāi Tahu tūpuna (ancestors).

The lagoon was a major mahinga kai (food-gathering area), providing a wide variety of reliable food resources for both nohoanga (seasonal hunter-gatherers) and permanent inhabitants. The wetlands also provided plant-based building, clothing, craft and cultural materials, including paru (black mud), particularly sought after for making dyes.[1]

Prior to the 1960s, farmland along the floodplains of the lower Waikato River was regularly inundated during these heavy rain events. To protect farms from damaging floods, the local drainage board devised a detailed engineering solution. It built a canal (the Pungarehu) from Lake Waikare to the Whangamarino wetland and partially drained the lake, creating capacity to hold water during floods. In extreme events, floodwaters from the Waikato can now be directed across the highway into Lake Waikare, to reduce the amount of water in the river. If the lake fills up and threatens to flood lakeside farmland, an outlet gate can be opened, sending floodwater directly into the Whangamarino wetland. The Waikare–Whangamarino flood control system can store up to 94.8 million cubic metres of water and lowers the flood peak in the Waikato River by around half a metre.

So, floodwater now often flows from Lake Waikare into the Whangamarino and ultimately back to the Waikato River, instead of backing up from the river into the wetland and lake. It's back to front, but the wetland still gets wet, so no biggie, right?

The problem is that the water in Waikare is not pretty. The lake has a muddy bed and continues to receive high sediment inputs from catchment inflows, and being so shallow (now 2 metres at its deepest), it's constantly stirred up by the wind-whipped water. Add to that a seething mass of invasive koi carp that love nothing better than grubbing around in the lake bed looking for tasty morsels, and you have a recipe for very turbid, muddy water. The resulting low light levels killed off the native macrophytes, but aquatic bacteria and algae loved the conditions and proliferated, sometimes turning the lake a startling orange — a rather good disguise for the vibrant koi carp. The macrophytes have never returned.

During floods, Lake Waikare's tainted water is now flushed down the Pungarehu Canal directly into the internationally significant wetland. Hugh is concerned about the effect of the sediments and nutrients from Lake Waikare on the special values of the Whangamarino: 'The hydrological system is completely turned around. Now [the

Whangamarino] gets much dirtier water coming out of Lake Waikare, because the Waikato River is quite clean compared to what's in Lake Waikare. So the water that is now inundating this wetland is much more nutrient-rich, much more sediment-rich. On average, around 20,000 tonnes of sediment are now discharged through the Waikare Gate into Whangamarino wetland every year.'

A major problem with all this excess sediment is how it affects the plants and animals that live there. Hang out around the margins of Whangamarino in spring, and you might be lucky enough to hear a low, resonant foghorn sound, booming gently from the raupō reeds — this is the territorial tune of a bachelor bittern looking for love.

However, Australasian bitterns (matuku) are masters at the art of disguise. They're enormous birds (bigger than a chicken), yet somehow manage to melt into the background to ensure that they aren't seen by predators. But being able to see their own prey is a big deal for them. They hunt by sight, looking for the slightest movement of a fish in the water before making a lethal stab with a beak you wouldn't want to mess with. For bitterns, the trouble with sediment is that it means they can't see their food — it's like wearing a blindfold at a buffet. This might make it sound like dirty water is great for native fish — there's less chance they'll be on the bittern's menu — but just imagine you're a fish swimming in gritty water. It clogs your gills, tearing your fine gill filaments,[2] and guess what — fish hunt by sight too. Not only that, but the food they eat, including underwater plants and the aquatic larvae of insects, also suffer and become less abundant. The effects of sediment in the wetland trickle right the way through the food chain, potentially leading to starvation all round.

In a Waikato Regional Council review of the Waikare discharge consent in 2015, Hugh and his fellow experts presented data showing that the flood scheme was affecting the wetland water quality, its vegetation and its inhabitants, including rare species such as the matuku. They used their scientific evidence to argue for a reduction in the level of sediment to match the 'background' level:

> We were able to use our data to say that the lake discharge is depositing sediment into the wetlands at a rate higher than what it used to be. In that situation, an RMA process* enabled positive action, using science to justify our recommendations. It led to a good outcome, with new efforts to reduce the amount of floodwater released into the wetland.

One example of the sort of scientific data used was a measure of sediment in the water called the nephelometric turbidity unit (NTU) — basically a gauge of how grungy the water looks. Hugh presented data showing that, at 95 NTU, the turbidity of the water in the Pungarehu Canal was four-and-a-half times greater than in the Whangamarino River where it enters the wetland (21 NTU), and more than twice the level where the water leaves the wetland (42 NTU). With this evidence he was able to recommend a target of

* A Resource Management Act process, in this case the review of a consent held by the Waikato Regional Council to operate the lower Waikato flood scheme.

Silt-laden floodwaters now flow from Lake Waikare into Whangamarino, annually discharging around 20,000 tonnes of sediment into the Ramsar wetland.

50 NTU be set, to improve water clarity to a state that approaches the average turbidity level recorded in the Whangamarino River as it exits the wetland.

The data collected from Whangamarino has been useful for other wetland policies. Hugh was on an expert panel for the Waikato Healthy Rivers/Wai Ora programme — a big overhaul of the regional rules to protect freshwater, and a rare opportunity to seek environmental wins for wetlands in the region. He elaborates:

> One of the things I observed is that there are all these water quality standards for streams, rivers and lakes, a target or environmental limit, but none for wetlands, because we often don't have enough information on wetlands to say that this is the water quality standard that these wetlands have to achieve. But with Whangamarino we actually did have enough information to propose a water quality standard for Whangamarino that the new plan could consider. By collating the scientific evidence and working within the policy frameworks that are in front of you, you can actually do more to protect wetlands.

Invisible drains — too much drainage

By definition, wetlands need water. The widespread destruction of wetlands across New Zealand has involved extensive drainage to make the land dry enough to build on,

INTERNATIONAL STANDING

Whangamarino is a 7000-hectare freshwater wetland complex with several wire rush bogs, surrounded by fens and mosaics of open water, river systems, low-lying swamps and marshes. Its large size and diversity mean that it provides habitat for many wetland species. It supports 143 native vascular plant species and 38 native bird species, and is a national stronghold for the swamp helmet orchid (*Corybas carsei*), the sedge *Schoenus carsei*, stout water milfoil (*Myriophyllum robustum*), black mudfish (*Neochanna diversus*) and matuku (Australasian bittern, *Botaurus poiciloptilus*). Tens of thousands of waterfowl use the wetland, including open-water diving ducks, shallow-water waders, dabbling waterfowl, predatory gulls and terns, swamp crakes and rails, and riparian wetland birds.

The site is a Ramsar wetland of international significance. It is also culturally significant to Waikato-Tainui as a mahinga kai. Local hapū (Ngā Muka) identify the protection, quantification and restoration of the mauri (life force) of the wetland and its surrounds as a key concern.

raise animals on or grow crops on. But there's more than one way to drain a wetland. The usual practice is to dig channels along the edge, and then through the wetland, to draw water down from its surface. At Kaimaumau, in the far north of Aotearoa, a less visible form of drainage may threaten a significant wetland. Here, rather than being drawn away from the surface of the legally protected wetland, water is being taken from aquifers underneath it to irrigate food crops. There seems to be abundant groundwater, but does extracting it pose any risks to the wetland?

Kaimaumau wetland is a gumland bog and fen system with outstanding ecological and cultural values. From the air, it's a Florida Everglades-like wetland wilderness. Vast expanses of water are filled with emergent reeds, while on higher ground spiky sedges indicate areas of peat that developed on impervious soils formed by long-gone kauri forests. Each vegetation type is positioned in sequence within a network of ancient dunes. This type of wetland is kept wet mostly through rainwater, or potentially by groundwater seeping up to the surface from below.

Kaimaumau is largely unstudied and its hydrology is a mystery. So is groundwater extraction for irrigation any cause for concern? Naturally, Hugh wanted to investigate. If the wetland starts to dry out because the groundwater underneath it decreases, wattles or other invasive species may potentially march in and take over the native plants. Wetland biodiversity can be modified even with a change in water levels of just 10–20 centimetres. Hugh explains:

> We didn't have any really strong information at the time about the impact that groundwater extraction may have on the hydrology of the wetland

> system, and whether that can have long-lasting impacts. So, we had to really break it down, and go firstly, do we need to worry about this? Secondly, what evidence do we have to actually understand the hydrology of that wetland? Does it even rely on groundwater? We needed to know, because if there was a connection between groundwater extraction and wetland water levels, it might take a long time for it to recover.

Hugh set out to determine whether the amount of water taken out of the ground for irrigation was a threat to Kaimaumau. His team of hydrologists and ecologists looked at whether there might be natural springs coming up into the wetland system. They examined hydrological data and vegetation patterns, and collected radon isotope samples, which are handy indicators of where groundwater is seeping up to the surface.*

Because of concern over the potential risks to the wetland and the lack of baseline monitoring to measure the effect of the increased water take, Hugh recommended a more precautionary approach, one that enables sustainable land use development with careful monitoring of the wetland. 'We got a range of specialist wetland scientists together to work out how much of a problem do we have here, and whether we needed to collate evidence to make a strong case for this wetland. Again this was positive, making sure there were clear limits for groundwater extraction. Now if there is any impact on wetland hydrology, the amount of groundwater take from pumps can easily be reduced.'

Joining the dots

A big part of Hugh's science-broker role is seeing how all the different bits of information need to come together. For instance, a research agency might be doing something specific with the use of sediment isotopes to understand the sources of sediment entering the Whangamarino wetland, and changes in sediment accumulation rates. According to Hugh, 'What I can do is work out how can we use that data to translate into a management action or policy change. I can draw that knowledge out from experts and then articulate it in a way that the policy makers can actually make sense of it and make use of it.'

There is much we need to learn about thresholds for wetlands. There is generally a good understanding of how nutrients affect lakes, rivers and estuaries, but we still don't really know how nutrient or sediment inputs affect the function of wetlands such as bogs or swamps. As Hugh says, 'We often promote wetlands as tools to help capture sediment and nutrient run-off, but research to date indicates that wetlands may also have a point where they become overloaded. We need to know more for this to translate into more sustainable catchment management.'

* Radon is abundant in groundwater but it rapidly dissipates into the atmosphere when it reaches the surface. Elevated concentrations of radon in surface water indicate a location where groundwater is seeping up into the surface water.

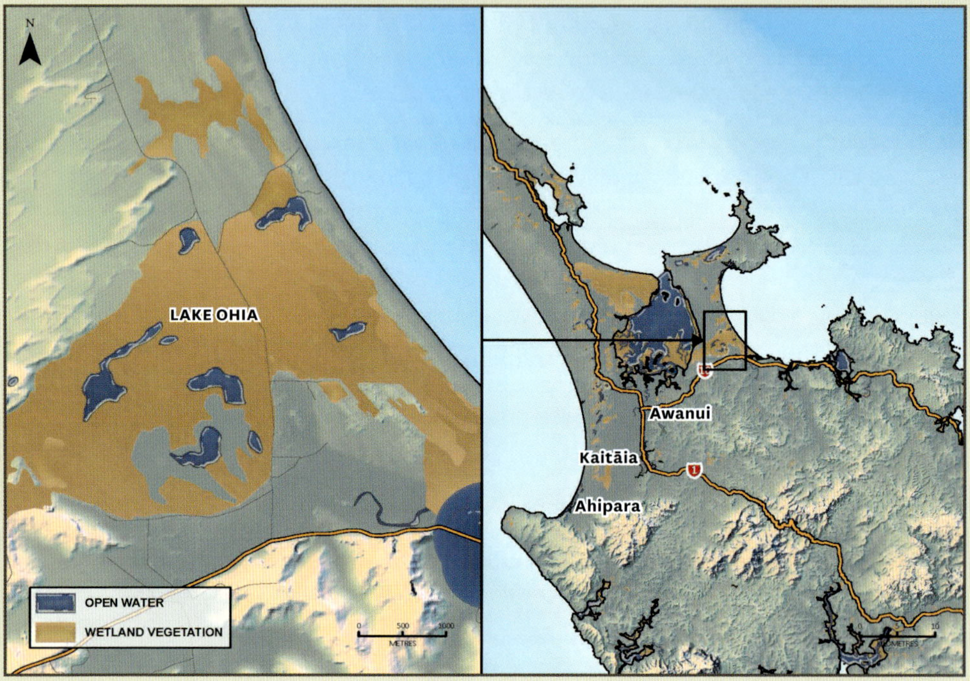

Lake Ohia

Ephemeral lake and gumland

Northland's Kaimaumau gumlands, where Hugh has worked, are protected in a scientific reserve and are not open to the public, but a similar type of wetland can be visited at nearby Lake Ohia.

Like a setting out of *Lord of the Rings*, Lake Ohia is an other-worldly site where you will see a bizarre landscape of charred stumps, sun-baked silt crusts and gumland scrub in a former lake bed. Sculpted by a landscape-altering sequence of events, it's an ideal place for a historical outing.

This was once the site of a dense kauri forest, but near the end of the last Ice Age, around 30,000 years ago, the water level rose and formed an ephemeral lake, drowning the trees and entombing their remains. The thick tangle of ancient kauri stumps was exposed in the early 1900s, when the lake was drained to aid gum diggers seeking the highly valued fossilised kauri resin left by the long-gone forest. The lake is now a fleeting feature, present for only a couple of months each winter. Summer visitors may well ask, 'Lake? What lake?'

Highlights

- Lake Ohia gumland is an important habitat for rare ferns (including the nationally critical *Phylloglossum drummondii*), and for mosses and orchids, including the mauve sun orchid (*Thelymitra malvina*)
- The surrounding wetlands support threatened black mudfish and bird species such as mātātā (fernbirds) and pūweto (spotless crake)
- The rare dwarf mistletoe (*Korthasella salicornioides*) can be found growing on mānuka shrubs.

Access

Access is via the base of Karikari Peninsula, a 40-minute drive northeast of Kaitāia. Turn off State Highway 10 onto Inland Road; a 300-metre sand track to the edge of Lake Ohia is 1 kilometre on the left. Continue another kilometre along Inland Road to the Gum Hole Reserve, where a short metalled track connected with boardwalks loops through stunted mānuka scrub and passes a series of holes excavated by gum diggers.

Drones on Duty

Can an eye in the sky give us a better picture of the state of wetlands? Marinus Boon deploys drones to explore wetlands in three dimensions and determine their values and needs.

Marinus Boon

Specialist field: Wetland ecological studies and remote sensing

Currently works for: Pattle Delamore Partners Ltd (PDP), Te Whanganui a Tara Wellington, as a Service Leader — Remote Sensing and Ecology

Marinus Boon has experience in wetland mapping and functional assessment in several countries, including Aotearoa New Zealand and South Africa. In addition to wetland mapping, his work includes vegetation and hydrological integrity assessments, wetland rehabilitation planning, constructed wetland monitoring, peat fire detection, and managing drone hardware and software. He is a certified drone pilot and holds a formal wetland assessment practitioner certification from Rhodes University, South Africa.

EVER HAD ONE OF THOSE flying dreams? Where you're swooping and gliding blissfully over beautiful landscapes in your sleep, free as a bird? While unaided flight eludes us as a species, we can still experience the world as a bird does. Drones (remote-controlled aircraft) have transformed how we view the landscape. They allow us to see a wetland from a bird's point of view, exactly as a matuku slowly flapping across swaying raupō reeds sees it.

From our usual human perspective, feet firmly on the ground, it's easy to see wetlands and their parts in isolation. But wetlands are intimately linked to their surroundings, affected by what's happening upslope and upstream, influencing what's going on downstream, connected by water and by air to forest, farmland, urban areas, the ocean and other wetlands. Satellite and aerial photographs have been standard tools of the trade for ecologists seeking to map and describe wetlands, but being able to hover above a wetland via a drone, look slowly around 360 degrees, rise up, dip down and zoom in, gives us a much clearer picture of a wetland in three dimensions and in real time.

Ecologist and drone pilot Marinus Boon recalls a saying: 'Wetlands are the expression of water within the landscape.' The water that nourishes a wetland may be in full view as still ponds or shallow pools, gently flowing seepages or meandering river braids. Or it may be hidden, tucked just beneath the ground surface or obscured by plants. It's often the plants that betray the water source, but as Marinus says, to really understand a wetland you need to see how it is positioned within the larger landscape as well as observe it up close — you need both the big picture and the detail to make sense of a wetland. Marinus explains:

> Drone technology gives you the ability to explore all of the different views of a wetland. It gives you an interactive feel. It's so much more than just looking at a map or air photograph . . . you get a better feel for the site, and you can see how the wetland lies within the larger landscape — is it in a valley, or on the side of a slope? What are the surrounding land uses? It gives you detail and context.

Sensing wetlands

A drone is so much more than a flying toy with a camera slung beneath. With a bit of scientific know-how and some lateral thinking, drones can be used in countless ways — many we have yet even to imagine. Already we can load them up with small amounts of herbicide to spot-spray a solitary willow that is impossibly out of reach for ground control. We can use them to scatter seeds of native plants to revegetate a degraded wetland. We can use them to swoop down and collect a water sample. We can use them to count waterbirds, monitor harmful algal blooms and erosion, map weeds and measure the success of planting programmes.

The camera possibilities and capabilities are also increasing. Thermal cameras can pinpoint non-native predators to allow targeted placement of traps or bait stations, or measure the success of pest control. We can use infrared sensors to see where plants are thriving or struggling, and thermal sensors to see where groundwater seeps up into a wetland. Lidar (light detection and ranging) sensors fire laser pulses from the camera lens down to the ground and use the time it takes for the light to bounce back to measure vegetation heights and draw 3D landscapes. Computer models can then virtually strip away overlying vegetation and buildings to reveal the hidden terrain beneath, including old river channels or drained waterbodies.

Reflectance sensors can pick up ultraviolet or infrared colours that we can't see with our eyes, allowing researchers to spot different plants in a wetland. Marinus gives an example: 'Look at kikuyu grass. When it's intermingled with other grasses it's hard to spot if you're doing a survey on the ground, but you can use the infrared reflectance value of kikuyu to work out the percentage cover of this weed compared with native grasses.'

Marinus also talks about the ability of drone images to ramp up the accuracy of wetland maps using landscape position and surface hydrology — the subtle bumps and dips in the ground that determine where water ends up. 'You can import drone photos into the computer and use a process called photogrammetry to convert the photos into a three-dimensional surface, allowing you to simulate which areas are likely to flood first. We can also use those images to simulate the impacts of things like various increased flooding or sea-level rise scenarios.'

A picture of wetland health

While you can use drone images to map the boundaries of a wetland, to go a bit further you need to understand what is driving the wetland. Where does its water come from? Is it a year-round wetland or just temporarily wet for a couple of weeks of the year? How well is it doing in terms of its health? What benefits is it providing for people? In South Africa, Marinus used a spider diagram tool[1] to portray scores on a wide range of things we value or need wetlands for — cultural significance, flood protection, water purification, recreation and the like — but he says he hasn't yet seen it used in Aotearoa New Zealand.

Here, ecologists use similar spiderweb-shaped diagrams, but generally assess wetland well-being using a narrower range of features that represent the ecological health and pressures on the wetland, without addressing wetland values. It is termed the Wetland Condition Index, a sort of warrant of fitness for wetlands that was developed by Dr Beverley Clarkson and others.[2] The index scores a site on features like the amount of the wetland covered by invasive plants, or the effect of human activities such as drains or dams, and the information is usually gathered during a visit to the wetland. In a large site, looking for these features on the ground can be impossibly time-consuming, and it can also be damaging to the wetland. But with the drone's high-resolution imagery, Marinus can easily see where there's a ditch or a drain, a weed infestation, or if farm animals are trampling the wetland. He can also see how healthy the vegetation is. He explains further:

> You can use drones with special processing software and image sensors to measure things like vegetation and the amount of 'green', so that would let you know where the vegetation is growing more vigorously than in other areas. If you have NDVI* reflectance values, you can see areas of dead or unhealthy vegetation, areas that are degraded, or where sedimentation is smothering the vegetation so it isn't growing. You can use that information to split a wetland into a few classes and then use machine learning** to train a computer to recognise healthy versus unhealthy wetland areas in an image. The computer can then do the mapping for you and we can use the information to apply the Wetland Condition Index.

---

* NDVI (Normalised Difference Vegetation Index) is a measure of the state of plant health based on how the plant reflects light at certain frequencies. Healthy vegetation will have a high NDVI value, while unvegetated areas such as built surfaces, bare soil and open water, and vegetation that is growing very slowly, will have lower NDVI values. These values are usually transformed into a colour-coded reflectance map.

** Machine learning is a form of artificial intelligence that involves training computers (via algorithms) to learn and improve their performance. For example, Marinus teaches his computer what a particular feature on a drone image looks like, such as kikuyu. He then instructs the computer to search for, recognise and signal to him other parts of the drone footage that also look like kikuyu. The more examples Marinus shows the computer, the better it becomes at detecting kikuyu in other parts of the drone footage.

DIY droning

Marinus started working with drones in 2014. Initially, he worked with the images that someone else had collected. Having previously relied on aerial or satellite photographs, he was amazed at the incredible detail he could see on drone photos. 'On an aerial photograph plants may look like indistinguishable green blobs; on a drone you can sometimes tell what species it is. It was wow! And also the speed — so quick to fly and process; no waiting for satellites or aircraft. You can do it when you need it.' He's now a certified drone pilot, having taught himself a lot of the required skills:

> It's mostly been learning by doing; trial and error; getting guidance from other people, taking courses, online training, and YouTube tutorials. I was initially more academic . . . but I wanted to bring the science and technology together — how can I apply this with something else or something newer? Some people in the science field might say, 'It's not proven yet' . . . but if we don't try it, we won't know if we can make it easier.

When Marinus started working with drones in South Africa, they weren't readily available to the average consumer, so his company employed a local manufacturer to build one.

> We then attached a Garmin GPS and a normal camera. You had to merge the photos and GPS coordinate data — it was time-consuming, but we got good-quality data. Then Chinese drones became commercial and so good that the local manufacturers couldn't compete. Now they're available to anyone. Computers and software processing programmes are also much better and faster, able to process huge amounts of data in a short time. It's come a long way quickly.

Safer surveying

Drones can be literally life-saving, reducing the need for researchers to physically get into wetlands for a close look. Fieldwork in New Zealand wetlands may have its health and safety challenges, but one thing Marinus doesn't have to worry about here is venomous snakes. He recalls a wetland survey in a remote rural area in Mpumalanga province in South Africa. Halfway through the survey, something told Marinus to stop. 'I looked up and there was a large puff adder ready to strike, just 2 metres away from me!' Puff adders account for more deaths than any other snake in Africa. Marinus instantly froze, knowing that it would not be wise to get bitten in such a remote area.

> I decided that I at least needed a photo of my encounter to share, so I took a photo and then slowly moved backwards, changed direction and continued with my survey. My colleague decided she had had enough after encountering the snake with me and decided to go and sit in the car. I unlocked the car for her and took the keys with me to continue on my own. Late in the afternoon I headed back to the car, and upon arriving realised I had lost the keys . . . it wasn't my favourite day in the field, or hers!

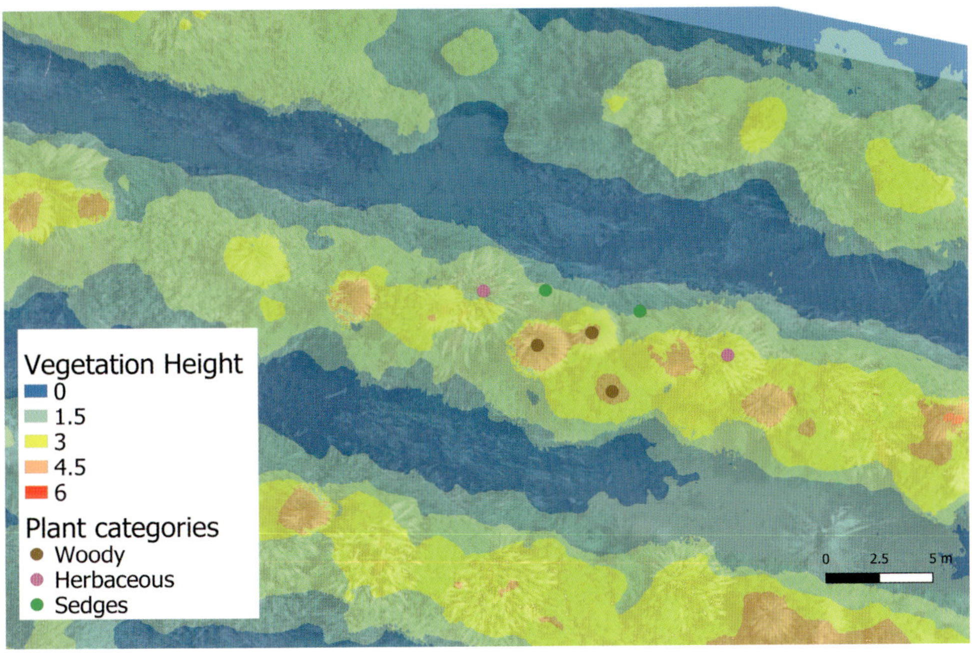

A data model is used to convert a photograph captured by a drone (above) into a map of vegetation height and types (below).

BUSY BEES — DRONES AT WORK

The owners of a dairy farm near Carterton who were restoring wetlands on their land wanted Marinus to track their progress and find out if the restored wetland was actively providing ecosystem services (the useful things that natural areas do that benefit people or the wider environment, such as trapping sediments or storing carbon).

The wetland was once a kahikatea and maire tawake swamp forest, burnt to clear land for grazing around 200 years ago. Today, it is a sedgeland with some areas of mānuka. Because the surrounding land use has changed substantially during the past two centuries, the values of the wetland have also changed.

Marinus has been monitoring not only the wetland's recovery but also its ability to perform important ecosystem services once again. He used an off-the-shelf drone to capture a series of overlapping photographs and video footage, geotagged so he could pinpoint the exact location of each photo or video segment. To ensure that the drone would follow the route Marinus wanted it to, he used a flight-planning application to pre-program its course.

Back in the office, Marinus stitched the images together, along with their location data, to create a fine-scale map (similar to the one shown below) of the wetland showing vegetation types, open water and saturated areas. The map also revealed the location of invasive trees and grass, zones of deep flooding (which can destroy wetland vegetation), excavations and erosion gullies. By adding information on slope, drainage channels and natural or unnatural blockages, it provided an insight into the movement of water through the wetland. The drone data gave Marinus high-resolution, accurate, up-to-the-minute information to help assess wetland health, hydrological functioning and biodiversity.

A better day at work for Marinus is when he is analysing and interpreting various images and field data to build a 'wetland puzzle', as he calls it:

> I love assessing what exactly makes a wetland 'tick' and what is important about a specific wetland, since it's not always obvious. A scenic wetland with some important species, for example, may attract the attention of most people, but it may not necessarily always be the most functionally important wetland. Whereas a not-so-aesthetically-pleasing wetland may in fact be of extremely high ecological importance, for things like slowly releasing water to maintain stream flows. Building a wetland puzzle can keep me going for some time. I also love spending time in the field looking at wetland soils, hydrological indicators and vegetation, and trying to understand why this wetland is located at its specific position in the landscape. And yes, I also enjoy flying drones over wetlands to gather data, whenever the opportunity presents itself!

Making connections

Sometimes, science relies on happy accidents to bring disciplines together and take research and technology in new directions. Marinus's father is a horticulturist and his mother was a town planner, so it was only natural that he developed an interest in vegetation, geography and environmental science. He says, 'My parents took me on hiking trips to nature and game reserves, from when I was a baby through my teens, and from that I developed a passion for nature conservation. Discovering more and more about wetlands was like opening a treasure chest, with no end of treasures.'

Drones will never completely replace humans in terms of wetland fieldwork — you need the human mind and experience to interpret what the drone can see, and there will always be features too small for the drone to spot, samples that need to be collected by hand or species that can be identified only by an expert in the field. But combining the power of the human brain and the power of drone technology can drastically improve our understating of how wetlands work, what they do for us and what they need from us.

The challenge, says Marinus, is convincing people in the wetland industry who are used to conventional approaches to embrace new technologies such as drones and machine learning. But he is optimistic about the future:

> Luckily there have been people along my career path that believe in my research, have collaborated and helped spread the message about drone potential, and we are now seeing drone and computer-vision technologies quite widely adopted, providing a wave of useful information to assist with wetland conservation. Learning is an everyday thing; learning should never stop, and science is constantly evolving. I am a strong advocate for thinking about the big picture when undertaking scientific studies.

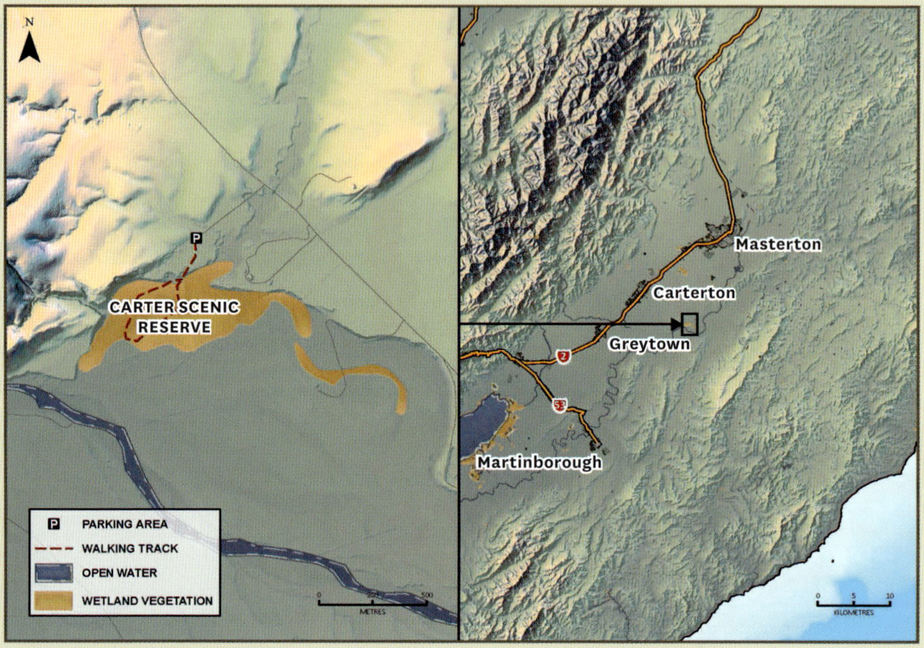

Carter Scenic Reserve

Lowland swamp and forest

Many of the wetlands Marinus works with are swamps. He says that one of his favourite lowland swamps is the Carter Scenic Reserve:

> This wetland and its interface with grassland, forest and shrubland was once typical of the Wairarapa landscape. There are very limited remnants of these patchwork landscapes remaining after the clearance of the Wairarapa plains. Walking in this reserve I cannot help but think how wonderful it would be to have kilometres of this landscape to walk through and experience. What makes it so special for me is the diversity of habitat types compared to some other wetlands.
>
> I am fascinated by the beautiful swamp maire tree with its aerial roots that help it cope with a wet and boggy habitat. Wading through a swamp with swamp maire is very special; these trees have very beautiful creamy-white flowers and red fruit. Although these trees are not yet endangered in New Zealand, they are not found everywhere anymore due to the clearance of swamp forest, and they are now under threat from myrtle rust.

Because the site is one of the rare remnants of flooded lowland forest, Greater Wellington Regional Council and the Department of Conservation have installed water monitoring equipment to regularly record the height of the water table, providing crucial information on which to base future water management decisions.

Highlights

- Mature lowland kahikatea swamp forest with tracks and boardwalks
- Uncommon native plants, including maire tawake (swamp maire) and narrow-leaved houhere
- Pools of water with rare brown mudfish and banded kōkopu.

Access

Carter Scenic Reserve is located at 340 Gladstone Road, a 10-minute drive southeast of Carterton, in the lower North Island. An easy half-hour stroll along a loop track winds through the remnants of a flooded kahikatea forest and back through a drier forest section.

www.doc.govt.nz/parks-and-recreation/places-to-go/wairarapa/places/carter-scenic-reserve/?tab-id=50578
https://carterreserve15.wordpress.com

Looking Back to See the Future

What can the past tell us about a wetland's future? Dr Olivia Burge's research used modern techniques to unravel the past vegetation cover of the Awarua wetland and predict its future.

Dr Olivia Burge

Specialist field: Vegetation succession, invasive species, ecosystem restoration, and disturbance

Currently works for: Manaaki Whenua — Landcare Research, Lincoln

Olivia Burge, PhD, is an ecologist with a particular interest in how changes in nutrients and hydrology affect wetland plant communities, both now and in the future. She collaborates with palaeoecologists to understand the long-term history of our wetlands, such that we might better understand and predict their future. Olivia's research has been recognised and supported by the National Wetland Trust's Golden Plover Award, the Miss E. L. Hellaby Indigenous Grasslands Research Trust and the Department of Conservation's Arawai Kākāriki programme.

OLIVIA BURGE HAS a confession to make — she hated science as a teenager. She avoided the optional science classes at high school, with too much focus on memorising random facts. It wasn't Olivia's preferred way to learn; she was looking for 'a knowledge system for discovery!'

Still, she was always passionate about the environment, and ended up doing a double major degree in law and science. She worked as a resource management lawyer for three years, cutting her teeth on cases involving biodiversity offsets or reverse sensitivity — those 'not in my *new* backyard' cases, such as when someone moves in next door to an existing gun club or composting facility and then starts complaining about the noise or the smell. Says Olivia, 'We worked on really interesting cases, but I found it was quite "reactive" work — we could only work on issues where a contested case arose, it was ambulance at the bottom of the cliff type stuff. So I decided to quit law and undertake post-graduate study.'

Olivia claims that she 'sort of fell into wetland research'. She was assessing different projects for what was intended to be an MSc degree but was later converted to a PhD, and one project on offer was about forest regeneration in one of New Zealand's Ramsar wetlands. She thought it sounded interesting. The essence of her thesis work was to answer the question: Could we, should we, put a native forest here? The 'here' in question was Awarua peatland, near the coast at the very bottom of New Zealand's South Island.

Awarua peatland is big.* Really big. About 12,500 hectares big. OK, it's not Pantanal big, or Okavango Delta big, but it's one of the biggest remaining areas of natural peatland vegetation left in Aotearoa New Zealand.[2] It's significant too. In 1995, ecologists Pam Cromarty and Dr Derek Scott compiled information on the nation's best wetlands and concluded that the coastal peatland was worthy of international recognition. Just over a decade later, in 2008, it was added to the Ramsar list of internationally significant wetlands, as part of the existing Waituna Lagoon Ramsar site.

When the Ramsar Convention was first enacted in 1971, it was all about the birds. It was the first, and as of 2021 remains the only, multilateral environmental treaty that focuses on a specific ecosystem type, and it was initially formed out of the need for international cooperation to protect the global habitat needs of migratory birds. That's how the shorebird-saturated Waituna Lagoon got on the list in 1976 — one of the first three wetlands in New Zealand to take the honour.

Fast-forward half a century, and the Ramsar criteria have widened to include a wide range of wetland ecosystem values. Good thing, because Awarua peatland isn't really *that* special for its birdlife. It is habitat for a handful of rare bird species, and a southern stronghold for some of them, but in the peatlands it's more about the plants (and moths and lizards). The Department of Conservation, which administers the Awarua peatland, says it 'supports an unusual array of plant communities . . . dominated by wirerush (*Empodisma minus*), with tangle fern, swamp inaka (*Dracophyllum oliveri*) and mānuka shrubland'.[3] It's also home to hardy cushion plants *Donatia novae-zelandiae* and *Oreobolus pectinatus*; these usually grow only in subalpine conditions, but in the cold, blustery conditions at Awarua they grow almost at sea level. In addition, native orchids flourish throughout the wetland.

The Awarua peatland is rather snugly described by botanist Dr Peter Johnson as an 'undulating blanket bog'.[4] It forms part of the Southland Plains, a massive area of flat land built from sediment sluiced off the flanks of the Southern Alps — pea-sized gravel carted down towards the sea by myriad rivers and spread out like a slightly crumpled tablecloth. Small streams entrapped by a gravel bar that built up along the coast kept the surface damp, and trowelled over the top was a smooth layer of peat 1.5–2.5 metres deep, built up from the poorly decomposed remains of sedges and rush-like plants. As the post-Ice Age climate warmed up, trees arrived: ancient southern conifers,[5] the 'fleshy-footed' podocarps — bog pine, kahikatea, mataī, lowland tōtara and rimu.

Then humans arrived. Humans, of course, brought with them one of their most important innovations — fire. The wetland has never been the same since.

After the fires came drains, and today the Southland Plains have been largely developed into farmland, although about a third of the original extent of the Awarua blanket bog remains.[6] Historically, the peatland would have been a mishmash of forest

* At 10,000 hectares, the North Island's Kopuatai Peat Dome (also a Ramsar wetland) is the largest single 'chunk' of peatland. While Awarua has a larger overall area of peatland, it comprises several discrete remnants.

Historically, Awarua peatland would have been a mishmash of forest on higher, drier ground, with sphagnum moss, sedges, rushes, ferns and a rush-like restiad in wetter areas.

on higher, drier ground, with sphagnum moss, sedges, rushes, ferns and a rush-like restiad* in wetter areas. We know this because in the 1930s scientists Dr Lucy Cranwell and Professor Lennart von Post examined soil cores bored metres deep into the ground, and identified pollen and spores that told them what species were abundant here through time.[7] Peat builds up in layers, forming a 'history sandwich' by locking in pollen and other material into a perfectly preserved time capsule, with the youngest layers at the top and the oldest at the bottom.

Today, just 2 per cent of the Awarua peatland (300 hectares) is in mature podocarp-broadleaved forest.[8] These stands, some with trees up to 20 metres tall and trunks almost a metre across, are likely hundreds of years old, and consist of kahikatea, rimu, mataī, miro and tōtara, and occasional mountain toatoa. The hardwood components

---

* Plants in the Restionaceae family. In Awarua these are represented by a single species, the southern wire rush (*Empodisma minus*).

are small-leaved trees, kāmahi and pōkākā , along with many smaller shrubs. According to Olivia:

> We know from buried pollen that forest was historically there. We just don't know where exactly in the wetland it was. If we had good lidar maps,* we could actually say for certain where some areas of forest were, because there's still a lot of standing tree spars in the wetland, the skeletal remains of past forests. My colleague Janet Wilmshurst aged one — a tōtara — by counting its rings, and it was at least 300 years old when it died, and based on radiocarbon dating we know it died around 1377 AD!

Olivia refers to a scientific paper written by Peter Johnson, who also found evidence of 'ghost forests' at Awarua.[9] First, he discovered a small island protected from fire by a surrounding tarn 'moat', upon which he found forest trees and shrubs. The second clue was a layer of wood buried knee-deep below the surface of the peat in a nearby quartz gravel quarry. Much of the wood was charred, but it was still recognisable as mānuka, rimu, lowland tōtara and Hall's tōtara. Johnson concluded that:

> It is probable that in the natural vegetation pattern of the study area herbaceous wetland communities would have occupied just the poorly drained hollows and level crests, as a mosaic within podocarp-broadleaved forest and scrub on the better-drained ground of gentle slopes . . . Even without a scattering of wind-shorn tōtara and rimu crowns overhead, it is easy to envisage a low forest, ferny inside, growing in today's climate on the same peat landform where today most people would accept the low bog and heath vegetation as the natural cover.[10]

So what happened to the rest of the Awarua podocarp forest? New Zealand is a temperate, rainy country — its ecosystems are not adapted to frequent fires. Fires lit by settlers to clear land for farming no doubt found their way into the wetland, destroying forest and setting in train a predictable sequence of natural regeneration: sedges and ferns, succeeded by shrubs, succeeded by forest.[11] The whole process can take many decades, and with every major fire, the vegetation succession clock is 'reset' at time zero — back to sedges, then mānuka, then maybe forest again — unless another untimely fire rewinds the process all over again.

Olivia provides further evidence: 'Janet [Wilmshurst] drilled out a core of peat by the dead tōtara that we found, and based on charcoal in the peat, we could see when fires occurred in the past. The first time we saw evidence of fires in the wetland was around the time that tōtara was killed, and then a big increase in fires associated with land clearance following European settlement (after 1800).'

---

* Lidar (light detection and ranging) is a form of laser-scanning across landscapes to build a three-dimensional map incorporating tree height, and is usually conducted from an aircraft. Lidar maps of Awarua would detect tall forest and standing dead trees.

APPRECIATING AWARUA

Olivia reflects on her doctorate field days:

> I spent a fair amount of time in Awarua alone, so I got to really spend uninterrupted time there.
>
> There are stunning freshwater pools in amongst the mānuka. The mānuka itself is amazing — it's like a natural maze and you could easily get lost in amongst it — in a flat land area, there are no distinguishing features!
>
> There was a forest remnant that was beautiful to take a moment in, particularly when the Southland sun came through. I also worked in tussocklands, where you could see Bluff hill and Tiwai smelter, and the tussock would wave up and down in concert with the prevailing wind.
>
> I visited areas that had been burnt, and they were dramatic and solemn and striking, with burnt mānuka stems, but also surrounded by new regrowth. I visited the site in hot summers, snowy winters and windy springs, so I really got to appreciate the wetland in its many climatic cloaks.

ABOUT FOREST SUCCESSION

A forest is a living entity. Like people, trees are born, they grow up, they produce offspring, they get old, they pass on.

Forests grow wherever the going is good. They're the Goldilocks of the vegetation world — if it's too hot, too cold, too dry or too wet, they turn up their roots. Other types of vegetation, such as scrub, grassland or reedland, grow in these harsher areas — deserts, mountain tops, sand dunes, lake margins, geothermal sites. Where the land is gentle, fertile and temperate, forests will muscle all other vegetation types out of their way and, by the sheer power of their size and the shade they cast, take over. In Aotearoa, they once covered more than 90 per cent of the land.

When an entire forest dies — perhaps obliterated by a volcanic blast, incinerated by a fire or washed away as smashed-up kindling by a massive flood — it opens the door for a whole new cycle of growth. This is a fleeting chance for the grasslands, the fernlands, the shrublands and the scrub. For a brief period, the newly bare land is not ideal for tree species. Without the shelter of a forest, the sunlight beats down, winds blow unabated and frosts can blanket the ground.

Hardy plants are the first to gain a foothold here. Perhaps sun-loving ground ferns like bracken arrive, their tiny spores blown in from a great distance. As they spread their fronds over the ground, creating dappled shade and trapping moisture, shrubs find a microcosm that suits them. Their seeds may have blown in on the wind too, and perhaps lain dormant in the ground for a few years, waiting for a little overhead protection. Or perhaps they were deposited there recently, along with a handy dose of nature's fertiliser, by a passing bird. The hardier shrubs survive and grow taller than the ferns, turning fernland into shrubland. As the shrubs grow taller and wider, nudging branches and foliage with their neighbours, they become scrub — no longer scattered among the ferns, they now form a solid canopy of their own. The forest plants, many of them happy in the shade, come next. Over the course of some decades, trees loom over their shorter shrub cousins, and the forest is back in charge.

Flammable field sites

Most of the Awarua peatland area is now covered by mānuka scrub or shrubland.[12] Prized for the antibacterial properties of its oil and honey products, this native species is common in many New Zealand wetlands. But these same oils and its dry, twiggy foliage and flaky bark make mānuka extremely flammable. While fires are less frequent now than they were when the Southland Plains were being converted to farmland, accidental fires still occur every few years, and with mānuka so dominant, there is an increased risk of larger and hotter blazes, and more damage. As Olivia's PhD thesis notes, two fires in 2009 and another in 2012 burnt an estimated 2029 hectares of the wetland. She says, 'My supervisor was always concerned about me working alone in the field — we were really conscious of the fires that have occurred and continue to occur down there. There was actually a fire on land neighbouring some of my experiments, and if you were working in there, with the dense mānuka over your head, you wouldn't even know!'

Most fires in recent times have been accidental, and each has sparked a rapid response from the regional fire brigade, who douse the flames with monsoon buckets to protect wildlife and neighbouring farmland. According to Olivia, 'We've been really lucky that there's been no, or very limited, underground fire, because when the peat catches fire underground, it's insulated from the aerial water drops and it's really difficult to control.'

Slow forest movement

Even where the wetland hasn't been burnt for many decades, forest regeneration has been extremely slow in the Awarua peatland. So slow, in fact, that Olivia set out to determine whether forest regeneration had actually 'stalled', and if so, why and can we/should we do something about it?

In dryland sites mānuka is typically a nurse crop, a temporary shelter under which native forest plants can become established and eventually take over. But this didn't seem to be happening in the drier parts of Awarua, and Olivia wanted to find out what was slowing down forest regeneration. Was it a lack of suitable seeds in the area? Were the seeds not being carried into the wetland? Or were seeds and seedlings being eaten by introduced mammals such as brushtail possums (*Trichosurus vulpecula*) and rodents? She also wanted to try out some tricks to give the forest a helping hand. Could seedlings be protected from possums? If you cleared away some of the competing vegetation, did seedlings grow faster? If you put artificial perches in the wetland, would birds bring in more seeds? Could you just directly plant native forest plants into the wetland, and if so, where were the best areas to do that?

Philosophically, there was also the big question: should we intervene or just leave nature to do its own thing? Olivia muses:

> Forest regeneration in suitable (drier) areas would return the full range of plant communities to the site as well as reduce the overall flammability of the wetland, and provide more food resources for birds. [But is Awarua on] a trajectory to an 'acceptable' state? Altered hydrological regimes caused by drainage and forest clearance might make natural regeneration [back] to the mosaic of forest that historically existed impossible; however, a slow return to forest would likely be more desirable than the system becoming dominated by invasive shrub species like gorse, or early-successional vegetation, both of which reinforce the frequency and extent of fire within the wetlands.

Seed rain and bird droppings

A key ingredient in any forest restoration is plant seeds. If there are no viable seeds remaining in the soil — and the seeds of most New Zealand tree species don't last long in the ground[13] — and no source of replenishment seeds within bird-pooping or wind-whipping distance, a forest will not be able to establish itself, even if all the soil and climate conditions are perfect.

To find out if the slow forest regeneration was down to a lack of seeds, Olivia set up a number of field experiments. Within 160 small vegetation plots across the wetland, she counted the number of seedlings on the ground. She also counted the number of seeds that landed in 'seed traps' she installed near a forest remnant. In addition, she installed some avian 'rest rooms' — tall perches she hoped would entice native birds to stop, relax and make a deposit (of seeds).

Olivia then compared her results with distance to the nearest natural seed source (nearby forest patches). She found that except within a very short distance of existing forest patches very few forest plant seeds were being dropped into the wetland. She also found that the bird perches were not worth the time and cost to install them — not a single bird was photographed using them on her motion-activated game cameras. Instead, the perches just seemed to attract possums, which are implicated in both seed and bird predation.

Wetland sediment cores — here being collected from the Amoeboid Mire on the Kepler Track by Manaaki Whenua — Landcare Research staff (from left) Chris Morse, Olivia Burge, Michelle McKeown and Jamie Wood — helped Olivia determine when fires occurred at Awarua.

Her results highlighted two problems facing this site: introduced predators like rats or stoats taking out natural seed dispersers (birds), and introduced omnivores like possums and rats eating seeds and succulent seedlings. These animals seemed to have a greater impact on forest regeneration than competition between seedlings and existing native plants for space, light, moisture and nutrients, because when Olivia sowed seeds directly into her plot under the protection of herbivore-exclusion cages, twice as many seedlings survived. As Olivia says, 'This indicates there are available microsites within the wetland into which forest species, if dispersed and not eaten, may grow.'

Nevertheless, forest seedlings were a bit fussy about where they were planted, and seeds (when protected from herbivores) grew much better in soil that was drier. But even if you control the pest animals, planting out a massive wetland is a daunting, if not impossible, task. Olivia's solution was to trial the concept of nuclei — small patches of suitable land where forest species can be deliberately planted to kick-start the process, providing some shade and luring in the few birds that are around. Based on Olivia's seedling trials, we now know the optimum soil moisture to encourage forest regeneration, and that even in a massive wetland like Awarua, introduced animals are playing havoc with natural ecosystem processes.

Model wetland

Although it might be possible to pinpoint where forest grew in Awarua in the past, it doesn't mean that these areas are the most suitable for forest restoration efforts today. Given the cost of hands-on forest restoration at Awarua, including pest control and active planting, Olivia spent a considerable part of her doctorate study on exploring future scenarios of forest regeneration within the wetland. This was no exercise in crystal ball gazing — Olivia is a skilled quantitative ecologist, or in her own words, 'I spend a lot of time mucking around on the computer'.

Olivia used a computer program called NetLogo to create a grid-based landscape (similar to Minecraft and other large-scale computer games). She then built various layers of map-based information into the model, including the location of existing native forest patches (seed sources), vegetation types within the wetland, and a map of highly flammable and less flammable vegetation. She explains how the model worked:

> We used the computer modelling as a big 'what would happen if?' experiment. We used the current vegetation cover and experimented with different scenarios — such as no more fires, lots of fires, lots of birds moving around the landscape, active restoration planting in different arrangements. We let each simulation run for 500 years — and we ran thousands of simulations! — and then looked at which elements in what combination led to successful forest regeneration.

The simulations indicated the importance of forest patches in the landscape as a seed source to generate new forests — but also that we need birds to disperse these seeds. The animated computer model showed forest spreading from the east to the west of the wetland, which reflects that there are more patches of remaining forest in the east of the wetland. This suggests that if birds could be attracted from those remaining patches into the wetland by planting some trees that produce attractive fruits, there is potential to speed up regeneration. Olivia says the experiments didn't stop there: 'The computer simulations let you test out multiple scenarios. We tested whether a better strategy for encouraging forest would be planting out new 'islands' of forest within the mānuka, or whether it would be better to increase the size of existing remnants.'

One interesting finding was that the ideal place to actively plant forest patches for restoration depends on the fire risk. In areas with a high risk of fires, it's best to plant around the outside of existing forest patches, because these have lower flammability, and are 'safer'. In areas with lower risk of fires, you will get better results by restoring scattered pockets among non-forest areas to provide a source of seeds to the non-forest areas around them.

Future forests?

Olivia discovered that there are multiple constraints to forest succession within Awarua: non-native herbivores, recurrent fire and a lack of bird dispersers. As she points out, the problem is not just within the wetland: 'You actually need mobile birds in the wider landscape and food and seed sources within the wider landscape. If you don't have enough habitat within the wetland to support a permanent bird population, then you need that habitat for a permanent population of birds elsewhere and nearby. And then you still need something to essentially to lure birds in: food, shelter, breeding habitat and so on.'

Questions remain about the future of Awarua. DOC freshwater ecologist Dr Hugh Robertson says there are signs that parts of the wetland are drying up: 'Do we look to a different future for this site? In some parts of Awarua drains have done considerable damage, lowering water levels. And with recent dry winters the wetland was noticeably drier. We've seen a decline in *Empodisma* and *Sphagnum* from one research site over the past few years, shaded out by encroaching mānuka. It's a challenge.'[14]

Olivia's plea is that we shouldn't forget about forest regeneration, even in wetlands that are already dominated by native species. Forests are less flammable than mānuka scrub, and will also provide canopy cover that will keep out shade-intolerant weed species such as gorse. With so little lowland forest left in many parts of New Zealand, there is a great opportunity at Awarua for forest restoration. Olivia is thinking about the next steps:

> One approach would be to use remote sensing to quantify both where the still-standing [tree] spars are, and also current ground levels, to infer where forest used to be, and also where it might best be restored to.

There's a human element and an ecological element, and there's no real easy answer in how you balance that, particularly when you have limited funds for enhancement, while you're also managing a wetland, right? . . . For instance, should we plant forest in areas of pasture grass in ex-leasehold land to create a seed source for natural regeneration in the wetland? Should we try to do that as a project that people can really see and think, wow, this has gone from pasture to regenerating forests and that's really cool. Or should we plant forest in the peatland mānuka — which is a current fire risk, so you'd be reducing that risk, but it would be a lot less accessible and less visible to the public.

So, yeah, we've got some thinking to do about how to take my PhD research the next step further.

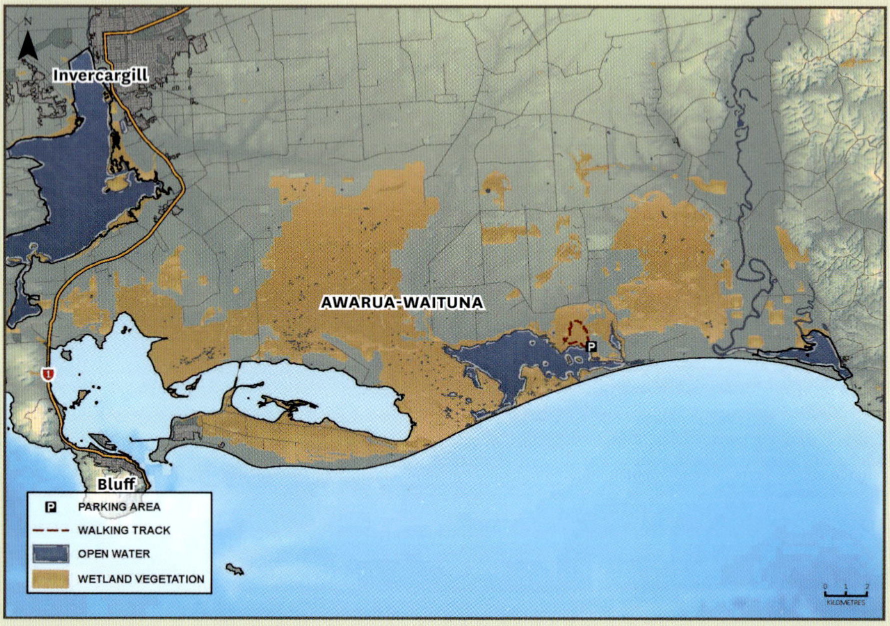

Awarua-Waituna

Lowland peatland

Awarua peatland was appended to the Waituna Lagoon in 2008, as part of a Ramsar wetland of international significance. The blanket mire and associated ecosystems are among the largest, most diverse and most intact examples in New Zealand. There are 80 moth species in the peatland, some found nowhere else in the world, while the streams, pools and lagoon support native fish such as the giant kōkopu, banded kōkopu, īnanga, tuna (freshwater eels) and kanakana (lamprey). Its flora is also fascinating with typically subalpine plant species growing here near sea level, and tiny insect-eating sundews and bladderwort plants.

Highlights

- Wide variety of peatland plant communities, including cushion bogs, coastal tussocks and low, windswept herb fields — keep an eye out for tiny ground orchids, colourful fungi and carnivorous sundews among the sphagnum moss and wire rush
- Threatened plant species include the Nationally Critical dwarf mistletoe (*Korthalsella salicornioides*), often found growing parasitically on mānuka shrubs
- Abundant small tarns and rivulets, home to giant kōkopu, banded kōkopu and species of tuna

- The site is a national stronghold for the Nationally Critical moth species *Asaphodes frivola*, found only in the Invercargill area
- A variety of native lizards, including a rare species of korero gecko (*Woodworthia* 'Otago large'), which is yet to be formally named, the green skink *Oligosoma chloronoton* and the cryptic skink *O. inconspicuum*
- Listen carefully for fernbirds (mātātā)
- Enjoy views over Waituna Lagoon from the information shelter.

Access

Access to the lagoon is well signposted from Invercargill. Just follow the road south to Awarua Bay and drive to the end (approximately 15 kilometres).

A 250-metre boardwalk winds through low-growing mānuka and shrubs, emerging at a viewing shelter that overlooks Waituna Lagoon. Interpretation panels in the shelter provide a wealth of information on the area. A 5-kilometre loop track starts near the shelter, with easy walking on both boardwalk and gravel tracks through mānuka wetland and past bog tarns. There are interpretation panels and rest stops along the way. The site is exposed and has a cool temperate climate, so be prepared for cold and blustery conditions any time of year.

www.wetlandtrust.org.nz/get-involved/ramsar-wetlands/awarua-waituna-lagoon

Green Invaders

How can we sort the 'good' plants from the 'bad'? Paul Champion's career has led to the development of new techniques to spot and manage invasive plants before they become unstoppable.

Paul Champion

Specialist field: Biosecurity, plant ecology and conservation of endangered plant species, especially in freshwater and other wetland habitats

Currently works for: NIWA (National Institute of Water and Atmosphere Taihoro Nukurangi), Hamilton, as Principal Scientist — Freshwater Ecology

Paul Champion's focus research areas include assessment of weed potential of introduced plants, management of alien aquatic weeds (including surveillance, control techniques and strategies), assessment of environmental impacts of both freshwater pest invasions and weed-control strategies, and restoration of habitats impacted by invasive weeds.

FOR SOME INEXPLICABLE REASON, aliens from outer space are frequently depicted as 'little green men'. One of Hollywood's most famous extraterrestrials, E.T., was a little human-like creature accidentally left behind by a group of alien botanists. Ironically, in real life Paul Champion is a botanist who advises governments on how to deal with little green aliens — alien plants.

In science-speak, alien is another word for non-native, foreign, exotic, introduced, not from round these parts. In short, it's a species growing and spreading where it shouldn't be. Most people simply refer to alien plants as weeds.

Of the plants that are always, or more often than not, found in Aotearoa New Zealand wetlands, 523 are native species and 226 are non-native.[1] So, for every two native wetland plants there's now an introduced species out there potentially competing for the same habitat. The worst of these weeds are collectively called invasives — plants that have escaped cultivation and are merrily spreading in the wild, smothering or competing with native plants and causing other problems.

Paul reckons that grey willow is one of our worst invasives, but he also mentions royal fern, which is encroaching upon fens at the edges of peat bogs; alligator weed, which is forming dense mats in swamps; spartina grass, which is taking over estuaries; and many aquatic plants that are clogging up lakes. 'Even [in] a beautiful lake with really good water

WHAT WEEDS DO — THE STORY OF SEBAEA

Sebaea ovata is a small herb that has nearly been driven to extinction in Aotearoa by invasive plants. At no more than 30 centimetres tall, this yellow-flowered gentian would be unlikely to grab the attention of anyone other than a botanist, but it has an interesting story of intrigue, disappearance, threatening bullies and an unlikely ally.

The little turf plant was once found in scattered dune lake margins and wet sandflats across the country, but by 1971 it had virtually disappeared. In his 1981 book *Rare and Endangered Plants of New Zealand*, botanist Dr David Given classified *Sebaea ovata* as 'indeterminate' — meaning there wasn't enough information about the species to say whether it was rare, vulnerable, endangered or even extinct.[2]

To the great joy of botanists, a small population was discovered in a wet dune hollow near Whanganui in January 1989 by members of a field excursion. Ecologist Colin Ogle was among them. He wrote:

> [The field site was] a damp sandy flat covered in short turf vegetation . . . Once gathered at the site, the entire party immediately fell to hands and knees. Some minutes and 20–30 metres from the start of our slow crawl of discovery we encountered some slender erect herbs with tiny, bright yellow flowers . . . None of us was sure as to whether we were

quality like Lake Taupō, look underwater [and] you've got native turf communities in the shallows, then you've got big walls of invasive lagarosiphon, and below that hornwort dominating all the way down to the bottom of the vegetation in quite a few places. Once you get below a couple of metres, the native underwater vegetation is completely excluded by weeds.'

The Ministry for Primary Industries Manatū Ahu Matua (MPI) has a hit list of nine invasive plant species it aims to eradicate from Aotearoa* — five of them are aquatic plants. They include water hyacinth, which sounds friendly enough and looks pretty in a garden pond, but has been rated the world's worst aquatic weed by the Food and Agriculture Organization of the United Nations.[5] All of the top nine plants are banned from sale, distribution and propagation in New Zealand, yet some people continue to grow and sell them illegally, or inadvertently spread them into natural waterways and wetlands, potentially undoing many years of hard work spent getting rid of them.

---

* These are referred to as National Interest Pest Response (NIPR) species.

among plants of the long-lost native species, because we lacked suitable reference books in the field. However, we took photographs, and also two of the 50 or more plants around us as voucher specimens. Their identity was later confirmed [as *Sebaea ovata*].[3]

Sebaea needs short, open vegetation with bare ground for its seeds to germinate successfully.[4] The arrival and spread of ground-smothering weeds wiped it out across much of its range. Ironically, at the Whanganui site the weeds had been kept in check by another introduced species — rabbits. Rabbits aren't much interested in sebaea and would rather eat the grasses and dandelion plants that take over its habitat.

Unfortunately for sebaea, the 1997 release of calicivirus almost eliminated rabbits in their habitat, allowing the weeds to run rampant. The Whanganui populations haven't been seen for more than a decade, but a population translocated by the Department of Conservation (DOC) to an area near the mouth of the Rakahuri Ashley River appeared to be thriving in 2020, says Paul, although it's likely still at risk of losing its habitat because of weeds.

Now that we have more information on sebaea, the plant has been formally classified as Nationally Critical. At least we know it's not extinct, but it's still a plant on life support because of invasive weeds.

Sorting the good from the bad

Invasive plants are like a disease. The best defence is to spot them early, go hard and eliminate them before they get out of control. The sooner you can find and deal to them, the cheaper and more successful an eradication programme is likely to be.

As a botanist with extensive field experience, Paul has a keen eye. He can spot a plant out of place a mile away — what may look like an amorphous green blob to the average person is like a screaming headline for Paul. Subtle clues are hard-wired into his mind . . . native raupō has a slight twist to the top of its strappy leaf . . . that plant over there doesn't have a twist . . . crikey, it's Manchurian wild rice!

But how do you know if a new plant that pops up in a wetland is a harmless herb or a potential pest ? To help government agencies predict which plants are likely to go rogue, Paul has developed a clever risk assessment model.[6] It ranks aquatic weeds based on knowledge of how a species, or its near relatives, has behaved overseas: 'What effects does it have on ecosystems, how versatile is it at invading a range of habitats, how vigorously does it grow, how does it reproduce and spread, where it could potentially end up and how hard is it to get rid of?'

HAERE RĀ, HORNWORT

Hornwort (*Ceratophyllum demersum*) is a submerged freshwater weed that looks like one of those classic fish-tank oxygen weeds. In 2013, MPI declared that the species had been eradicated from the South Island.[7]

The plant first turned up in the South Island in 2002, in a stream and several ornamental ponds in Motueka. Those populations were removed, but four years later it was found in a Timaru lake. MPI called in the cavalry — NIWA — who drafted a plan to eradicate the species from the lake. NIWA scientist Rohan Wells says, 'It was a difficult site and conventional methods of weed removal were not successful. We tried the new herbicide endothall and only one treatment was needed to successfully eradicate it.'[8]

Endothall is broken down naturally to carbon, hydrogen, oxygen and organic acids. NIWA trials showed that it is very specific to certain targeted weeds and that native plant species present in the trial treatments were flourishing a year later.[9]

Once an alien plant has been detected in New Zealand, the first step is to collect information on its extent, how big the infestation is, how many sites it has infested and how much of a risk it poses. From this, the most appropriate management intervention option can be selected.

The first option is to do nothing. This doesn't necessarily mean taking no action — you might monitor the plant, keeping an eye on it to see if it looks likely to take off. At that point you can then switch strategies and quickly jump on it.

The next option is exclusion, keeping the plant out of certain areas. For instance, you may see that it's moved in down the road from a highly valued wetland, and you want to stop it before it reaches that site. Exclusion is a very cost-effective strategy. As Paul says, 'It's not like an oil spill, where you're dealing with a massive sudden shock to the system, but in 10 years you probably wouldn't even know it had been there. If you have a really invasive species turn up and you do nothing, in 10 years' time you're going to have 100 times the problem.'

The third option is eradication. But it requires a huge amount of effort to get rid of every last bit of an invasive plant. Says Paul, 'Often, half of the cost of eradication is spent doing all the follow-up to make sure that you've completely got rid of your target — every last seed, every last bit of live underground root or rhizome. It's an incredibly difficult thing to do, but New Zealand has been really successful managing species in that way.'[10] NIWA is trialling new detection methods that could be game changers. One of these involves programming waterproof computers to recognise and record the locations of specific underwater weeds.[11] Sniffer dogs and environmental DNA are also being trialled to detect weed species in low abundance.[12]

If you can't exclude or eradicate the weed and don't want to let it run amok, the next option is progressive containment. In this strategy you start with outliers — the areas furthest from the main infestation site that are just starting to be colonised. For plants that are spread by water, you'd start at the most downstream infestation and work your way upstream, as Paul explains: 'So, you're actually squeezing the plant back to the core areas that it occurs in, and hopefully in the long term, you can then switch to an eradication programme.'

If the plant has become widespread or too expensive to eradicate, the final option is sustained control. In this strategy we accept that it's here to stay, but we keep the pressure on to reduce its spread or impact. An example is the way in which aquatic weeds are managed in the Rotorua Lakes district, by regular control around boat ramps to reduce the risk of boats picking them up and spreading them around the lake or to another site.

Paul and his colleagues have pulled all this advice together into a best-practice guide for managing aquatic weeds.[13]

Being picky

Haloxyfop, triclopyr, endothall. The names of the chemicals used to kill weeds certainly sound alien and sinister, but herbicides are sometimes the only practical option if you want to control or eradicate an invasive plant effectively. The key, says Paul, is to look for the right herbicide for the job and apply it in the most effective way, using the least amount to achieve your goal and minimise 'off-target' effects.

While glyphosate is the most widely used herbicide ingredient in the world, it's a broad-spectrum chemical, so will destroy just about any plant. Another option is to use selective herbicides, ones that will target a certain species or type of plant and not harm those you want to protect. Some are developed to kill broadleaved plants, and are used by farmers to maintain pasture grass. Other herbicides take out only grasses, which is handy when you are trying to grow leafy crops and don't want grass competing with them for water, space or nutrients. Although they have mainly been developed for the agricultural industry, herbicides are increasingly being turned to for control of invasive plants in natural areas.

Until 2005, glyphosate and diquat were the only herbicides that could legally be used over waterways in Aotearoa, which severely limited the options for wetland managers. Today, there are a few more herbicides to choose from, thanks to Paul's research and experiments:

> We were looking at selective herbicides, as a potential way of controlling weeds while protecting threatened plant species. I managed to get a whole range of threatened wetland plants, other native wetland species, and weed species. We grew them all in pots in our experimental facility. Then we tested out a range of non-selective and selective herbicides — so your haloxyfops, your triclopyr and clethodims . . . I think it was eight herbicides in total.

The researchers monitored the fate of each plant/herbicide combination, rating each as healthy, damaged or dead at the end of the trial.[14] Four of the eight herbicides (haloxyfop, clethodim, clopyralid and triclopyr) showed promise in terms of effectively killing the target species with minimal by-kill.

The NIWA team then took their trials out into the field, to a mostly dry arm of Lake Lyndon and the shallow Barker tarns near Lake Coleridge in the South Island, and two North Island sites, Lake Whangape near Huntly and South Taupō Wetland near Turangi. They set up quadrats (square plots), measured the amount of cover per species and then applied each herbicide to a separate plot. All four of the selective herbicides gave good control of their target species. Grass-specific herbicides had very little impact on non-grass species. The broadleaf herbicide triclopyr amine gave good control of willow without touching sedges, grasses and rushes, but tenacious grey willow showed a lot of regrowth the following year, and some non-target broadleaf plants such as coprosma shrubs were damaged or killed.

Their research findings formed the basis of successful applications to the Environmental Protection Authority to approve triclopyr amine and haloxyfop, plus two other herbicides for use over water,[15] subject to specific conditions.* Paul recounts subsequent successes with selective herbicides:

> The first trial we did with haloxyfop was on spartina, a highly invasive grass that smothers tidal mudflats. This is the tool that has really turned the tide, if you like, on spartina management, and it's now been eradicated from huge parts of the country, even heavily infested areas like Havelock in the Marlborough Sounds.

Bring in the sheep

While Paul's research has helped improve the effective use of herbicides, they have their detractors. And although herbicides are one of the cheapest ways to get rid of a plant in the wrong place, they are not the only option. Manual control — elbow grease, as it's known — can get rid of a weed from a small area, but sometimes you need a helping hand from another species.

One example Paul recalls was a series of shallow tarns on the South Island's Molesworth Station:

---

* These herbicides are restricted to use by central and regional biosecurity authorities and a number of conditions apply.

Jan Clayton-Green and Shannel Courtney of DOC investigate the re-establishment of rare native turf plants following the control of weedy oval sedge on Molesworth Station.

> There are a whole series of tarns and there was one that was being invaded by oval sedge, *Carex leporina*. It was a concern because there were rare plants growing among the turfs in the shallow areas of the tarns. There was a species of *Craspedia*, a little daisy thing that was only known from that tarn — it hadn't even been given a formal scientific name yet — and a rare cardamine, *Cardamine mutabilis*, and, yeah, this oval sedge was threatening to take over the entire tarn.

Paul knew there were no selective herbicides that would kill only the sedge, and that the most effective herbicide, glyphosate, would also take out valued native plants. This is where DOC ranger Jan Clayton-Green started thinking outside the box. Molesworth is a high-country sheep station, with no shortage of woolly weed-eaters, so the team started trialling different stocking densities of sheep to graze the bulk of the oval sedge down to a manageable level. They were then able to bring in human volunteers to mop up the rest, until they eventually eradicated the weed from the area. As far as Paul is aware, all of the threatened species are now thriving in the tarns.

On a much larger scale, an ambitious attempt to control weeds without chemicals was undertaken by members of Waikato-Tainui in the North Island. In 2010, as part of a wide-ranging settlement with the government, a series of swampy islands in the Waikato River were returned to the tribe. The islands had been largely neglected under Pākehā ownership and were returned in a much-tattered state, overrun with a medley of invasive weeds. The haukāinga (local people) wanted to remove the weeds but were not keen on using herbicides.

Waikato-Tainui embarked on a two-year experimental project to trial non-toxic weed control options alongside herbicide control carried out by contractors. They brought in scientists from NIWA and Manaaki Whenua — Landcare Research to help guide the project, including a plan to eradicate weeds from a portion of the largest (10-hectare) island. Among other things, they wanted to test whether native plants could be used to outcompete weed plants. The members put in a huge amount of effort and developed some new techniques for controlling yellow flag iris, which formed extensive thickets across the island. They spent 6041 hours trying to control the weeds manually, but it was tough going. For every 5 hectares that were sprayed to control pest plants, just over 1 hectare was successfully controlled by hard labour.[16]

Despite the setbacks and eventual acceptance that judicious use of herbicides may have a place in some circumstances, the many hours spent on manual labour enabled iwi members to make detailed observations and develop a much deeper understanding of the river islands' ecology. This information will be invaluable for their ongoing restoration efforts and is documented in the excellent handbook *Te Reo o Te Repo: The Voice of the Wetland*.[17]

Natural enemies

While sheep, young cattle, goats and even chickens have all been used in the fight to knock back weed infestations prior to hand clearing, there are also many smaller allies that can be left to do the job in their own time — biocontrol agents.

One of the reasons alien plants have done so well in Aotearoa is because the natural predators or diseases that keep them in balance in their country of origin are not present here. So a logical solution is to visit their native country, find out what their natural enemies are and bring these to New Zealand. It seems risky to bring in one new organism to control another, and besides the fictional old lady who swallowed a fly, there are many real examples from the past where such a move turned out to be a massive fail: the diurnal mongoose released in Hawaii to control nocturnal rodents, cane toads released in Australia to control a sugar cane beetle, stoats brought to New Zealand to control rabbits. All of these biocontrol agents are now pests in their own right.

Having learnt that lesson, the Environmental Protection Authority now requires any new organism entering New Zealand to be put through thorough testing. As Paul says, 'You have to show that the target plant is the only species that's going to be attacked by this thing. So you do starvation trials, which means you force your insect or disease onto other related plants, native plants, whatever. If there's any damage to native species, I don't think you pass go, so there's pretty rigorous screening.'

One biocontrol agent that targets a wetland invader is the South American alligator weed flea beetle (*Agasicles hygrophila*), introduced to Aotearoa in the early 1980s.[18] According to Paul, it didn't seem to be having much of an impact in the Waikato region where he lives. It was during a conference field trip in Florida that the penny dropped:

> We were on Lake Okeechobee, a big shallow lake full of the weed hydrilla, but there were also tiny little shoots of a plant, and I said to the trip leader, 'That's alligator weed isn't it?' He said, 'Yeah.' I asked why it wasn't taking over the whole area, and he said, 'Biocontrol.' So in Florida you've got high temperatures, and the insect really hammers the invasive plant.

In the cooler Waikato region, the beetle doesn't overwinter as well as the plant does, so the plant has three months of unchecked early-season growth, then the beetle emerges to have another go at it. As Paul notes, 'It's about understanding the biology of the plant and the biology of the control agent.'

Some of the most successful international biocontrol examples are the ones targeted at free-floating aquatic plants such as water hyacinth and salvinia, because the insect can devour all of the productive parts of the plant and they don't have any roots that can survive and re-sprout. But it's not the solution for all invasive plants. Paul says that there haven't been any successful examples of biocontrol for submerged weeds despite a lot of research investment overseas, so for species such as hydrilla, selective herbicides are currently the only option to eradicate it.

Fighting fire with fire

Can we let our native wetland plants do the heavy lifting? Paul thinks it's worth trying in the right locations. Most of New Zealand's wetland plants are small herbs or shrubs, easily overcome by taller invasive wetland plants such as willow or alder, but there are a few native species that might just have the edge on these woody invaders.

Kahikatea emerging above invasive willows on a Waikato River island.

Healthy kahikatea forest, Ōmaha, north of Auckland.

Paul's master's thesis at the University of Waikato was on kahikatea, our tallest native tree, capable of growing in swamps and shallow water throughout much of our lowlands. His work revealed that kahikatea would actually spread out from remnant patches into areas infested by willows. He thinks the spread would probably be faster if the willow wasn't there, but it presented a possibility in terms of restoration ecology: 'Essentially, what you'd try to do is plant at a density that very quickly forms a canopy that sits over the top of these invasives.'

Weed management is as much a social issue as it is a biological one. Although many people are spending hours of volunteer or work time eradicating weeds, a lot of invasive plants are still being spread by others. This could be through lack of awareness, such as dumping the contents of fish ponds in places where they can get into waterways, or through inadvertent actions, such as when weeds are picked up on boats or fishing gear, diggers or other machinery. Some plants are also spread deliberately, because someone considers them attractive or of medicinal or food value, or in the misguided belief that the weed would benefit fishery and wildlife values of waterways.

Paul notes how close New Zealand came to eradicating water hyacinth and salvinia in the early 1990s, but even though it's illegal to propagate and spread these plants, people still keep them to raise tropical fish. 'So, essentially what MPI has been doing since that time is just mopping up new sites as they turn up, rather than dealing with the people that are the source of these plants. You know, it requires education and behaviour change.'

One programme Paul feels really privileged to have helped initiate and be part of is MPI's National Pest Plant Accord. The aim of this is to stop the deliberate spread of plants that are sold in the nursery and aquarium trades by working with the industry to agree to a list of species that are banned. Paul recalls going to a nursery one day and recognising a plant that stopped him in his tracks:

> I said to the nursery worker, 'Oh, where'd you get that from?' And he said, 'Oh, it came in with some stuff I was importing.' So I said, 'That's a terrible weed in North America. Can I have it?' And he happily gave me all his stock and it's never been seen in New Zealand since! That was flowering rush, *Butomus umbellatus*, . . . a species of invasive aquatic plant currently spreading across the northern US and southern Canada displacing many plant species and affecting fish spawning habitat.

If only it were that easy to wipe out the many other weeds that manage to sneak into Aotearoa! Paul and his biosecurity colleagues will never run out of work. However, it's not always hard yards that are keeping weeds out of wetlands; sometimes, that initial conversation with a person makes a real difference, as Paul points out: 'Actually, the same nurseryman got in touch with us again and he had another plant that he saw was having quite a major impact on his nursery. So, yeah, he got the message. Education really worked in this instance.'

INVASIVE IMPOSTERS

Lookalike plants cause all manner of headaches for folk restoring wetlands. Is that plant a rare species I should nurture, or is it the vanguard of a nasty weed species that I should nip in the bud right now?

Wetland plants are hard enough to identify at the best of times. Rushes look pretty much the same to the untrained eye, and even the most experienced pro needs flowers or seeds to distinguish some species. So when native and invasive plants are closely related or look similar, it adds another layer of head-scratching for wetland managers. As a final tip: check your prey before you spray!

Some of the tricky lookalikes are listed below.

Native species	Invasive lookalike
Toetoe (*Austroderia* species) • tough midrib, hard to rip the leaf in half • base of leaf has a bluish appearance • seed head curves attractively 	**Pampas (*Cortaderia* species)** • weak midrib, leaf easily ripped across • usually has curled-up dead leaves at base of plant • flowers are more upright 
Pink bindweed (*Calystegia sepium* subsp. *roseata*) • bright pink flowers • slender green bracts at base of flower (in *Convolvulus* species, these green petal-like structures are on the flower stem) 	**Great bindweed (*Calystegia sylvatica*)** • has white flowers with swollen bracts that overlap one another • cross-breeds with the native species; the hybrids have pale pink flowers with overlapping bracts 

Native species	Invasive lookalike
Stout water milfoil (*Myriophyllum robustum*) • rare native species • tends to look bronze 	**Parrot's feather (*Myriophyllum aquaticum*)** • aggressive weed • bushier than the native species 
Swamp millet (*Isachne globose*) • sward-forming wetland grass • dainty seedheads like those of millet • little row of hairs in leaf base 	**Mercer grass (*Paspalum distichum*)** • also creeping bent (*Agrostis stolonifera*), reed sweet grass (*Glyceria* species) • seedheads sticky or feathery (not like millet) • papery membrane (ligule) in leaf base 
Giant umbrella sedge (*Cyperus ustulatus*) • large, dark-coloured seedheads 	**Exotic umbrella sedges (*Cyperus* species)** • 12 exotic species (not all found in wetlands) • yellow, green, red or purple seedheads 

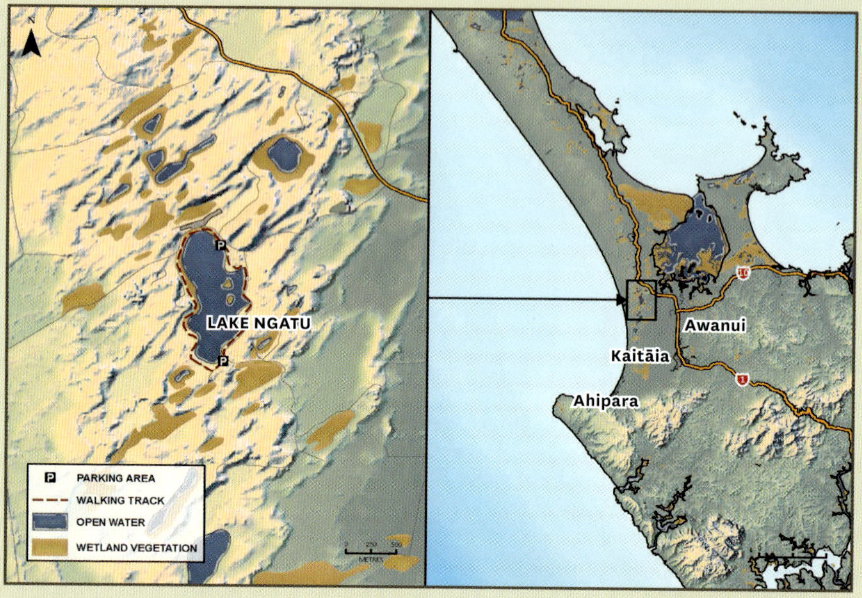

Lake Ngatu

Dune lake

While weeds are a scourge in many wetlands across Aotearoa New Zealand, there is one area that Paul is pleased to report has very few weeds — the Northland dune lakes.

Fifteen per cent of New Zealand's larger lakes are dune lakes, formed by shifting sands that block off streams or create depressions that trap rainwater.[19] Northland Regional Council describes them on their website as 'one of the rarest and most threatened aquatic habitats in the world' and says they support an array of threatened plants and animals.[20] NIWA has been monitoring Northland's dune lakes annually since 2004.[21] Paul says that while most of them are generally still in very good condition, they are vulnerable to nutrient enrichment, declining water quality, pest fish and invasive weeds.

One lake NIWA has been concerned about is Lake Ngatu, north of Kaitāia. Ngatu is among the best dune lakes in Northland because of its diverse aquatic plant community, supporting 20 native plant species. But there are also five exotic species, including the invasive oxygen weed *Lagarosiphon major*, first recorded at Lake Ngatu in 1988.[22] This species is highly invasive and capable of occupying the entire lake bed, but until recently it was kept in check by low nutrient levels. However, when the nutrient status began increasing to high (eutrophic) levels in 2015 and the native plants started declining, Northland Regional Council took action. Following the successful eradication

of the same weed from nearby Lake Ngakapua in 2015 through the application of endothall,[23] the council used the same herbicide over Lake Ngatu in late 2020. Endothall breaks down into natural elements, it doesn't affect other native plants, fish or invertebrates, and the treated water is safe to swim in after 24 hours. Within two weeks, the *Lagarosiphon* weed beds had collapsed. It will take five years of all-clear surveys to officially declare the lake free of the weed, but Paul says there was no sign of it during NIWA's March 2021 underwater survey and the lake vegetation index value at the time was the highest recorded since their annual monitoring began.

Highlights

- The lake has a diverse aquatic plant community, including the rare fern *Cyclosorus interruptus*[24]
- Supports a large population of the Nationally Critical *Trithuria inconspicua*, a tiny rush-like submerged plant that has become extinct in most of the Northland lakes[25]
- Keep an eye out for matuku (Australasian bittern), weweia (New Zealand dabchick) and taranui (Caspian tern).

Access

Lake Ngatu is 16 kilometres north of Kaitāia. Follow State Highway 1, then turn left onto West Coast Road. Access is signposted 1.9 kilometres along on the left. There is parking at several different spots around the lake, but the most common access sites are at the northern and southern ends, where there are toilet facilities.

There is a 4-kilometre loop walking track around the perimeter of the lake, which can also be explored by kayak. The lake bed is owned by Ngāi Takoto, who have installed bollards to keep power boats out of the lake, and information signs to encourage responsible use of the lake.

www.wetlandtrust.org.nz/get-involved/visit-wetlands/wetlands-you-can-visit-in-the-northland-region

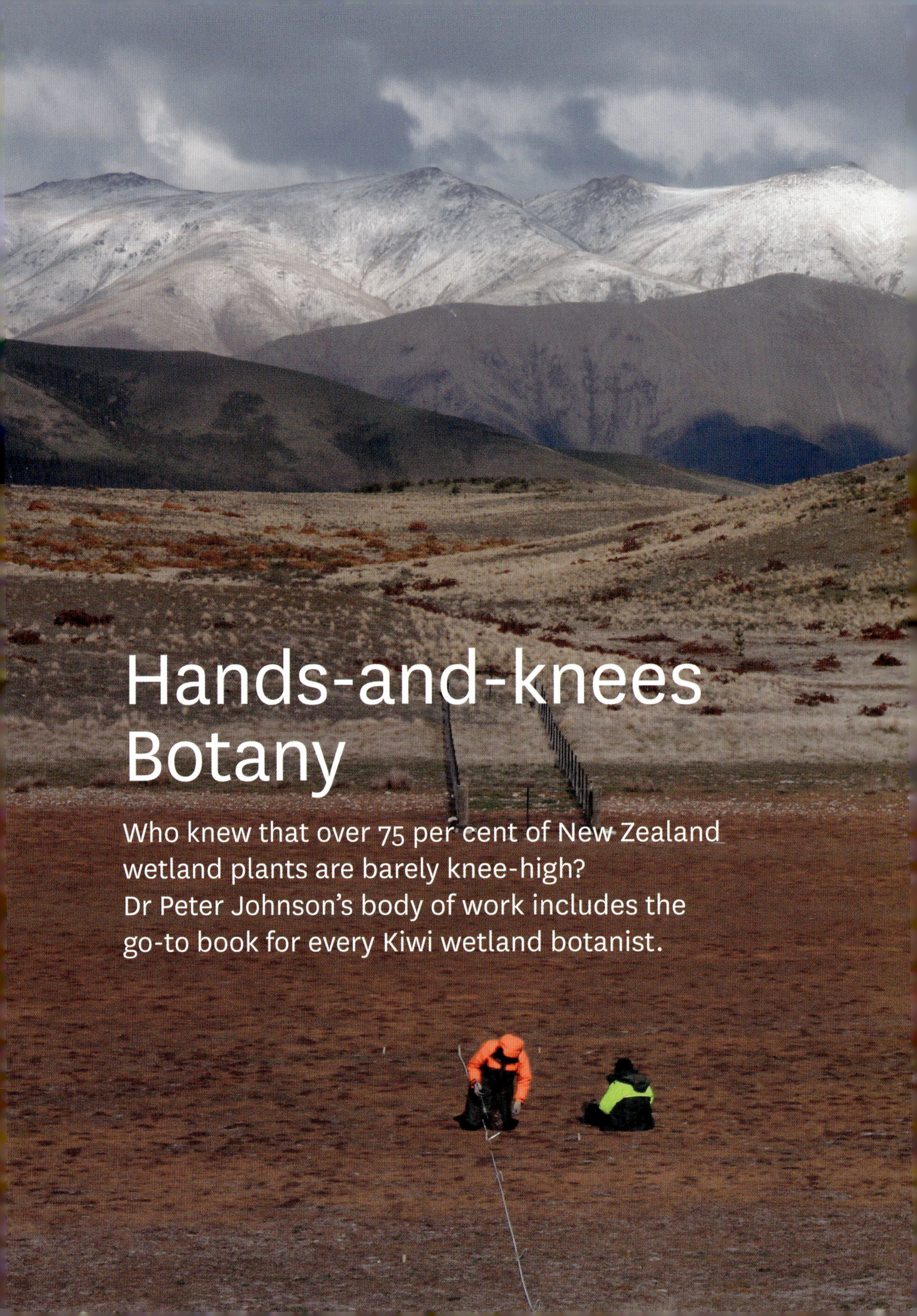

Hands-and-knees Botany

Who knew that over 75 per cent of New Zealand wetland plants are barely knee-high? Dr Peter Johnson's body of work includes the go-to book for every Kiwi wetland botanist.

Dr Peter Johnson

Specialist field: Botany, plant ecology, wetlands

Currently works for: Manaaki Whenua — Landcare Research, Ōtepoti Dunedin, as an emeritus research associate

Peter Johnson PhD is a semi-retired botanist who has contributed his expertise to growing knowledge and understanding of wetlands, their flora and their management. He has authored and co-authored seminal works on wetland plants and the diverse wetland systems found in New Zealand. He has had a particular interest in kettle-hole wetlands (surface depressions in glacial moraines) and their tiny turf vegetation and diverse flora, along with other ephemeral and dune wetlands. Peter has shared his expertise on wetlands with the Fiordland National Park and Chatham Islands Conservation boards, and is a trustee of the Te Nohoaka o Tukiauau Sinclair Wetlands Trust, a long-term restoration initiative on the Taieri Plain near Dunedin.

FEW PEOPLE WOULD HAVE experienced such an intimate connection between politics and cushion bogs as Peter Johnson and his field companions.

It was the night of the New Zealand general election, 29 November 1975. Most New Zealanders were packed into draughty town halls or clustered around their neighbour's black-and-white television. Peter and his colleagues were a million miles away.

Picture howling wind and damp, low clouds scudding across the 700-metre-high plateau of Ajax Hill, in the blustery southeast corner of the South Island. Two tents perched on a hummocky *Oreobolus* cushion bog, semi-sheltered behind bushes of pink pine and stunted kaikawaka, 'all buffeted, bonsai, and flagform,' as Peter recalls. He admits that it is a place 'where nobody but wetland botanists would ever likely choose to camp'.

In one tent, botanists Geoff Baylis and Alan Mark were cheerfully cooking dinner, while in the other tent, Peter shared the cramped space with palynologist Matt McGlone and his sprawl of peat-coring gear, pottles of samples for pollen analysis and grubby boots. Over the simple camping fare, Alan had his ear to a tiny transistor radio, yelling election news updates tent to tent through the mist: 'there's another seat gone, there's another one . . . it's all over'. It was a gloomy night as the conservative National Party 'Rob's Mob' raced into election victory bolstered by an infamous television advertising campaign featuring dancing Cossacks.[1]

Power plant threatens wetland plants

Their gloom was triggered by recollections of the National Party's grand plans to build a vast hydropower scheme during its previous term in government — a scheme that threatened to destroy many hectares of rare lake shoreline turf vegetation in Fiordland National Park (which now lies within a UNESCO World Heritage site). The clash between economics and these tiny lake-edge plants began Peter's distinguished career as a wetland ecologist.

Planning for the scheme had begun in the 1960s and proposed building dams to raise the water level of two natural South Island lakes by a whopping 30 feet,[2] creating the water storage and energy to generate electricity. It's hard to imagine these days, but carrying out a routine ecological assessment prior to a landscape-scale development simply wasn't considered at the time — not even a development that would involve excavating 1.4 million tonnes of rock, drowning wooded islands and merging two iconic Fiordland lakes. Such was

THE MINIATURE MARVEL: *VERONICA LILLIPUTIANA*

Reminiscing about his favourite plant species, tarn speedwell, brings out Peter's inner 'poet'.

> See me on a rainy day in a tent above a little sandy cove at Lake Manapouri, too wet for fieldwork. Trying to catch up with plant identifications, under trying conditions. A wee turf plant with tiny leaves and an oversize pale blue flower. No clue as to what it is, but Allan's *Flora* is balanced on a food box, a candle offers its one candlepower, and I juxtapose specimen and hand lens for a systematic keying out . . . the whole way, to family (Scrophulariaceae: fair enough); to genus (*Parahebe*: well, maybe, but keep going); to *Parahebe canescens*. And yes, the description fits! Wow, satisfaction.
>
> And in later years, further satisfaction, as the 'accepted' name reverts to *Veronica lilliputiana* — so much more descriptive (Lilliput is the island home to a race of tiny people in [the novel] *Gulliver's Travels*). By which time I have been able to marvel at the tiny flowers, en masse, creating their own blue zone at a particular ponding elevation around kettle holes in Canterbury moraines. And then, more recently and not far from home, on Otago Peninsula, to be shown in an ephemeral dune hollow the same plant, that had not been seen here for 114 years!

the political backdrop for a series of shoreline vegetation surveys initiated by far-sighted botanists Geoff Baylis and Alan Mark at Lakes Manapouri and Te Anau in 1969–1971.

The two University of Otago scientists had managed to secure funding from the New Zealand Electricity Department to study the shoreline vegetation and gain some understanding of it before it became permanently flooded by the rising water levels. They were joined by Peter, then an undergraduate who attended the field survey as part of a summer work programme. Meticulous investigation and mapping of the shores of both lakes revealed the special relationship between shoreline vegetation patterns and natural water-level fluctuations.[3] The environmental damage was going to be much worse than anticipated by the project's engineers or promulgated by the assurances of government departments.

Peter's early days surveying the shores of Manapouri were not only part of an in-depth study to fill huge gaps in our knowledge; they also added scientific heft to the Save Manapouri campaign, which ran from 1969 to 1972. The campaign is regarded as one of New Zealand's longest and hardest-fought environmental campaigns, with almost 10 per cent of the nation's population adding their signature in opposition to raising the lake levels. The resounding success of the campaign became a deciding factor in the 1972 general election, which ousted the National Party after 12 years in power. The incoming Labour government immediately invited the campaign leaders, including Alan Mark, to become members of a new statutory body — Guardians of Lakes Manapouri, Monowai and Te Anau — to make recommendations on management of these lakes for both hydroelectricity generation and protection of national park values. The power scheme was developed, but the lakes weren't merged, and the fragile flora and shorelines were protected from wide-scale inundation and slumping. What's more, the environmentally exploitative mindset of the time began to shift towards one of more sustainable use, eventually culminating in the Resource Management Act 1991.

From fieldwork to field guide

A particular challenge of the fieldwork at Lakes Manapouri and Te Anau lay in identifying the very large number of tiny turf plants. There were no illustrated manuals at the time, and only the highly technical, rice-paper-paged Volume I of H. H. Allan's *Flora of New Zealand* series to guide them. Peter was very reliant on giving his herbarium specimens 'tag names' (such as 'white-flowered spoon-shaped mudwort') and then sending them off to get accurately identified. This task fell to a well-known trinity of botanical experts, Dr Lucy Moore, Dr Elizabeth Edgar and Ruth Mason, based at what was then the Botany Division of the Department of Scientific and Industrial Research (DSIR) at Lincoln.

In Peter's mind, the experience crystallised into the need for a simple guidebook to help identify our diminutive lake-edge turf species. However, it soon became clear that a book showcasing all of New Zealand's wetland plants would be much more useful. The DSIR Botany Division (Peter's employer following his university studies) agreed and allowed Peter to travel throughout New Zealand, liaising with colleagues and collecting plant specimens — surely *the* dream job for any botanist.

Although Peter is an accomplished sketcher, Pat Brooke, an illustrator, was employed to create the detailed line drawings critical for identifying each species. Brooke's drawings were based firstly on freshly collected or cultivated specimens, secondly on dried and pressed herbarium specimens, thirdly on Peter's photos and drawings, and lastly on Peter's draft sketches for each page, which drew attention to the diagnostic features Brooke needed to emphasise in her illustrations. These could be the location of a tiny sickle-shaped mucro — a short, pointed projection from the stem of a rush — or the length of a ligule — a tiny flange of hairs or tissue found in a grass plant's 'armpit'.

Owing to the enormity of the task, the production of *Wetland Plants in New Zealand* spanned eight years, with the first edition published in 1989. The book features more than 530 native and naturalised (non-native species that have become established in the wild) plants of bogs, swamps, estuaries and lakes. It covers the entire New Zealand botanical region, including the subtropical Rangitāhua Kermadec Islands and Manawatāwhi Three Kings Islands groups to the north, Rēkohu Chatham Islands to the east, and the subantarctic islands to the south. The result is a landmark for sharing knowledge of our nation's wetland plant species, and the book continues to be the 'go-to' identification guide for New Zealand's field botanists. Many cherish their own slightly muddied, well-thumbed copies, typically filled with scrawled field notes and dried plant specimens spilling out from between the pages. Peter modestly states: 'It is nice to think that the several editions have been useful to users, and are still sought after.'

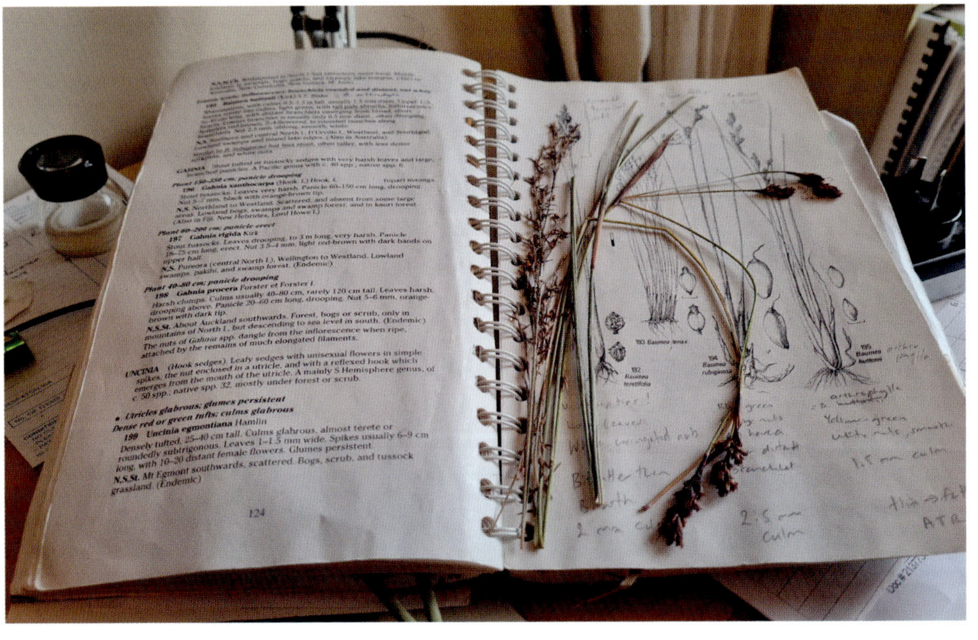

Peter's book Wetland Plants in New Zealand *is a well-used resource for wetland botanists.*

To think that the suite of delicately hand-drawn plant images were once considered lost! All reprints of *Wetland Plants in New Zealand* by Manaaki Whenua Press today are facsimile editions — created from high-quality photographs of each page of the original publication. The drawings were thought never to have returned from the printer in Hong Kong. Yet three decades later the precious originals resurfaced — they had lain completely forgotten in storage at a research facility in the South Island!

The evolution of plant identification

The process of working out the identity of a plant has evolved dramatically during Peter's career. The traditional method was to follow a dichotomous key — usually at the front of a botanical guide. These keys are a series of paired technical questions about various features of the plant, perhaps whether the stem is bright green or dull brown, or whether the leaves are flat or rolled. Each answer directs you to the next relevant pair of questions, until you reach the ultimate question that leads to the plant's name. You can then turn to the relevant species in the guidebook to confirm that you have arrived at the correct answer . . . or not! Dichotomous keys can be incredibly frustrating — if you answer a question incorrectly, it can send you off on a wild goose chase and it can be difficult to backtrack. If the first few questions relate to flowering parts and your plant isn't yet flowering, you're stumped right from the get-go.

With the advent of computers came the user-friendly lucid key, an interactive tool that allows you to plug in as many diagnostic features as you can, based on the sample you have. No flowers? No problem. Just type in the length of the leaf, whether it is hairy, whether it is sharp-edged, and so on. The more features you add, the shorter the list of potential candidates becomes. Having whittled the options down to a few, you can then check the photos and descriptions offered to decide which species your sample is. In some cases, scientists will send their pressed and dried plant specimen to the experts at a herbarium (a library of dried plant specimens) for final confirmation.

The ultimate in easy plant identification is the smartphone app. In the remarkable way that science fiction often appears to predict future technology — or perhaps the imagination of science fiction writers inspires technologists to bring those ideas to life — the *Star Trek* series seems to have predicted the advent of smartphone plant identification. Where once Dr Beverly Crusher merely pointed her tricorder at a patient to determine their ailment, today's field botanist can point their smartphone camera at a plant to identify it. Applications such as iNaturalist compare your photo with a massive image library to suggest species that look similar to your specimen, and a bevy of keen botanists provide a form of crowd-sourced verification (in their spare time).

However, smartphones are not likely to displace human botanical skills any time soon, because many diagnostic features of wetland species require a hand lens or even a microscope to see, and verification by experts is still crucial. Peter's own journey into the meticulous world of accurate plant identification has relied strongly on others: 'And here lies an important point: that these botanical experts willingly shared their considerable expertise, and time — for careful identifications do take a lot of time, which illustrates how important it is to do the same for others.' Now with many decades of experience and an expert himself, Peter can reciprocate: 'I like to think that I have repaid the same with my own hours spent with someone else's specimens with microscope, dissecting needles and reference herbarium specimens.'

Putting wetlands in their place

A further collaboration, with wetland expert Dr Philippe Gerbeaux, led to the development of another vital publication in the wetland management toolbox. *Wetland Types in New Zealand* sets out a detailed system for classifying all of the different wetland ecosystems found nationally — from common to uncommon, freshwater to saline, coastal to inland, subterranean to montane, geothermal to frozen.[4]

A standardised national classification system for wetlands means everyone is speaking the same 'language', and each type of wetland can be better understood, managed and monitored. *Wetland Types in New Zealand* provides a multi-layered 'family tree' of wetland classification, from broadly defined hydrosystems (where the water comes from, and how salty, fresh, hot or cold it is) and wetland classes (the terms we are most familiar with: swamps, bogs and so on), to structural classes of vegetation (forest, scrub, sedgeland and the like) and wetland composition (a shortlist of the dominant plants, such as the iconic kahikatea generally found in swamps). Together, the combination of categories from each level in the classification makes up the wetland 'type'.

Classifying and mapping different types of wetland can be very tricky. Changes from wetland to dryland, or from one type of wetland to another, can be so gradual that it's hard to know where one type ends and another begins. Or they can be a mishmash of several types rapidly changing over the space of a few paces. In coastal systems, for example, the part of the wetland furthest from the shoreline might be freshwater, with species that survive best where there is little or no saltwater influence. Closer to the ocean, the species will change to those that can survive and thrive in a very different set of conditions — think of the hardy mangrove, with its leathery leaves and intriguing system of upright pneumatophores (breathing roots) to help it cope with sand, silt, saltwater and sun.

A career in wetlands

Since Peter's botanical career began in the 1970s, his research has spanned all aspects of wetlands and their management, from individual plants to whole ecosystems. One of his first jobs following university was carrying out vegetation surveys of scenic, scientific and recreational reserves throughout Aotearoa as botanist for the DSIR. He recalls his time in Otago and Southland with fellow Dunedin-based scientific colleagues Dr Ralph Allen and Dr Bill Lee as 'excellent training'. Together, they trudged through bogs, marshes and swamps — 'places you wouldn't have otherwise thought to go yourself' — surveying the local flora, describing the vegetation patterns and assessing management needs such as fencing, fire management, pest control and nutrient management. Their methodical (and muddy) work in these legally protected areas called out where better physical protection and recognition were needed.

Notebooks have always been a feature of Peter's life: 'The ideal, for me, is A6, case-bound, side-opening with a partly red cover so it is visible among green vegetation, opens flat and fits in a back trouser pocket. Hard cover allows it to be sat upon and also held between the teeth when standing in a lake or climbing a bank.' Decades of observations are documented as scrawled text, diagrams and drawings. This takes both diligence and discipline — as a child, he kept 'nature books', and then at 13 he made his first vegetation map of an island in the upper Mata-Au Clutha River.

Looking back, it was almost inevitable that Peter would become a leading expert on wetland plants. Studying botany at Otago University was a natural choice for a boy growing up with a garden and home next to Dunedin's forested Town Belt. There, native trees could be identified and used for hut-building. A small stormwater stream provided a site for attempting (unsuccessfully) to translocate kōura — small freshwater crayfish. Family holidays were spent exploring many wetland locations, including coasts, lakes and riverbanks — places he is still drawn to as an adult.

> Wetlands have for me been a career — an annual weekly activity, and in addition to that, I like going fishing, camping, swimming — those things take me to wet places and natural places . . . A wetland isn't just sitting there . . . it's a dynamic functioning thing . . . the passion part is trying to work out how a wetland works — observe and understand patterns and processes that I could never quite work out before.

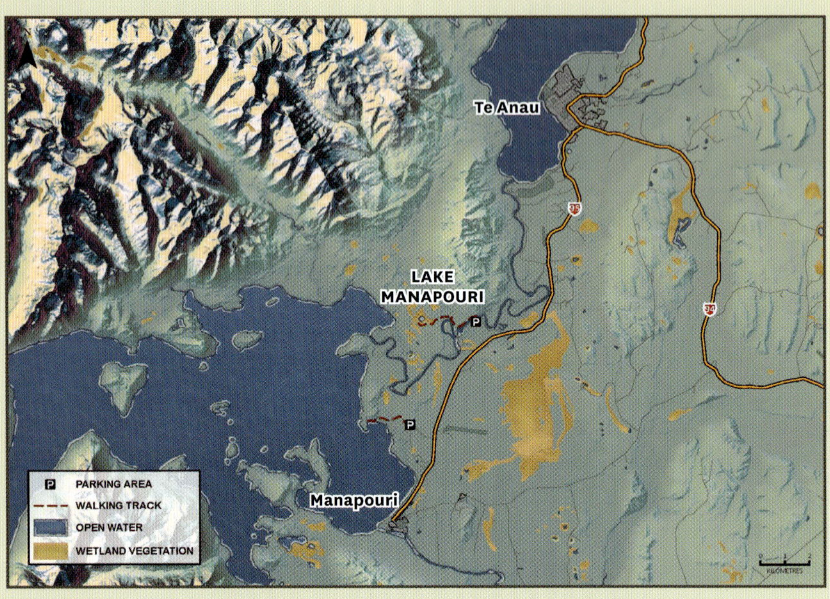

Lake Manapouri and surrounds

Glacial lakes and peatland

One of Peter's favourite places is undoubtedly Lake Manapouri in the South Island's Fiordland National Park, as he explains:

> Because it is an old haunt. Because it is familiar, an old friend. Because whenever I visit, I always like to again touch the waters, paddle or go right in, to sit and look, to keep wondering, afresh and incrementally, how all the lake processes work, how the sediments keep cycling, how the plants are disturbed and reinvigorated, how something never fully understood before can suddenly, years later, seem to explain itself, and all this on the shore of one of the huge Fiordland lakes, such as I once described as having 'disparate moods: ecstatically idyllic; overpoweringly sullen; unforgivingly nasty'.

Examples of Peter's beloved lakeshore turfs can be seen on the shoreline at Frasers Beach beside Manapouri township — be prepared to spend time on your knees, and pack your best macro lens to capture the delicate beauty of these tiny plants. An even better way to enjoy them is by kayaking slowly around the lake margin.

A must-see is the curiously named Amoeboid Mire, nestled between Lakes

Manapouri and Te Anau. It was formed when the Manapouri Glacier scraped across the landscape and then slowly melted around 14,000 years ago, leaving unevenly furrowed terraces. Peat gradually began to accumulate in the low-lying areas and was colonised by a suite of mosses, rushes, sedges, herbs and insect-eating sundews, all tolerant of the acidic, permanently wet ground. It's sobering to think that this near-pristine mire and its collection of small lakes were almost drowned in the quest to raise the level of Lake Manapouri for power generation.

Highlights

- Stunning mountain and lake scenery
- The lake margins at Manapouri and Te Anau are a national stronghold for many turf species and among the best examples of this habitat in New Zealand. Look for the highly threatened herbs *Tetrachondra hamiltonii* and *Gratiola concinna*, the tiny mud buttercup (*Ranunculus limosella*), its rare cousin *R. ranceorum* and the non-threatened mudwort (*Limosella lineata*)
- Information panel at Amoeboid Mire and views over the hauntingly named Spirit Lake
- Stunted patches of bog pine and pygmy pine, both uncommon at such low altitudes.

Access

Lake Manapouri is a 20-minute drive from the town of Te Anau along the Manapouri–Te Anau Highway (State Highway 95). For the Ameboid Mire walk (an easy 45–60-minute walk each way), drive from Te Anau along the highway for 10.6 kilometres, then turn right at the signposted road to the Rainbow Reach car park. From Rainbow Reach you can also walk the Kepler Track along the full length of the Waiau River to Lake Te Anau (9 kilometres — allow two hours).

While you are in the area, plan to visit nearby Rakatu wetlands, restored and enhanced to mitigate the environmental effects of the Manapouri hydroelectric power scheme.

www.wetlandtrust.org.nz/get-involved/visit-wetlands

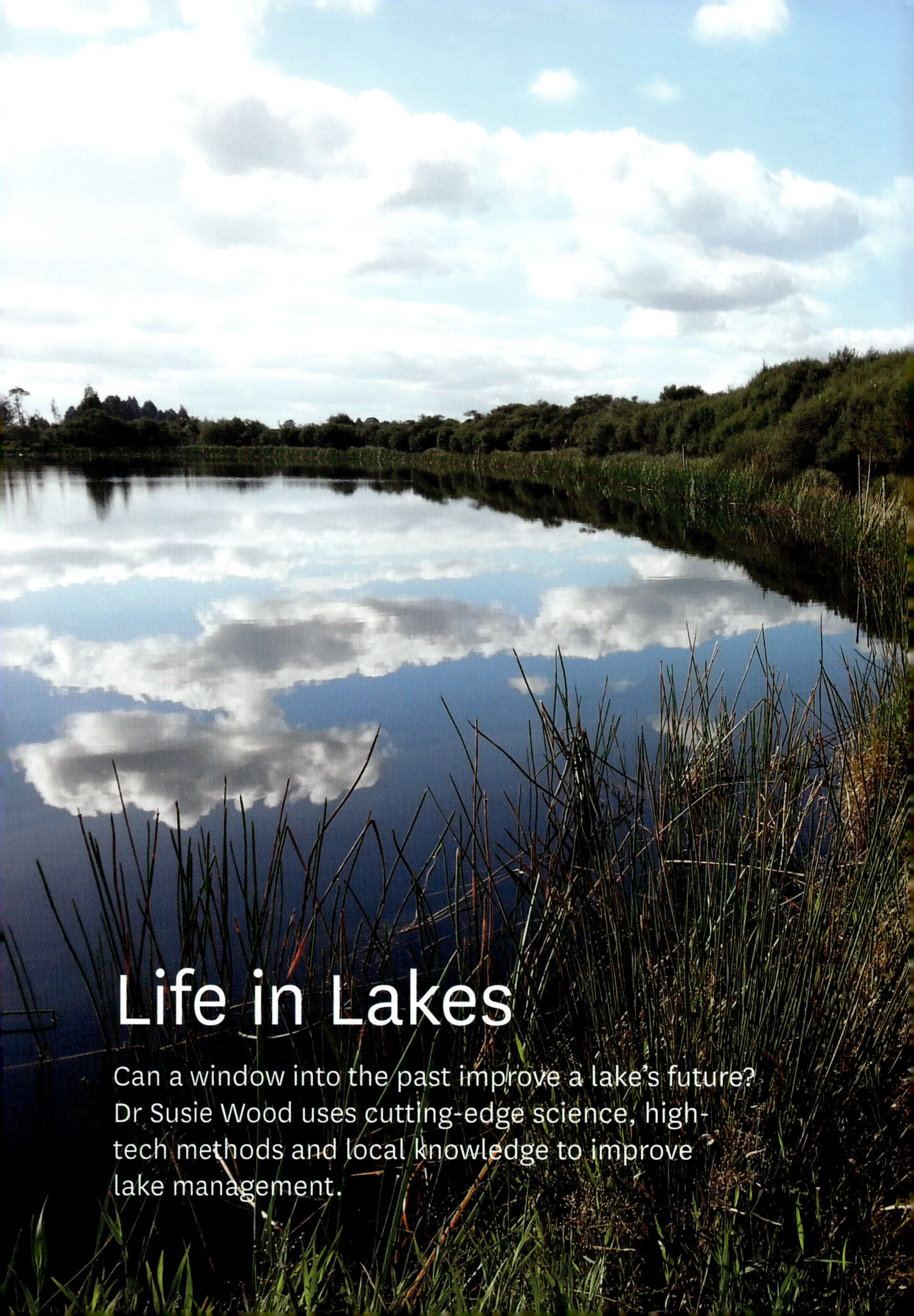

Life in Lakes

Can a window into the past improve a lake's future? Dr Susie Wood uses cutting-edge science, high-tech methods and local knowledge to improve lake management.

Dr Susie Wood

Specialist field: Lake ecology, toxic algae, environmental DNA, molecular ecology

Currently works for: Cawthron Institute, Whakatū Nelson

Susie Wood, PhD, is a widely respected freshwater scientist who works collaboratively with other scientists in Aotearoa New Zealand and around the world, and is regularly consulted for her expertise, particularly on cyanobacteria. Susie co-leads a five-year national lakes research programme that involves collecting and analysing lake sediments and water samples from approximately 10 per cent of New Zealand's 3800 lakes. In 2019, she was the winner of the New Zealand Freshwater Sciences Society Medal, 'For her outstanding contribution to freshwater science and management, and her leadership of women in science'.

BEING BORN IN AOTEAROA New Zealand offers a unique and privileged opportunity to grow up in the outdoors. The iconic Kiwi childhood includes days spent roaming nearby native bush, playing in clean rivers, hiking in the hills, lazy summer picnics at the beach, hunting, fishing and gathering kaimoana (seafood). If you talk to any Kiwi scientist, a childhood like this is often a common feature of their backstory, and Susie Wood is no exception.

Susie says her passion for the environment stems from her personal experiences of camping trips by lakes and rivers with her family. 'My parents were amazing at introducing and showing us the beauty of New Zealand's outdoors, and it really instilled a desire in me to enjoy and preserve our amazing environment. My siblings and I would dare each other to swim in the coldest water, despite the large lurking eels!'

Susie was also an avid reader, and a major fan of 'choose your own adventure' books. 'I was actually a bit of a weird kid — when I read those adventure books, I would go straight to the ending and weave my way backwards to try to understand how each decision had led me to my final destination.' It's precisely this interest in cause and effect, or action and consequence, that led to her pursuing a career as a research scientist.

She says her absolute passion in life is working on New Zealand's lakes, and that the country is incredibly special because we have a huge diversity of different kinds of lakes within a relatively small geographical area. 'We have over 3800 lakes bigger than 1 hectare, from our largest, Lake Taupō, which is over 600 square kilometres in area, to our deepest alpine lake, Hauroko, which is about 460 metres deep.'

Unfortunately, many of New Zealand's lowland lakes are in poor condition. Of the 200 or so lakes that are regularly monitored, more than half are classified as being in poor or very poor condition in terms of water quality and plant communities.[1] Susie says that most of the monitored lakes are in lowland areas, more likely to be impacted by human activities, and that understanding the causes of this degradation is vital to making a difference. She adds, 'There's not a single example of a lake restoration project in New Zealand that has been successful without some kind of ongoing intervention. To me, that indicates that it's so important to try to prevent lakes from reaching that tipping point, because it's so much harder to reverse the damage that has already been done than it is to stop it from happening in the first place.'

A love for water

They say that art imitates life, and maybe that's the case for work too. In her day-to-day life Susie spends an enormous amount of time enjoying being in the water, and her career reflects that passion.

Through university, Susie followed her interests, taking just about every biology course offered, and graduated with a degree in marine biology from Victoria University of Wellington. Showing drive in both her personal and professional pursuits, she began her PhD in 2005 while training to represent New Zealand in cycling at the 2006 Melbourne Commonwealth Games.

During her PhD, Susie stumbled upon the wonderful world of freshwater algae and was immediately captivated by the diversity and intricate structures of these organisms, as well as the crucial role they play in freshwater systems. 'You're always going to find algae in a lake as they are a naturally occurring part of the ecosystem, but when the balance of that ecosystem is disturbed, they can take over and cause problems. At university I was really excited by the opportunity to understand the role of these tiny organisms and to identify how human actions could cause them to grow out of control.'

In the end, it was Susie's expertise in algae and her love of working backwards through time to explore the causes of an effect that led her to study lakes. Because so many of New Zealand's lowland lakes are in a poor condition, and algae are a major indicator of ecosystem disturbance, Susie can apply her unique skills to help figure out why and how a lake has degraded, and whether there is anything that can be done to restore it.

Past, present, future

In 2016, after years spent researching the health of lakes throughout Aotearoa New Zealand, Susie was becoming increasingly concerned about their continual decline. She and fellow researcher Dr Marcus Vandergoes at GNS Science decided that more needed to be done to understand the current and historical health of lakes at a national scale and to get communities involved with helping to restore them. They agreed that if just 10 per cent of New Zealand's 3800 lakes were studied, that would provide an enormous amount of baseline data that could transform how lakes and their surrounding land are managed and protected. And so the Lakes380 Research Project was born.* Marcus and Susie submitted a proposal to the government's Endeavour Fund.** They were elated when they were notified a few months later that they'd been given the green light and the project would receive five years of funding.

When the project began, only about 200 lakes in Aotearoa were being actively monitored by regional councils and the Department of Conservation. The team set out to survey a diverse range of lakes, from lowland to alpine, including those that were culturally significant and important to local communities. The team wanted to use a suite of cutting-edge scientific techniques to build a rich and detailed understanding of each lake's health and history. Understanding the current state of New Zealand's lakes was a key part of the project, but Susie also wanted to apply her 'adventure book' technique and work backwards to figure out what the lakes used to be like, how they had changed and what had caused these changes. She says, 'We knew the technology existed, and that collecting and publishing this level of detailed information about our lakes and their histories would be a massive step forward for New Zealand.'

However, Susie and Marcus and their research partners had even bigger aspirations for Lakes380, beyond analytical science. The large group involved in the project were all in agreement that raising public awareness about lake health, and getting communities involved, was central to the project's success. As Susie puts it:

> We were absolutely committed to incorporating mātauranga Māori, Māori knowledge, and the oral histories of these lakes that belong to iwi to ensure that we had a rich, full picture of the lake's past, but also to guide the development of future aspirations for restoring and enhancing the health of that lake.
>
> Lakes380 isn't just about using traditional scientific methods, but also bringing in social science, mātauranga Māori and science communications tools, things like storytelling, infographics, animation and art, to help the public understand and use our findings.

One way the team raised local awareness was through a number of smaller projects, such as the Lake Stories Aotearoa New Zealand website.*** This presents a number of short documentaries and recorded stories, including 'Wairarapa Moana Kete Pūrākau, Wairarapa Lake Stories', compiled by Lakes380 researcher Dr Charlotte Šunde. Susie says that this approach has been hugely successful in raising local awareness and inspiring people to get

---

* See www.lakes380.com.

** The Endeavour Fund was established to support research, science, technology and related activities that have high potential to positively transform New Zealand's economic, environmental and social outcomes, and give effect to the government's Vision Mātauranga policy (a policy to unlock the innovation potential of Māori knowledge, resources and people). See www.mbie.govt.nz/science-and-technology/science-and-innovation/funding-information-and-opportunities/investment-funds/endeavour-fund.

*** See www.lakestoriesnz.org.

IWI ENGAGEMENT

Relationships and engagement with iwi have always been a top priority within the Lakes380 project. Susie, Marcus and the wider Lakes380 team have worked hard to connect with iwi and include them in the project. They will never sample a lake without first talking to iwi to make sure that they are not working on wāhi tapu (sacred sites), and that iwi, hapū and whānau perspectives are included in the approach.

Usually, before any sampling takes place, the Lakes380 team will hold a knowledge-sharing event with the local iwi and community, where the researchers will explain their sampling approach. The attendees are then invited to share their knowledge and insights about the lake's health and history. Ultimately, by drawing upon the principle of whakawhanaungatanga (making connections and relationships), the Lakes380 team have enabled the integration of mātauranga Māori with the project and initiated a shared approach to assisting with long-term lake protection and enhancement of lake health.

involved in freshwater environmental issues in their area: 'Projects like Wairarapa Moana Kete Pūrākau are ideal because they tap into people's emotional connection to their local lakes, rather than just hitting them with scientific data — Lakes380 is about using both to help people to connect to their moana, their local waterbodies.'

Cutting-edge techniques

Every researcher has a slightly different method and approach to research. Susie says there is a consistent method to sampling in the Lakes380 project, but she has her own personal, intuitive approach at the start of each survey. 'One of the first things I do when I arrive at a lake is go down to the edge and look into the water. I'm looking at the clarity of the water, and immediately my scientific brain is kicking into action and I'm thinking about what organisms live in the lake and why.' Susie says that after she has studied the water, her eyes start moving upwards and she'll scan the land around the lake. 'By visually exploring the landscape around the lake, I'm looking for clues that might help me to understand why the lake could be in its current condition.'

Once an initial assessment has been completed, the sampling begins. The team load their gear onto boats and head out into the middle of the lake, where they collect water samples and sediment from its bed.

First, they collect data to tell them about the lake's current health. It's standard stuff, a bit like a medical check-up — pulse, blood pressure and so on. They collect surface sediment samples (scoops off the top of the lake bed) using a device called a ponar grab, which is like a big set of metal jaws. It's held open by a spring-loaded pin and lowered on

Lake Wainamu, a Lakes380 site on the west coast near Auckland, is a popular recreational lake. The lake is a drowned valley blocked by a large dune of black iron sand at its north-western end and is surrounded by native bush.

Susie Wood and Marcus Vandergoes extract a sediment core to examine the history of Lake Rotonuiaha (Wairoa, Hawke's Bay).

a rope to the lake bed. The pin fires when the ponar hits the bottom and snaps its jaws closed, collecting a scoop of sediment in the process. The team also takes a series of water samples. Along with the sediment samples, these help Susie and the team figure out the present-day state of the lake and what organisms live there.

The next step is to delve deeper into the lake's past by taking a 2-metre-long sediment core sample from its bed. This is basically a tube of mud, but oh, the stories that mud can tell! To extract the core, tubes attached to a sediment corer are hammered into the bottom of the lake. A vacuum seal at the top of the corer stops the sediment slipping out of the tube as it is pulled up.

Lake sediment cores are like the pages of a history book, with continuous layers laid one on top of the other recording environmental history, 24 hours a day, 7 days a week, year after year, going back thousands of years. Everything that happens in or around the lake leaves a trace in the sediment — dead lake organisms, seeds, pollen, nutrients and toxins from the surrounding landscape all end up on the lake bed. By analysing each layer with scientific tools, researchers can start to understand how and why organisms and environments within and around a lake have changed over the course of thousands of years.

Data analysis

When the fieldwork is finished, the Lakes380 team take their spoils back to the lab to analyse the samples using a range of scientific techniques.

To unlock the secrets contained within a lake sediment core, it is first split in half

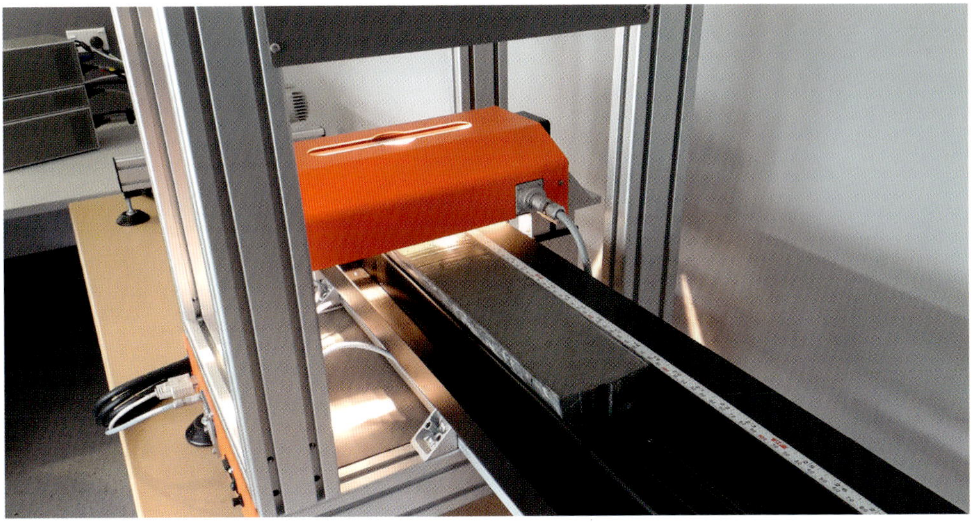

Sediment cores are scanned on a high-resolution hyperspectral scanner, which uses many wavelengths of light to identify shifts in algae communities in the lake over time.

lengthways. Small plugs of sediment are extracted along the entire length of one half of the core, providing a range of samples to assess what organisms lived in the lake and the lake's health. The other half of the core is kept intact as a reference.

There are some challenges in making sure that the samples collected date back as far as the team's goal of 1000 years. The older the sediment layer, the more likely it is that anything biological has been obliterated, but Susie says it's remarkable the things that have been preserved for hundreds of years within the mud where there is no oxygen to break them down. One example is pollen grains. The outer walls of pollen are very tough and can remain preserved in the bed of a lake for thousands of years. Following some initial sample preparation to remove the surrounding sediment, scientists can look at the size and shape of pollen under a microscope and work out what sort of plant it came from. The types and amount of pollen found in the sample can tell us how the plants within and around a lake have changed.

Diatoms are another feature the team look for in a sediment core. These are single-celled algae, which sounds mundane, but when viewed under a microscope they look like the escaped contents of a kaleidoscope or a bizarre button collection. For scientists, though, diatoms are phenomenally useful for figuring out how lake health has changed. They live in most aquatic environments and are sensitive to temperature, pH (acidity or alkalinity), light, nutrients and salinity. Their cell wall is made of silica (a compound of sand that is used to make glass) and they preserve well in lake sediments. After laboratory preparation to remove sediment, the diatoms are analysed under a microscope to identify their species. Different species prefer specific environmental conditions, so the team can use the data they gather to help paint a picture of the state of the lake in its past.

MAGIC MACHINES AND METHODS

The Lakes380 researchers rely on hi-tech laboratory equipment and analytical methods to unlock the secrets of our lakes.

Machines

Hyperspectral scanner This machine has an amazing camera that uses both visible and infared light to analyse a sediment core and figure out when there were changes in the amount of algae that lived in the lake over time.

ITRAX scanner The ITRAX scans the surface of the split core and generates geochemical data — for example, the amount of metals such as arsenic, cadmium, chromium, copper, nickel, lead and mercury in the core at different points. It can reveal how lake conditions have changed over time and the patterns of erosion on the land around the lake.

Methods

Radiocarbon analysis Radiocarbon dating is used to determine the age of a layer in a sediment core. Lakes380 scientists age leaves, twigs and seeds taken from sediment cores and record the depths at which they were found. This allows them to work backwards to figure out how old the sample is.

Environmental DNA (eDNA) In lakes, traces of DNA can be found in sediment and water samples long after the living things that shed them have died. From that information scientists can establish what species were living in and around the lake, how many and when.

Pigment analysis The natural colourings in plant and animal tissues are called pigments. By testing the pigments found in sediment cores, scientists can identify the types and amounts of phytoplankton (algae and cyanobacteria) that have lived in a lake. Because different types of phytoplankton thrive in different water-quality conditions, this helps researchers understand how the water has changed over time.

Surface sediment Nutrients are commonly mixed in with the surface sediment on a lake bed. Although they are sitting at the bottom of the lake, some processes occasionally cause them to be released into the water, such as when oxygen levels drop or the water becomes more acidic. Too many nutrients in the water fuel the growth of algae and bacteria, tipping the healthy balance of the ecosystem and causing an environmental problem by putting pressure on other species and processes. To analyse the nutrient content of the sediments, researchers collect samples with a ponar grab and take these back to the lab for analysis.

Water quality Water samples help researchers understand the chemical conditions in a lake. The samples are analysed for chlorophyll-a, nutrients, organisms and dissolved metals.

One of Susie's key areas of expertise is in the use of environmental DNA (eDNA). All organisms that live in and around lakes shed DNA. This can come from hair, scales, feathers, skin and saliva, or even entire bodies of small organisms such as bacteria. Using a range of techniques, the team can detect DNA in a sediment core, and by working out the organisms that were present and the age of the core segment they can establish the past health of the lake and the historical distribution of taonga species such as kākahi (freshwater mussels) and tuna (freshwater eels) at a frequency and resolution not possible with other approaches.

Beyond Lakes380

Lakes380 is the biggest scientific study of New Zealand's lakes in our history, and it's been a massive part of Susie's life for the past five years. She hopes to continue the work beyond the first phase of funding as there is still so much more to learn and understand:

> The Lakes380 data will help set realistic restoration aspirations and identify lakes in need of protection. In an ideal world we'd be able to sample all 3800 lakes, but that's not likely — however, it would be great to continue to work on those we have sampled and expand the pool of monitored lakes, because the more information we have, the more we can do to improve the health of our lakes throughout Aotearoa.

Susie says that many people become discouraged when faced with the scale of the challenges confronting our environment, but she is unwilling to give up and is a passionate ambassador for community engagement in freshwater restoration initiatives.

> A few years ago, I gave a speech at the Cawthron Institute Annual Lecture about my work in Lakes380, and I quoted a song that I really love by a band called Jack and the Weatherman called 'Rock Bottom'. It really sums up how I feel about how we need to respond to the challenges we are facing.
>
> *Voices whisper that I'm not to blame*
> *They say everyone's afraid of change*
> *Sounds to me like they've given up*
> *Well they don't know the heart I've got . . .*
> *I'm willing to change, I just don't know how*
> *But I hit rock bottom so I'll start right now.**
>
> I've rewritten that last line for myself, so it's kind of a mantra that I live by — 'I'm willing to change and the time is now' — because it's never too late to make a difference.

* Courtesy of Jack and the Weatherman, www.jackandtheweatherman.nl.

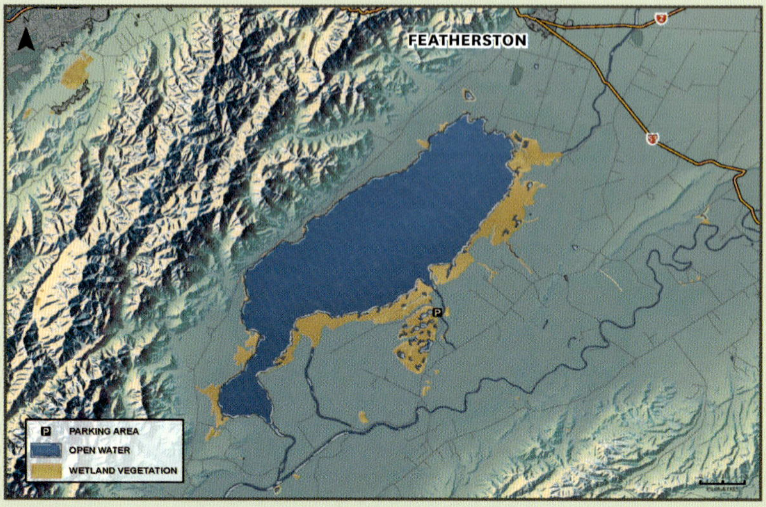

Wairarapa Moana

Lowland lake and swamp complex

One of the lakes that Susie has worked in is New Zealand's newest Ramsar site, a wetland of international significance. Wairarapa Moana, meaning 'Sea of Glistening Water', has been a place of significant customary and spiritual importance for Māori for some 800 years. The cultural significance of the site stems in part from its history as an Indigenous settlement and rich fishing ground, in particular for tuna (freshwater eel), which could be preserved and therefore traded.

The Ramsar site is the largest wetland complex in the lower North Island, and comprises Lake Wairarapa itself, its associated freshwater swamps and marshes, brackish Lake Onoke, coastal marshes, river and streams, and extensive coastal shore habitats. It extends for 30 kilometres from Wairarapa Lake Domain on the northern shore of Lake Wairarapa, south via the Ruamahunga River to Lake Onoke, which is impounded from the sea by Onoke Spit at the edge of Palliser Bay.

Lake Wairarapa and the surrounding wetlands retain a high degree of natural character, as well as important cultural heritage, recreation and other values. The diversity of habitat types present provides a home for a wide range of species. More than 50 rare and threatened species are found here, along with over 100 native birds and 20 native fish.[2]

The edge of Lake Wairarapa is regularly inundated by water and then exposed, supporting a submerged turf community of small native plants, greater in area than at any other North Island lake. Scattered stands of kahikatea and cabbage trees, and isolated tōtara, ribbonwood, kōwhai and lacebark, are remnants of the extensive areas of native forest that once existed in the area.

Highlights

- An important breeding, feeding and roosting habitat for resident birds, including matuku (Australasian bittern), koitareke (marsh crake), and pūweto (spotless crake), and a significant wintering site for populations of the nationally critical tarāpunga (black-billed gull) and the nationally vulnerable tūturiwhatu (banded dotterel)
- Habitat for native fish including threatened brown mudfish, giant kōkopu; panoko/papangoko (torrent fish); and tuna (longfin eel)
- Largest community of shoreline turf plants in the North Island, with many rare species such as the herb *Crassula ruamahanga*
- Rare native plants, including water brome (*Amphibromus fluitans*), dwarf mistletoe (*Korthalsella salicornioides*) and the moss *Fissidens berteroi*.

Access

Wairarapa Moana offers a range of visitor activities, including picnicking, gamebird hunting in season, fishing, boating, kayaking, birdwatching and walking.

Lake Domain is around 8 kilometres from Featherston, about an hour's drive from Wellington. Turn right off State Highway 53 onto Murphys Line, right onto Lake Domain Road, then left onto South Soldiers Settlement Road.

Onoke Spit is near the end of Western Lake Road, Palliser Bay. From here, turn onto Beach Road and follow this short gravel road to the small car park at the end. There is a small tidal stream to cross to get to the spit, so be prepared to get wet feet. Note that when the gap through the spit is closed, this stream does flood and you may be unable to cross. At such times, alternative access is along the coast at Ocean Beach.

https://lakes380.com/wairarapa-moana-iwi-rohe-study

Working Wetlands

Can reconstructed wetlands perform the many jobs that natural wetlands are credited with? Dr Suzanne Lambie's research reveals the machinations of microbes in designer wetlands.

Dr Suzanne Lambie

Specialist field: Soils and microbial function and diversity in wetlands

Currently works for: Manaaki Whenua — Landcare Research, Kirikiriroa Hamilton

Suzanne Lambie, MSc, PhD, is a soil biochemist investigating how microbial communities function in restored and constructed wetlands. This is an uncommon field where there is still much to be learnt. She has carried out research on numerous wetland restoration and wetland construction projects in the Waikato, with a focus on their effectiveness for treating farm run-off and pollutants from mining and roading. Suzanne works closely with farmers, iwi and science agencies on projects on both private and public lands.

Ko Taupiri te maunga e rū nei taku ngākau.
Ko Mangawara te awa e mahea nei aku māharahara.
E mihi ana ki ngā tohu o nehe, o Ngāruawāhia e noho nei au.
Ko Suzanne Lambie tāku ingoa.
Nō Taupiri ahau.

Taupiri is the mountain that speaks to my heart.
Mangawara is the river that alleviates my worries.
I recognise the ancestral and spiritual landmarks of Ngāruawāhia where I live.
My name is Suzanne Lambie.
I'm from Taupiri.

IT'S OBVIOUS THAT Suzanne Lambie gets into wetlands a lot. Her office is crammed with small plastic bags filled with soil samples. A mud-splattered backpack and well-worn gumboots are tossed in one corner, along with a soil-coring device and various well-used spades. Most of the shelf space is taken up with an ordered chaos of reference material — and what looks like several boxes of homebrew beer kits turns out to be empty flagons for collecting water samples.

Despite growing up in a small town next to a willow-filled swamp remnant, Suzanne took little notice of wetlands until later in her career, when, she says, she just sort of 'fell into them'. At the University of Waikato, under the tutelage of soil scientists Dr Megan Balks and Louis Schipper, she developed a fascination for soils — what Suzanne calls 'the great ignored world beneath our feet'. Upon returning to Waikato after many years working in Palmerston North, two areas of research interest coalesced in Suzanne's mind: soil biochemistry and natural wastewater treatment systems. Wetlands are often described by Māori as ngā whatukuhu o Papatūānuku — the kidneys of the Earth — acting like giant sponges and filters across the whenua. Healthy wetlands have an extraordinary ability to strip water of excess nutrients and other contaminants, trap sediments, control floodwater and remove carbon dioxide from the atmosphere.[1]

This is all well and good if you have plenty of wetlands to perform these jobs, but in Aotearoa New Zealand only 10 per cent of the original extent of wetlands remains. With hundreds of individuals, groups and agencies restoring or rebuilding wetlands for conservation reasons or recreational use, Suzanne wondered if newly reconstructed wetlands could perform these ecosystem services as effectively as naturally occurring healthy wetlands:

> I was interested in how restoring wetlands affected soil processes and ecosystem services — things like nutrient cycling and water treatment. There is so little information on this, and because wetland soils can act as sponges for nutrients, they are pivotal for meeting clean water aspirations in New Zealand. There are so few natural wetlands left in New Zealand, I figured that anything that I could to do to help restore or protect wetlands would be a good contribution to my community.

Microbes at work

A wetland is a marriage of water, soils, plants and animals, but so many of the important processes that occur are invisible to the naked eye.

When water enters a healthy wetland — be it through natural waterways, in a flood or via a drain or channel — not only can the wetland soak up the excess (provided the area is large enough) but it can also trap and process nutrients, sediments and contaminants such as heavy metals. Sediments can be trapped by plants, helping to keep the water clean and clear, but although plants and their roots help to physically sieve the water, the real cleansing action takes place in the microbial realm.

Microbes (microscopic organisms) are everywhere: on our bodies, in clouds, at the bottom of the ocean, in geothermal pools and in ancient ice. The microbes Suzanne studies are in the soil, in the build-up of dead plant material, in the sediments that have accumulated on the floor of the wetland and in the water itself. The main groups of microbes that interest her are fungi and bacteria. Most people think of fungi as mushrooms — the sometimes elaborate but ephemeral structures that burst out of the soil after rain. Microbial fungi are much more basic — but although they're single-celled organisms, they still break down plant and animal matter like their showy mushroom counterparts.

Bacteria are variously single-celled or have cells in pairs, chains or clusters, but all are

WHAT IS SOIL BIOCHEMISTRY?

Biochemistry is the study of the invisible chemical processes that take place in living beings, such as how our cells communicate with one another during illness or growth. It's the science of transformation at scales that are barely comprehensible, but even though the action takes place at a cellular level, we see the results daily: growth and decay, the cycle of life.

Applied to soil, biochemistry looks at the processes and chemical reactions that create soil and how factors like temperature, water acidity and nutrients affect those processes. Carbon, organic matter, plant material and the microbial community work together to make litter from plant material, which then breaks down to become organic matter, nutrients and gases — mostly carbon dioxide in drier soils, but also methane in saturated wetland soils.[2] Biochemistry teases out these processes, giving us a better understanding of how nutrients are cycled.

invisible to the naked eye. Like fungi, they're found just about everywhere. Those that live in wetlands play an important role in cycling nutrients. Actinomycetes are a particular type of bacteria that are good at breaking down complex compounds — in other words, the bigger things that fall into or are carried into a wetland, such as twigs, branches and leaves. Other species of bacteria tackle smaller things. Together, all these microbes are part of the first step in a food web, where smaller organisms are eaten by bigger ones — the microbes are eaten by zooplankton (tiny aquatic animals), which are eaten by insect larvae, which feed fish that are then eaten by larger fish or birds.

Soil fungi and bacteria function quite differently from each other. The fungi are slow-growing but also slow to respond to changes in environmental conditions such as heat and drought. Bacteria are generally the opposite — they're fast-growing, they like plenty of nutrients and they can respond very quickly to nutrient additions or changes in the environment. Knowing which conditions suit each type of microbe helps scientists like Suzanne take the pulse of a wetland. The question of whether nutrients are being cycled quickly or slowly lies at the heart of her investigations, because this indicates how effective the wetland is at cleansing and purifying water. In other words, are the wetlands doing one of the key jobs they were designed for?

Out of the field and into the lab

Studying wetland microbiology is an involved process that needs specialist expertise and equipment. To begin with, Suzanne collects samples from four physically distinct parts of a wetland — water, soil, sediment and humus (partially broken-down plant material). From here on, the analyses take place in the laboratory. She filters water samples using

FRESHENING UP OUR FRESHWATER

In 2004, the Parliamentary Commissioner for the Environment released a landmark report on intensive farming, sustainability and New Zealand's environment.[3] This really put the spotlight on our declining freshwater quality. However, given the role that wetlands play in water treatment, it's a bit surprising that detailed research into how they actually process nutrient run-off and leaching from farms is still quite new. Suzanne's investigations are cutting-edge science — the more information we have, the more effectively wetlands can be designed to absorb and process nutrients before they flow into waterways or leach into aquifers.

a special paper with minuscule pores (0.22 microns in size), capturing a film of bacteria. She sieves soil, sediment and humus samples separately to remove roots or stones and make the material in each sample a uniform size. Freeze-drying the samples puts the microbes in suspended animation and halts changes to the community structure to get a snapshot of who is there.

The next step is to extract the phospholipids (fatty acids) from each sample. Phospholipids make up the cell wall, or 'skin', of the microbes. Gas chromatography is used to separate out the different phospholipids, which allows the types and abundance of microbes in the samples to be measured. The result is a line graph with peaks, each of which represents a different phospholipid, which is assigned to a bacterial group or fungi. From here, it's a matter of adding up phospholipids to get a measure of the abundance of each of the microbial groups. Knowing the type of microbe and how many there are allows Suzanne to better understand nutrient cycling occurring in the wetland. This method is by no means unique to wetland research — it's also used in crime-solving forensics, quality control in cars and pharmaceuticals, and even researching the composition of meteorites.

Working with farmers

Wetlands and farming have had a troubled relationship in Aotearoa. On the one hand, many farmers enjoy the vistas and duck-shooting opportunities provided by wetlands on their properties, but in low-lying areas many of the level green grazing paddocks we see today have replaced once-extensive tracts of natural wetland. A glance at the Hauraki Plains in the Waikato, Northland's Hikurangi Swamp, the South Canterbury basin bordering Te Waihora Lake Ellesmere and many other places throughout the country shows how quickly one ecosystem can be obliterated and converted to livestock and crop production.

Many now think the pendulum has swung too far — that we have removed too many wetlands from the landscape, and are constantly battling nature, digging ever-deeper drains in sinking peatlands to farm areas that are far too wet and would have been better off

Suzanne Lambie discusses the performance of an integrated constructed wetland with dairy farm manager Ivan Vickers.

left as wetlands. This makes a compelling case for not only restoring what remains but also re-creating 'new' wetlands in areas that can be retired from farming.

Suzanne is helping to make this happen. She returned to the Waikato region after several years working for science agencies in Rotorua (Scion) and Palmerston North (Manaaki Whenua — Landcare Research), and currently lives very close to where she grew up — the small North Island town of Ngāruawāhia, on the banks of Waikato Te Awa Waikato River. Today, Suzanne works with dairy farmers and local iwi to restore their degraded wetlands, as well as designing new treatment wetlands to help cleanse polluted water.

A wetland restoration or construction on a farm may begin with a landowner saying, 'I've got this bit of a paddock I want to retire', but as Suzanne has witnessed, there are other drivers:

> Often, there's a history of family involvement . . . in planting areas of native plants . . . duck shooting is a big motivator for many New Zealand farmers: 'I'm going to have a maimai in my own wetland — how awesome!' However, wastewater treatment stuff is definitely starting to get on the radar. Farmers have talked to me about this pathogen thing — faecal coliforms. Is it from ducks? Is it my cows? . . . I think the National Policy Statement for Freshwater Management[4] and the push towards farm environment plans[5] is starting to bring all of this to the fore.

Designing a wetland to carry out water treatment functions on a farm is a tricky business — there are so many variables, trade-offs and compromises. Essentially, the aim is to fast-track the evolution of a whole ecosystem, from the plants, animals and insects right down to the unseen life forms, plus all their interactions with one another and with the physical environment.

In an ideal world, the minimum size for a wetland to function best is around 2–5 per cent of the size of the whole catchment it sits within. In reality, this hardly ever occurs in present-day New Zealand. For reference, the nearly 10,000-hectare expanse of the Kopuatai Peat Dome (an internationally recognised wetland in the Hauraki Plains, Waikato) is still just under 7 per cent of the whole catchment area. In Canterbury, however, the size of Te Waihora is tiny relative to the catchment area — less than 1 per cent. Suzanne highlights that, in a farming context, the size of the catchment isn't the first consideration — it's most often the amount of pasture the farmer is willing to 'give up'. This could be a wet piece of grazing land that needs retiring, as 'having cows buried up to their knees in mud isn't great for animal health'. There are other considerations too, including the fact that soggy soils don't grow quality animal feed.

The design of a new wetland, or the modification of an existing remnant, to make it a super-efficient water filter depends on many factors. How much water comes into the wetland from drains and across the surface? What's the main nutrient in the water and soils? How much sediment is likely to travel into the area?

Of the possible nutrients present in farmed areas, nitrogen and phosphorus are the main culprits for declines in water quality, encouraging the growth of nuisance algae as well as making the water unfit to drink, fish in or play in. Each nutrient enters water in a different way: nitrogen dissolves easily, while phosphorus makes its way through the environment bound to soil particles. Once the nutrient type and concentrations are known, important design decisions can be made. For example, managing sediment-bound phosphorus may mean incorporating deep ponds into the wetland so that the sediments can eventually fall and accumulate at the bottom of these. The nutrient-rich sludge can later be dug out and reapplied to areas of the farm where phosphorus is needed, saving the farmer the cost of buying more fertiliser. Heavily planted shallow ponds can be very effective at removing excess nitrogen, a design that is being used as part of a project Suzanne is leading just north of Kirikiriroa Hamilton.

Working collaboratively with local farmers and scientists from NIWA (National Institute of Water and Atmospheric Research Taihoro Nukurangi), Suzanne designed a series of four shallow ponds that snake through an area of land long retired from dairy farming. The aim? To test the best plant species for the job of removing nutrients. And the goal? To lighten the footprint of dairy farming in the area. But there's more to this experiment: a local mountain sacred to Māori, Taupiri, gives its name to the area. The stream the constructed wetland feeds into, the Whangamaire, is culturally significant to the local iwi. The stream wends its way to the mighty Waikato Te Awa via the Mangawara Stream. It was important to Suzanne to bring the science right into the community.

At Taupiri School, Suzanne and her science colleagues provided the rangatahi (young people) with hands-on learning about wetlands: the rangatahi constructed their own treatment wetland — in miniature — and learnt about different species of wetland plants, along with their cultural uses. They were also given a crash course on the insects, fish and birds that live in and around wetlands. They got their feet wet, too, planting flaxes on the banks of each pond.

Raupō (*Typha orientalis*), with its distinctive tubular, velvety brown seedheads, filled the first pond; wīwī (jointed twig rush, *Machaerina articulata*) filled the second; swards of purua grass (*Bolboschoenus fluviatilis*) filled the third; and the last pond was filled with the clubrush kūwāwā (*Schoenoplectus tabernaemontani*). The first pond may need to be dug out when it eventually fills with sediment from the incoming farm drain, but raupō is hardy and will easily regrow from the chunky pieces of root left behind.

Although the wetland was constructed and planted as recently as 2019, the initial results are promising — the ponds, plants and their active microbes are doing a great job of removing both nitrogen and phosphorus. It will take a number of years to really see how the system functions as the plants mature and seasons continue to pass. And maintenance is needed to ensure the ponds don't become havens for weeds such as willow or other fast-growing exotics.

While funding in this case was available to develop the Taupiri wetland, it's much harder to find money to cover the works needed to maintain wetlands over the long term. Constructed treatment wetlands can be effective at stripping excess nutrients, but they're not a silver bullet for all of our freshwater woes. Whether constructed or restored, wetlands are just one part of a broad sweep of measures needed to improve water quality, from the effective management of farm animals and soil, right through to the regionally specific rules and regulations that govern activities on rural land.

Suzanne Lambie's integrated constructed wetland for treating dairy farm run-off in Waikato.

Pūkeko can play havoc with wetland restoration efforts, uprooting newly planted seedlings and feeding on the stems, shoots, leaves and seeds of wetland plants.

Funding constraints aren't the only problem wetland projects have to contend with. Sometimes the challenges come from nature itself!

Problematic pūkeko

Pūkeko really like pulling out plants, especially newly planted seedlings. This large, loud and distinctively coloured native species is generally considered to be a wetland bird, but it thrives on agricultural land throughout Aotearoa. Being mostly vegetarian, pūkeko forage the stems, shoots, leaves and seeds of grasses, sedges and rushes, making them generally unpopular with livestock farmers and wetland restorers.

Seedlings are easy pickings: the pūkeko use their stout red bill, a veritable multitool, to uproot, cut or tear plants, which the birds grasp parrot-style with their foot while they eat them. Nest-making also leaves a trail of shredded vegetation. Sometimes the birds seem to pull plants out of the ground for pure amusement. That's not good news for restoration projects. Having to replant a wetland four times (as Suzanne has had to do) is a frustrating business given the cost of plants and the back-breaking labour needed to put them in the ground. She exclaims: 'I would have thought curiosity would stop after a few plants . . . but we're talking 200 to 300 plants pulled out overnight!'

There is a thriving folklore on how to dissuade pūkeko from decimating new plantings. Suzanne recalls being told to flood new plantings with 'a foot of water' because 'they don't like to get their bums wet'. The trial was unsuccessful. Spoiler alert — pūkeko are a wading bird! Another somewhat gruesome approach, also unsuccessful, is to string up a pūkeko carcass as a warning to others. Protecting each seedling with a plastic cylinder pegged into the soil works in some settings, but it's challenging for larger projects and for those in wet environments. Unfortunately, problem pūkeko often end up being shot as a last resort (despite being a native species, they can be hunted under licence). For a bird with such well-known habits, there is surprisingly little research on how best to outwit them.

From regional to local

Suzanne is clearly passionate about wetlands and their function, both at landscape and local scales. When asked what she would like to see happen in the Waikato region, her home territory, she responds: 'I would like to see land being used appropriately . . . I want to see wetlands restored, fenced off, and intensive agriculture kept on the soils and on the landscapes to which it is most appropriate, to minimise air and water pollution, and to protect our soils.'

Down at the local level, she sees the fruits of her collaborative work with farmers in a more personal way: 'One of my farmers has a little chair by their pond . . . she sits there and enjoys the vista. The other one loves ambling down his farm paths and seeing the wetland in the distance . . . so there is a really good, nice mental health benefit around it.'

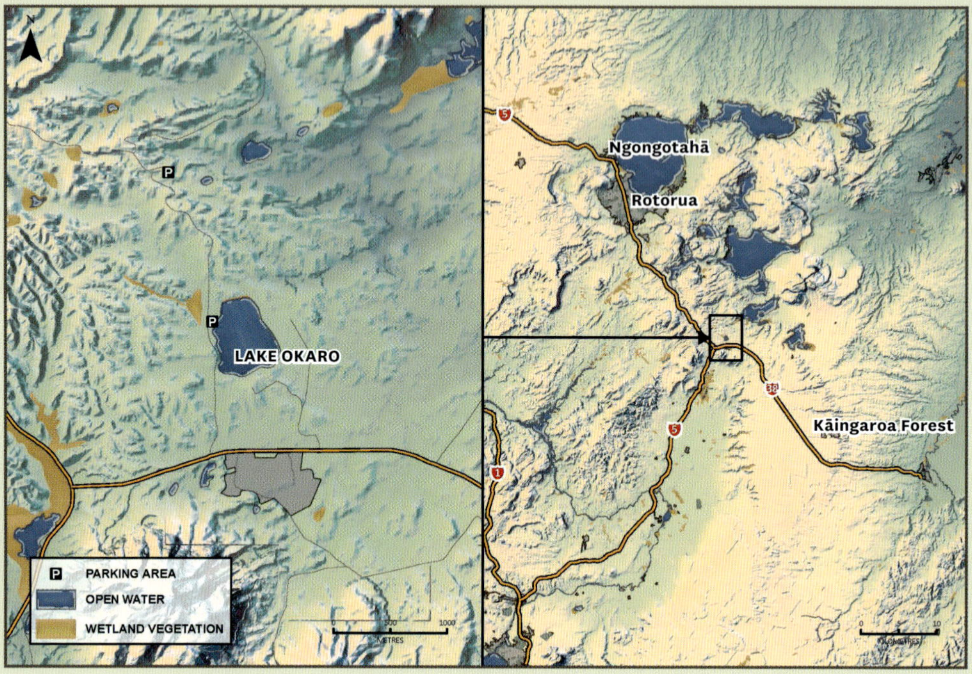

Lake Okaro

Constructed treatment wetland

Suzanne has worked on many constructed wetland projects on privately owned farmland, but to visit a well-established example on public land, head to Lake Okaro south of Rotorua. Adjacent to Lake Okaro are two shallow, plant-filled ponds built in 2006. The largest of these was built on private land, but the smaller pond is publicly accessible from the small car park on the shores of Lake Okaro. Both ponds were designed specifically to help reduce nutrient loads heading into the lake from the catchment.

There are many other wetlands to visit in the vicinity of Lake Rotorua. Examples of accessible restored lakeside wetlands include Lake Ōkareka and Otauira (Hannahs Bay wetlands). Hamurana Springs Recreation Reserve has crystal-clear, icy-cold waters, while Sulphur Point is home to geothermal wetlands and can be accessed through the Rotorua Government Gardens.

Highlights

- Look for the distinctive narrow arms leading into the ponds. These peninsula-like features create a meandering pathway for water entering the wetland, slowing it down so that any sediment can drop out before entering the lake
- Waterbirds such as tētē moroiti (grey teal) and pūtangitangi (paradise shelduck).

Access

Lake Okaro is a 25-minute drive from Rotorua. Head south along State Highway 5 (the Thermal Explorer Highway) and turn left onto Waimangu Road. Continue past (or stop for a look at) the geothermal wetlands at the Waimangu Volcanic Valley. Beyond here, Waimangu Road becomes Okaro Road and skirts the lake.

Although small, Lake Okaro is a popular waterskiing and kayaking destination. Facilities include a small campsite, toilets, picnic tables, a barbecue area and a boat ramp. There is a 1.8-kilometre walkway around the lake, from where waterbirds can be seen.

www.rotorualakescouncil.nz/our-city/lakes/lake-okaro
https://lakeswaterquality.co.nz/lake-okaro

Weaving Knowledge Systems

How can we blend the realms of cultural and Western ecological knowledge to understand and protect wetland values? Kairangahau Māori Yvonne Taura wades between both worlds.

Yvonne Taura

Specialist field: Kaupapa Māori research (application of Māori values and principles), freshwater science, iwi environmental management, and science communication

Currently works for: Manaaki Whenua — Landcare Research, Kirikiriroa Hamilton, as a kairangahau Māori (researcher) within Manaaki Taiao

Yvonne Taura descends from the iwi of Ngāi Te Rangi, Ngāti Ranginui, Ngāti Hauā, Ngāti Uenuku and Ngāti Tūwharetoa. She works within a dynamic kairangahau rōpū Māori (Māori research team), which is often called on to provide cultural and ecological expertise, and to work collaboratively with Māori and Indigenous communities at local, regional and international scales. In 2017, Yvonne co-edited and released the cultural wetland handbook *Te Reo o Te Repo: The Voice of the Wetland*, a collaborative project between Manaaki Whenua and Waikato-Tainui. This was followed by a second volume in 2021, *Te Reo o Te Repo — Kei Konei Tonu Au: Voice of the Wetland — I Am Still Here*. The cultural wetland handbook series includes processes for facilitating renewed and vibrant connections between people and their wetlands; understanding cultural resources; and learnings from on-the-ground case studies of wetland restoration and monitoring that have used cultural indicators.

HE MAIMAI AROHA ki a kōrua e aku poutokomanawa, e Te Rangituamātotoru Tamaira (Ngāti Tūwharetoa) kōrua ko Rakato Te Rangiita (Ngāti Tūwharetoa).

This chapter is dedicated to my dearest uncle, Te Rangituamātotoru Tamaira, and my koroua Rakato Te Rangiita, who sadly both passed away early in 2021 within days of each other. Together, they encouraged the active participation of wāhine of Ngāti Tūwharetoa in environmental matters, acknowledging the mana of wāhine across the whenua. Over the years they ensured the elevation of wāhine by supporting many to realise their potential in nurturing whenua and whānau. On a personal level, they guided and mentored me to pursue my career in science, more specifically freshwater wetland ecology, to play a role in raising the awareness of freshwater issues. I feel honoured to have been supported and encouraged, loved and cared for, and empowered by them both. They will be greatly missed by all who knew them.

> He paroha na te wāhine e ora ai te taiao
> It is with heartfelt weaving, as with the finely woven mat of a woman, that the environment is nurtured and made well.

Ka nui te aroha ki a kōrua mō ake tonu atu, nā Yvonne.

At the age of 21, Yvonne experienced a sudden moment of inspiration on the shores of Taupō-nui-a-Tia (Taupō Moana, Lake Taupō). She had just returned home to Aotearoa New Zealand after living in Australia since childhood — a move prompted by a growing sense of cultural disconnection as a young Māori woman. She based herself in the small central North Island settlement of Tūrangi to stay with her whāngai (foster) parents. Set in a geothermally active area beside the enormous volcanic lake, and surrounded by majestic mountains, ancient podocarp forests and rugged farmland, it was a far cry from the hustle and bustle of Melbourne, where Yvonne had spent most of her childhood and youth.

Determined to help her start her new life, Yvonne's uncle Toru said, 'Get in the car. We're going to go see our kaumātua [tribal elder].' They arrived at the marae, located on the whānau farm near the southern end of the lake, with the volcanic cones of Pihanga and Tongariro rising behind them and the clear, snow-fed Tongariro and Waiotaka Rivers flowing swiftly past. Speaking poetically, her koroua (male elder) Rakato brought the surrounding landscape to life. He shared tribal pūrākau (ancient stories) that explained the formation of each feature — the interaction of ngā atua Māori (founding elements, domains or deities) in shaping the ancient land and the effect of a romantic love rival upon the ancestral landscape — and the significant events of her tūpuna that gave each feature its name. Her koroua went on to describe the concept of kaitiakitanga (guardianship) and he taonga tuku iho o ngā tūpuna of the area (the rich culture handed down through generations by their ancestors), woven in karakia (prayer), pūrākau and waiata (songs). As an added special treat, he showed her where the taniwha (water spirit) Owheao resides.

Although Yvonne had never heard these concepts before, surprisingly she understood — her whakapapa (genealogy) connected her to the tribal landscape of Ngāti Tūwharetoa, the iwi of the area. It wasn't just a picturesque vista; it was a cultural landscape of ancestral mountains (maunga), rivers (awa) and lake (moana). Each physical feature was a being, acknowledged by the interaction of tūpuna, and each being formed part of a complex web of relationships spanning millennia.

Until that point, Yvonne had known only city life. Her habitat had been thoroughly urban. She rarely visited wild places and saw wetlands as just 'dirty, ugly swamps and dumping grounds'. The seismic cultural shift that took place on the shores of Taupō Moana occurred in a heartbeat:

> Thanks to my koroua, I'm seeing my surroundings through the eyes of my tūpuna . . . I'm drawn into this new world . . . it just came alive right in front of me! They're not just our maunga, awa or moana. They're our ancestors, that share special relationships with each other, and our tūpuna respectfully interacted with this tribal landscape. There's so much to learn from our mātauranga [traditional knowledge].

Although she wasn't clear exactly what pathway she was going to take, Yvonne knew she wanted greater insight and cultural connection to the landscape and te ao tūroa (the

Yvonne experienced a profound cultural connection on the shore of New Zealand's largest lake, Taupō-nui-a-Tia.

natural environment). With the support of her whānau, she embarked on a life journey to discover her Māoritanga, learning about whakapapa, te reo, kawa (protocols) and tikanga (customary values and practices). At the same time, she began to build a foundation in conventional science through higher education by completing university degrees. Weaving mātauranga and science together is not just a matter of contributing to research, says Yvonne, 'it's learning about our own tribal landscape through the eyes of our tūpuna'.

Wading into Western science

Yvonne didn't study science during her senior years at college — largely because she didn't believe she was smart enough and also because she was never encouraged to do so by her teachers. However, after her experience on the shores of Taupō Moana, she

WOODY INVADERS

Owheao Reporepo is typical of many degraded swamp complexes across Aotearoa New Zealand — a mixture of native and exotic plants under a canopy of crack willow (*Salix × fragilis*, below left) or grey willow (*S. cinerea*, below right). Willows were brought to Aotearoa in the late 1800s — like many other introduced species, to remind the colonists of the landscapes they left behind. However, willows are aggressive spreaders that can soon change the natural character of our wetlands and crowd out native plants. Their woody, deciduous habit (an uncommon feature among native wetland plants) creates an unnatural regime of dense shade in the summer, while the autumn leaf-fall results in a seasonal pulse of unwanted excess nutrients. They are also enormously thirsty, taking up water while blocking natural channels with their extensive mat-like root systems and fallen branches.

Crack willow spreads mostly by stem fragments, a few centimetres of which can rapidly sprout roots and become a new plant. Grey willow (sometimes also called pussy willow) has a different strategy for spreading — countless fluffy seeds are produced that are widely distributed by the wind. This guarantees that at least a few will find a damp spot and take hold. Today, the march of both willow species across the country's wetlands is mostly controlled by aerial applications of herbicide. Laborious ground-based methods are also used, which means clambering and kayaking through the wetland to drill holes in myriads of willow trunks and then injecting herbicide.

realised she needed to understand Western scientific principles and practices alongside mātauranga Māori to put herself in a stronger position to restore the tribal landscape and the culturally important features within them: 'The advantages of having our mātauranga alongside Western science knowledge is that it revitalises the rich narrative that enhances the data from your science experiments but is also verified or validated by what our tūpuna experienced in their lifetime.' Supported by her whānau, she first enrolled in a Bachelor of Environmental Studies course at Te Whare Wānanga o Āwanuiārangi in Whakatāne.*

After completing her undergraduate studies, in 2008–2010 Yvonne led an environmental enhancement programme (funded by the Ngāti Tūwharetoa Genesis Committee) with her hapū, Ngā Runuku, back in Tūrangi. As mana whenua (authority over land), Ngā Runuku are the kaitiaki (guardians) for the Owheao Reporepo, a wetland that forms a buffer along the eastern edge of Taupō Moana and is part of the larger South Taupō Wetland complex.

Taupō Moana is the largest lake in Aotearoa — almost 200 metres deep and nearly 200 kilometres in circumference. It's a natural crater lake, formed by a massive volcanic eruption some 26,000 years ago. Its single outlet is the nation's longest river, the Waikato. Since the 1940s, the lake's water levels have been controlled by large gates at the outlet, holding back water in winter to continuously feed eight hydroelectric power stations along the length of the Waikato River.

To be effective kaitiaki, Ngā Runuku wanted to know more about the environmental impacts of human-induced changes to their reporepo (wetland), including the ongoing control of the water level of Taupō Moana. The environmental enhancement programme enabled the hapū to inspect and investigate a range of environmental issues. Tasks included monitoring water quality, observing how erosion affects the Tongariro and Waiotaka Rivers, and the effects of periodic inundation on lake-edge farmlands. There are many unknowns — for instance, what effect does the heat from the tarsealed State Highway 1 have on the streams and gravelly fans that pass through Owheao Reporepo? Of major concern is how management decisions that influence environmental quality and condition within the rohe (tribal area) were being made without input from Ngā Runuku.

At the time the hapū-led environmental enhancement programme was under way, the Department of Conservation began an intensive grey willow (*Salix cinerea*) control programme across the South Taupō Wetland. Hapū members were deeply concerned. They wanted to know how the willows, and the herbicides used to kill them, were potentially affecting the aquatic life in their reporepo.

Investigating these concerns for her hapū was the motivator for Yvonne's Master of Science research, at the University of Waikato. She chose to focus on microfauna — zooplankton (tiny aquatic animals, including copepods, cladocerans, ostracods and rotifers) — which are so small you can suck them up with an eyedropper. These minuscule creatures might seem insignificant, but they can be useful indicators of environmental condition.

* Te Whare Wānanga o Āwanuiārangi is a publicly owned tertiary institution that provides education in a Māori cultural context.

Yvonne Taura takes field notes in South Taupō Wetland.

Microfauna are widespread throughout all parts of the aquatic ecosystem, from the sediments at the bottom, to the water column, to floating on the surface. Rotifers even survive in the thin films of water surrounding soil particles. Copepods and ostracods (seed shrimps) and cladocerans (water fleas) are all types of crustacean (distant cousins of crabs and lobsters). Rotifers (named after their wheel-like mouthparts) form their own special group of microscopic organisms. Microfauna are an important part of the food web, helping to move nutrients from one source to another. By consuming algae, dead bacteria and organic detritus, they become a food source for other organisms, both large and small.

Microfauna typically have short lives and are very sensitive to changes in their environment, including water temperature and pH. There was plenty for Yvonne to study, because much more is known about the microfauna in streams, rivers, lakes and terrestrial ecosystems than those in wetlands. She was grateful to her supervisor Dr Ian Duggan (now Associate Professor at the University of Waikato) for his in-depth knowledge and passion for freshwater microfauna communities.

Yvonne wanted to find out which microfauna species inhabit wetlands dominated by grey willow, how the composition of the microfauna community varies between these and areas of native wetland vegetation, and how willow control (using the herbicide metsulfuron) affects the abundance and diversity of the microfauna community. Her work formed part of a larger programme under way in the vast Whangamarino Wetland in the Waikato,* where studies were examining the impacts of grey willow control on both aquatic and terrestrial invertebrates.

She selected seven sampling sites in the Waiotaka Reserve that covered three different plant communities: plant species native to Aotearoa New Zealand, as these systems could potentially be invaded by grey willow; live grey willow; and dead (sprayed) grey willow. She sampled the sites over three seasons — February, July and December. Yvonne's field kit comprised a 10-litre bucket, a plastic jug and a bottle of alcohol (95 per cent ethanol, used to euthanise and preserve the microfauna). Her method was simple enough — climb into a pair of waders and then head through the reserve to each sampling site, dip a plastic jug in the water to collect 10 litres, pour it through a fine sieve and immediately add ethanol.

Back in the lab, Yvonne patiently sorted each sample into groups of microfauna and then, where possible, identified the main species in each group. The animals were so tiny (virtually invisible to the naked eye) that she sucked them up in a few drops of water and placed them under a microscope to identify and count them. The higher the number of different species and genera in a sample, the greater the species richness.

Overall, Yvonne's analyses showed no significant differences in species richness between native and live and dead willow sites over any of the three seasons — but this could also be because willow stands in the study area consisted of scattered trees that didn't form a dense canopy. To the relief of Ngā Runuku, it also appears that metsulfuron and the way it is applied (injected into holes drilled in the willow trunks) doesn't remain in the environment or detrimentally affect the native aquatic diversity. In fact, Yvonne found that fluctuating lake levels, the flow of the Waiotaka River and seasonal rainfall played a far greater role in regulating microfauna richness and abundance than either willows or weed control.

Following her MSc, Yvonne embarked on a two-year placement with Waikato Raupatu River Trust,** which included a secondment to the Crown Research Institute Manaaki Whenua — Landcare Research. She is now a full-time kairangahau (researcher) within

---

* Whangamarino is one of the seven internationally significant wetlands in Aotearoa. See www.wetlandtrust.org.nz/get-involved/ramsar-wetlands.

** The Waikato Raupatu River Trust was established in 2008 by the River Trust Deed and represents the Waikato-Tainui iwi following the land claims settlement in 1995. Waikato-Tainui was the first iwi to reach such a settlement with the Crown for injustices stemming from wars and land confiscations (raupatu) in the 1860s. See www.legislation.govt.nz/act/public/1995/0058/latest/DLM370553.html.

Manaaki Taiao, the institute's Māori research team, and is the Vision Mātauranga* lead for the restoring wetland ecosystem functioning programme led by Manaaki Whenua. Yvonne currently works on various collaborative projects to restore freshwater wetlands throughout Aotearoa, concentrating on Māori-led, culturally focused wetland research.

For Yvonne, the term 'culturally focused' means considering the many dimensions needed to heal and reconnect whānau with degraded environments through a Te Ao Māori (the Māori worldview) lens. In this context, when it comes to restoring wetlands (and any other ecosystem) it's not just the biophysical elements (plants, animals, water) that matter, but also how people experience, use and value the site and its features over time. This approach is rooted in tikanga and kaupapa Māori — traditional customs and practices that weave together both the physical and the spiritual realms. These elements inform the restoration journey, starting with 'ngā atua Māori and how our tūpuna navigated our tribal landscape'.

All of these narratives are vital — they are the foundation that connects people to their past and strengthens their connections to their cultural landscape. At the same time, they provide reference points to guide restoration priorities, aspirations and goals. Many Western-led ecological restoration projects start with the question 'What was it like in pre-European times?', with an aim of turning back the clock to the start of colonisation in the early nineteenth century, as a yardstick of the 'natural' condition of the ecosystem.

Māori have lived in Aotearoa since at least 1300 CE, so their knowledge and connection to the vast landscape has been critical for their survival. The importance of restoration actions underpinned by mātauranga Māori is becoming increasingly recognised and valued in Aotearoa conservation circles. This multifaceted knowledge system is built on understanding the world from an Indigenous cultural perspective. Fundamental to mātauranga Māori is whakapapa — being able to trace the intimate connection to specific places and the natural features within them.

Te Reo o Te Repo

Te Reo o Te Repo: The Voice of the Wetland is definitely a career highlight for Yvonne.[1] The illustrated 20-chapter practical handbook promotes Indigenous cultural wetland values and approaches to all wetland restoration projects. The handbook was a team effort, with Yvonne as lead editor, working closely with co-editors and contributors Cheri van Schravendijk-Goodman (Swamp Frog Environmental) and Dr Beverley Clarkson (Manaaki Whenua — Landcare Research). Other researchers and restoration practitioners contributed their expertise and appear as chapter authors.

* Vision Mātauranga is a government-led programme that recognises Māori as important partners in science and innovation. A key purpose is to build the capability of Māori to engage across multiple sectors (environment, health and economic development) by supporting research and mātauranga Māori. See www.mbie.govt.nz/assets/9916d28d7b/vision-matauranga-booklet.pdf.

The handbook had 'brewed' for a very long time before coming to fruition. Yvonne elaborates: 'The partnership between Manaaki Whenua and Waikato-Tainui have all of these amazing initiatives identified by the tribe — iwi-led projects, training and scholarships for tribal members, and the support of my two-year secondment . . . but there was no space for whānau to share their kōrero [narratives]. There wasn't an existing "hub" to house these exciting initiatives, so we thought, let's create one!'

The handbook showcases the work of kairangahau Māori (including Master's and PhD research) and iwi-led projects from around Aotearoa. Although there is plenty of technical material from the research, each chapter is written in everyday language (te reo Māori and English) to ensure that the content is accessible to whānau as well as non-Māori readers. Yvonne explains: 'We definitely wanted our whānau to see that we have dedicated Māori scientists and tribal scientists undertaking important mahi in te taiao — *Te Reo o Te Repo* provides a space to showcase the voices of the repo through the voices of our people.'

In the introduction to the handbook, the co-editors write: 'Environmental restoration work in Aotearoa cannot be undertaken without involving tangata whenua, as the Indigenous people whose culture and identity come from the land (hence "people of the land"), and who have existed within the local environment for many generations.' Questions directed at the reader, along with the step-by-step processes

Co-editors of* Te Reo o Te Repo: The Voice of the Wetland*; from left: Cheri van Schravendijk-Goodman, Yvonne Taura, Dr Beverley Clarkson.

for carrying out wetland restoration, are broadly framed so that iwi can use the handbook to guide how they design and direct restoration efforts in their own rohe. Following the success of the first volume, a second was published in 2021, *Te Reo o Te Repo — Kei Konei Tonu Au: The Voice of the Wetland — I Am Still Here.*[2]

The *Te Reo o Te Repo* cultural wetland handbook series has now been adapted for classroom use through a collaboration with the government-funded University of Waikato Science Learning Hub — an extensive online repository of tools and information designed for teachers. Yvonne worked closely with the Hub team and Māori advisors to create *Tuihonoa Te Reo o Te Repo*, a suite of online resources that includes more than 30 articles (including 24 in te reo Māori) illustrating Māori-led wetland research, teaching and learning guides for the classroom, interactive image maps to navigate the site, and a collection of short video clips featuring key wetland experts — kaitiaki and kairangahau Māori who supported the development of these resources.

Giving back to future generations

Restoring wetlands takes time. Revitalising a degraded site so that the flora, fauna, people and their cultural practices can thrive once more means taking a long-term view. According to Yvonne, everybody can — and should — be involved in setting the goals when it comes to restoring wetlands. In her view, it's an intergenerational process that includes tamariki (children) and rangatahi (youth), pakeke (adults) and kaumātua. Each generation has a distinct role to play: 'It's teaching our tamariki how to guide, plan, and take charge of the restoration work, while our kaumātua play an active role sharing traditional landscape narratives . . . it's far more enriching and the entire whānau feels empowered from the whole experience.'

Over the course of Yvonne's two-decade career, her role has shifted from field-based waders-through-the-swamp research to developing resources that enable others (including tamariki) to restore wetlands and better understand the intricacies of wetland ecology. She feels both lucky and privileged to have had her uncle and koroua guiding her from a young age. Not surprisingly, mentoring rangatahi and instilling the confidence in them to succeed remains her biggest passion: 'My uncle and koroua saw a clear pathway for me all those years ago when we gathered at Owheao . . . their guidance and mentorship has encouraged me to do the same for others over the years, especially for our rangatahi with a drive to work in te taiao.'

It's clear that Yvonne will stay engaged in wetlands for the long haul. Restoring her cultural connections to her tribal landscape built a strong foundation for her mahi (work) and ethos to share her growing knowledge with others, giving back what she herself was given by her whānau, cultural advisors and science mentors.

Her advice for budding scientists is to connect with networks in your chosen discipline. If it's wetland research that you're excited about, volunteer or intern with staff from research institutes, agencies or community groups engaging in wetland

TOHU: INDICATORS OF WETLAND HEALTH

Tohu are ecological signals that help Māori understand and interpret the health and well-being of taonga values (culturally important species) within wetlands. In science speak, indicators can be measured and monitored over space and time to show changes or trends (both positive and negative). Ongoing observation and monitoring over many seasons and hundreds of years have shaped the tikanga and kawa, along with rules (ritenga, tapu, noa) that guide Māori towards sustainable environmental practices, such as when, how much and in which way a species can be harvested.

While there are many synergies between locally relevant indicators and science-based ecological indicators, tohu originate from Te Ao Māori. As native and/or exotic plants and/or animals, tohu may be specific to katiaiki, whānau, marae, hapū, iwi and rohe — there's no real one-size-fits-all scenario that's relevant to all places and situations.

The *Te Reo o Te Repo* wetland handbook series showcases the diversity of tohu culturally significant to Māori, with chapters on matamata (whitebait), kōura (freshwater crayfish, below left), ruru (morepork), tuna (freshwater eels), wātakirihi (watercress), maire tawake (swamp maire, below right) and giant cane rush (*Sporadanthus*), among others. Each species has its own stories and history of cultural use. According to Garth Harmsworth, Toi Rangahau Māori (Principal Scientist) at Manaaki Whenua — Landcare Research, culturally specific tohu are conversation starters between tangata whenua and others interested in restoring and reviving degraded wetlands around Aotearoa. Incorporating locally relevant indicators 'is not only an appropriate recognition of the value of that local knowledge and expertise, but, also generates a new level of ownership . . . and can generate new innovations and techniques that can further improve the way we interact with the systems and taonga species we treasure'.[48]

RAUPŌ: MULTIPURPOSE WETLAND WONDER

Raupō (bulrush, *Typha orientalis*) belongs to a genus found in wet places all over the world. While raupō can be used in wetland restoration projects to help purify water by trapping sediments and absorbing excess nitrogen, historically the plant has a mulitude of uses for Māori. Every single part of the plant was used: the roots were eaten; the stems and leaves lined and insulated whare (as shown below, at Howick Historical Village), or were bundled together to make mōkihi or mōkī (rafts) for crossing rivers; the leaves were used as fishing floats, for sail-making and to make poi (light balls, swung or twirled to accompany dancing and singing); the downy seeds were stuffed into poi, bedding and pillows, and plugged holes in waka; and the pollen was baked into cakes or bread. In a modern context, this plant offers intriguing possibilities for paludiculture — an environmentally friendly alternative use for land that has been drained for intensive farming.*

---

* To learn more about raupō, go to https://maoriplantuse.landcareresearch.co.nz.

FROM WASTELAND TO WATERLOGGED WILDERNESS OF PLENTY

Colonial settlers arriving in Aotearoa viewed wetlands as wastelands. Over a mere 150 years, wholesale drainage and clearing occurred, until only 10 per cent of the original extent of the country's wetlands remained. The lost 90 per cent had been replaced with several of our major cities and towns, as well as extensive areas of farmland. For tangata whenua, wetlands were highly valued landscapes in themselves — they were mahinga kai (or hauanga kai for Waikato-Tainui, customary food-gathering sites) providing kōura, matamata and kōnehu raupō (raupō pollen for cakes); they were a source of rongoā Māori (traditional medicine); they provided fibres for raranga (weaving), and wood, rushes and sedges for construction; and they were a source of paru, a special mud used to dye muka (flax fibres). Culturally important items such as waka and carvings could be concealed and preserved in wetland sediments during times of warfare. For Māori, reinstating the ability to harvest in such areas once more, or just restoring the species and features, is of vital importance. When a wetland is lost or degraded, or access is impeded, so too is the tikanga associated with harvesting and other wetland uses.

restoration or freshwater science. Don't be shy to ask loads of questions, and reach out to mentors, academics and practitioners who inspire you to continue your work. Most importantly, look for those within your own whānau who can support you emotionally and spiritually. Yvonne says:

> I am delighted that I visited my koroua that day, inspired by his mātauranga [knowledge] and sharing knowledge of our tūpuna with me. That day of standing on the whānau farm, where Owheao resides, gazing at the repo submerging the farm edges and connecting to our tūpuna moana, hearing the rush of our tūpuna awa flowing into the moana, and flanked behind us by our tūpuna maunga, changed my course of direction almost instantly. I took the plunge and learnt conventional science and became a freshwater ecologist, but most importantly I draw upon the mātauranga of iwi and hapū, which plays the most crucial role in all of my rangahau [research].

Surprisingly, learning both knowledge systems simultaneously came quite easily to Yvonne, as if she was always meant to be on this path.

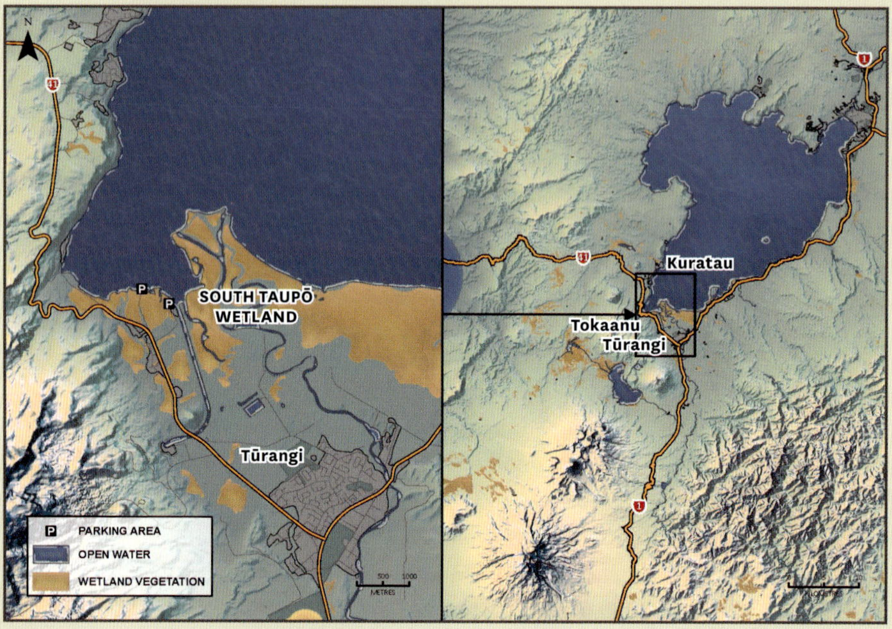

South Taupō Wetland

Freshwater swamp

Covering some 1500 hectares, this reporepo is one of the largest remaining in Aotearoa New Zealand. It was formed around 1800 years ago when the last major Taupō eruption hurled vast quantities of tephra onto the surrounding landscape. Over time, the Tongariro, Waiotaka and Waimarino Rivers eroded the piles of tephra and carried them to the lake, creating the Tongariro River delta, oxbow lakes and a series of distinctive ridges and hollows running parallel to the lake edge. Dryland plants colonised the ridges, while wetland plants took hold where their roots could remain seasonally or permanently wet. Introduced willows (*Salix* species) remain an ongoing challenge to manage and control.

Highlights

Thirty-six different species of native birds have been spotted in the wetlands. You may see weweia (New Zealand dabchicks), poaka (pied stilts), kawau (shags) and matuku (Australasian bittern).

The flora is diverse and includes broad

swathes of sedges such as soft twig rush (*Machaerina rubiginosa*) and rush-like plants such as oioi (jointed wire rush, *Apodasmia similis*), harakeke (flax), and shrublands with mingimingi (*Coprosma propinqua*), twiggy tree daisy (*Olearia virgata*), tī kōuka (cabbage tree) and mānuka. Raupō (bulrush) dominates shallow pools, while kuta (tall spike sedge, *Eleocharis sphacelata*) fringes the deeper pools.

Access

For a view of the wetland from Taupō Moana, walk along the 200-metre-long historic Tokaanu Wharf on Wharf Road (accessed off State Highway 41). The repo can also be explored by kayak or paddleboard as there are no formed walkways. Launch from the Tokaanu tailrace boat ramp (Kaiwaka Street) and head into the bay towards the Tongariro River delta, where hundreds of swans and other waterfowl gather, or along the Tongariro River. Lake Taupō is a large lake and conditions can change rapidly — wear a lifejacket, go with a buddy, and let others know your itinerary and when you plan to return, or join a guided tour.

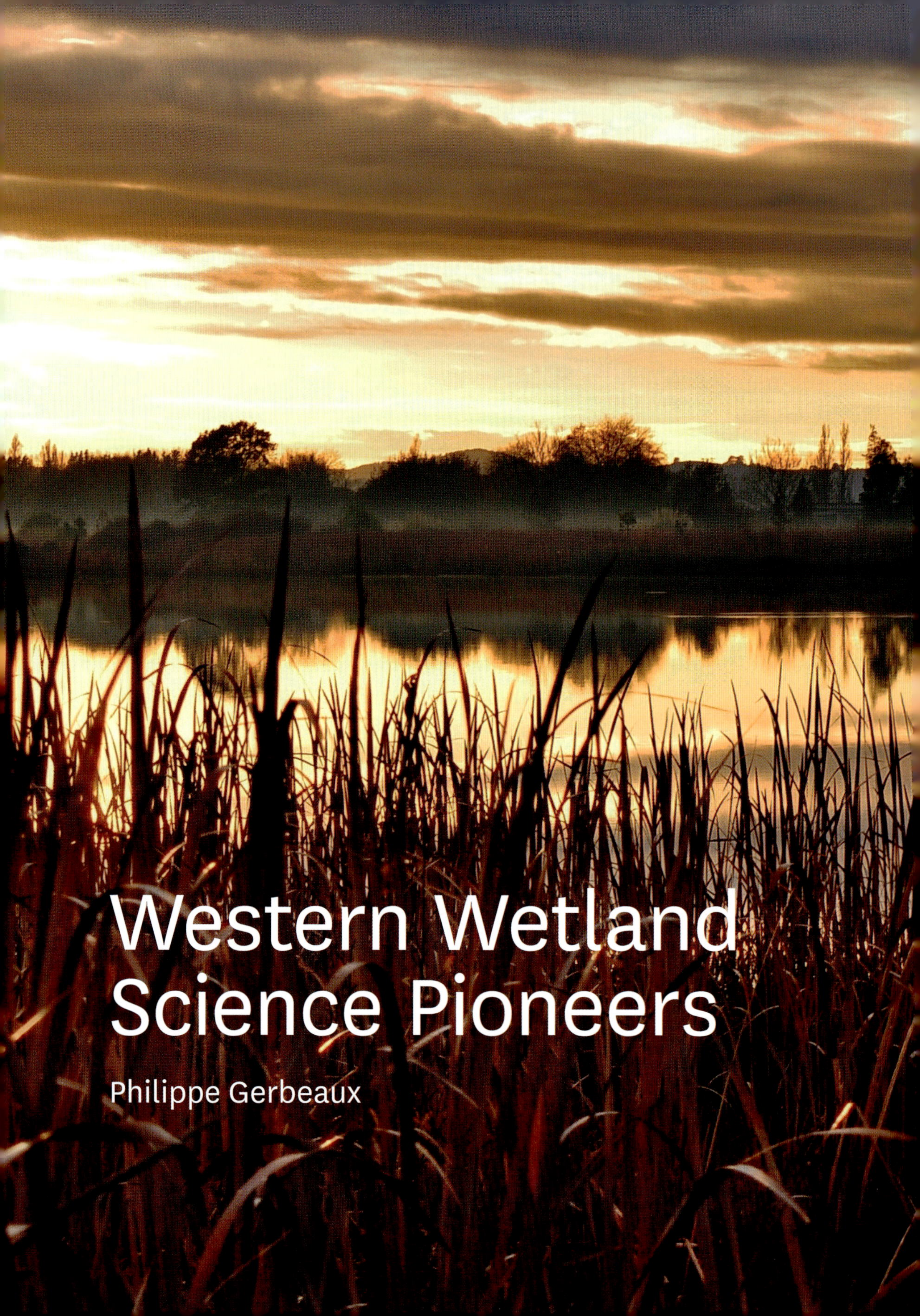
Western Wetland
Science Pioneers
Philippe Gerbeaux

IN 2021, NOT ONLY DID AOTEAROA NEW ZEALAND host the eleventh INTECOL International Wetlands Conference, but we also celebrated the fiftieth anniversary of the international Ramsar Convention on Wetlands. During this significant year, the National Wetland Trust suggested that it would be fitting to celebrate the many people who have progressed our scientific knowledge of New Zealand wetlands — both contemporary and early scientists, Māori and non-Māori.

When Māori first arrived in Aotearoa, they would have brought with them knowledge of the aquatic life and resources associated with tropical wetlands of the Pacific Islands. New Zealand's temperate wetlands, alpine tarns, vast peat bogs and geothermal wetlands would have seemed very different. In time, they developed a rich body of knowledge and understanding of how temperate wetlands function and change over the seasons, and which foods and fibres these wetlands could provide. We have no written records of the individuals who developed this work, because in te ao Māori it is the combined contributions of many generations that have created mātauranga Māori (traditional knowledge), and this is passed down through oral history.

Over the past century, Western scientific knowledge of New Zealand's wetlands has been added to by a body of Pākehā researchers, either born here or, in a few cases, attracted to the natural environment of our archipelago. Many have made a significant contribution to the field of wetland science, which today is being increasingly combined with ever-evolving mātauranga Māori for the betterment of our wetlands.

Interest in our wetlands has certainly grown steadily over the past 50 years, and today's scientists have benefited greatly from the work of our wetland pioneers, both Māori and Pākehā. My colleagues Dr Garth Harmsworth and Yvonne Taura expand on the contribution of Māori to our understanding of wetland values and processes, while my task is to reflect on New Zealand's Western scientists of the past century.

WHEN I ARRIVED in New Zealand in the mid-1980s as a young student, it struck me that most scientists interested in freshwaters and wetlands were conservationists at heart. They were very often actively engaged outside their normal day-to-day research activities through their contribution on various boards and councils concerned with the conservation and protection of the natural environment. They were also often active in their local community, carrying out whatever conservation work needed to be done on the ground. They were part of a generation of gifted scientists, passionate and tireless advocates for the protection of the ecosystems they studied, good mentors to their students, and often very humble. Many of today's scientists are equally gifted and passionate about the environment they study.

Many of those early scientists were women. All of them would readily and diligently share their knowledge and were highly respected within the wide range of national and international scientific societies they belonged to. That knowledge was mostly related, but not limited to, their specific field of expertise, whether in freshwater or terrestrial areas, in inland or coastal/marine ecosystems, or in macroscopic or microscopic plant or animal organisms. Most were effectively natural scientists who recognised that New Zealand's wetland ecosystems could have the same strong aesthetic and cultural appeal as forests, and that wetlands, lakes and rivers are part of what it is to be a Kiwi.

Inevitably, I have my own interest: freshwater and wetland ecology, with perhaps a slight preference for plant rather than animal ecology. It is therefore possible that some bias will creep into my choice of those Western scientists who have made a significant contribution in their respective fields to progressing the cause of wetland knowledge and conservation in Aotearoa.

In the first part of the twentieth century, wetlands* were the focus of mostly botanists and plant ecologists, and sometimes of zoologists, who worked across universities and government science and management agencies such as the Department of Scientific and Industrial Research (DSIR, founded in 1926), the Ministry of Works and Development (in particular, its Water and Soil Division) and the New Zealand Wildlife Service (established in 1945 and a division of the Department of Internal Affairs). New Zealand flora experts of the nineteenth century, such as Thomas Cheeseman (1845–1923) and Leonard Cockayne (1855–1934), would also no doubt have had an interest in plants from all habitats and have been attracted by the uniqueness and diversity found in wetlands.

THE FIRST WESTERN wetland scientist I would like to focus on is Dr Lucy May Cranwell (1907–2000). An internationally renowned botanist at the Auckland Institute and Museum, she established the science of palynology in New Zealand.** As it happens, peatlands are excellent environments for preserving fossil pollen and spores, and Cranwell wrote about peat deposits in both New Zealand and Hawaii (after she moved permanently to the United States in 1944). Professor Lennart von Post of Stockholm, Sweden, who founded pollen analysis (and created the von Post index, still used in New Zealand to assess peat decomposition), invited Cranwell to study post-glacial peat samples collected in Otago and Southland by Swedish glaciologist Dr Carl Caldenius.

Following in Cranwell's footsteps was Dr Neville Moar (1926–2016), who firmly established palynology in New Zealand (with the help of his soil colleague Dr William Harris). He also provided the foundation work that enabled one of his younger colleagues, Dr Matt McGlone, to re-create and publish the post-glacial history of New Zealand wetlands in 2009.[1]

I was fortunate when I arrived at Lincoln College in 1984 to be introduced to some

---

* 'Wetlands' is a relatively modern term that was hardly used in New Zealand until the country signed the Ramsar Convention on Wetlands in 1976. As Gerard Hindmarsh put it in his 2006 book *Swamp Fever*, the word was probably 'invented by the ecologically correct to categorise everything from bogs to mires, fens to flushes, marshes to seeps'. But the term 'swamp' was certainly commonly used, often as a synonym for 'wetland'!

** Palynology is the study of plant pollen, spores and certain microscopic plankton in both living and fossil form.

of those scientists and several other botanists interested in wetlands at DSIR Lincoln's Botany Division, which was only a few hundred metres from my office. In particular, I had great pleasure in meeting three wonderful women who are still highly regarded today for their eminent careers and contributions: Lucy Moore, Dr Elizabeth Edgar and Ruth Mason.

Lucy Moore (1906–1987) is well known for her contribution to the *Flora of New Zealand*. Soon after the death in 1957 of Dr Harry Allan, who was a key contributor to Volume I of the *Flora*, Moore was given editorial responsibility for the whole project. Originally from Auckland, she was a close friend of Lucy Cranwell and the two of them travelled together to Europe in 1935, attending botanical congresses in London and Amsterdam. They also went on many tramping trips, carrying out pioneering fieldwork in remote areas of New Zealand in the 1930s. On one particularly memorable trip, which they called their 'Bog Expedition', they visited all the Otago and Southland sites from which Carl Caldenius had collected the peat core samples on which Cranwell had worked with von Post in Sweden. Moore, who had moved down to Lincoln in 1960, wrote Volume II of the *Flora* (a volume on monocotyledons, published in 1970) with Dr Elizabeth Edgar (1929–2019), who had an interest in rushes and sedges.

Ruth Mason (1913–1990) had been retired from DSIR for ten years when I was first introduced to her. She used to keep in touch with her ex-colleagues at DSIR Lincoln Herbarium and was very excited to learn about my PhD research on the *Ruppia* beds of Te Waihora Lake Ellesmere, having contributed herself to reporting on the past fluctuations of those macrophyte beds in several unpublished reports. I still have those reports and own a few books from her library that I discovered one day for sale in an old bookshop in Christchurch after she passed away.

Dr Elizabeth Flint (1909–2011) was another significant female figure in the field. She lectured in botany at Victoria University of Wellington and then in England, before coming back to New Zealand to work part-time at DSIR Lincoln until her retirement in 1974. She returned as a research associate to the DSIR Botany Division in 1987 to co-write the three-volume series *Flora of New Zealand Desmids* with Hannah Croasdale and Marilyn Racine.* She continued research in a voluntary capacity at Lincoln University and Landcare Research (now called Manaaki Whenua — Landcare Research) two days a week, until she was 100 years old.

Another of New Zealand's leading experts on marine phytoplankton and freshwater diatoms was Dr Vivienne Cassie Cooper (1926–2021), who I met through the New Zealand Limnological Society (now the New Zealand Freshwater Sciences Society). Cooper worked at DSIR Auckland's Botany Division, and after she retired in 1986 she published the popular *Micro Algae: Microscopic Marvels*.

Other early New Zealand plant scientists have made important contributions to wetland botany and ecology, even though they are perhaps not always regarded as wetland specialists.

Dame Ella Orr Campbell (1910–2003), who in 1945 became the first female faculty member at Massey Agricultural College (now Massey University), studied at the University

---

* Desmids are single-celled microscopic green algae.

of Otago under Professor John Ernest Holloway. First and foremost an expert on bryophytes and orchids, she wrote 'Mires of Australasia' and 'The Water Relations of Heathlands' in the distinguished book series *Ecosystems of the World*,[2] a reflection of her strong interest in wetland ecosystems and peatlands in particular. She also visited many peat deposits of northern New Zealand, leading to the publication of several reports and papers.

Sir Alan Mark has spent the majority of his career at the University of Otago, where he is now Emeritus Professor. He is publicly renowned for the Save Manapouri campaign in 1969–1972, and his ecological studies of shoreline vegetation at Lakes Manapouri and Te Anau, and those of forest-wetland vegetation patterns in south Westland, are often cited in New Zealand wetland ecology papers. His main interest, however, is in alpine botany and mountain ecology, and he has also made a significant botanical contribution in highlighting the importance of wetland species to our alpine flora.

Dr Peter Wardle (1931–2018), who was an MSc student alongside Alan Mark at the University of Otago in the early 1950s, helped us better understand Westland podocarp swamp forests. He published research on the relationship of our tallest native tree, the kahikatea, with swamp communities on post-glacial surfaces. He was also one of the first New Zealand scientists interested in climate change studies. As early as 1954 he highlighted the strong imprint of recent climate change on our forests, and developed ideas about how the extreme cold of the Little Ice Age of the sixteenth to eighteenth centuries continues to affect the distribution of current vegetation. His book *Vegetation of New Zealand*, published in 1991, contains an important chapter on wetlands.

Dr Helen Hughes, the daughter of agricultural scientist and Cawthron Institute director Theodore Rigg, discovered a passion for botany during her education at Nelson College for Girls. From there, she went on to complete an MSc with first-class honours in botany at Canterbury University College, graduating in 1952. Her thesis, 'An Ecological Survey of the Pakihi Lands of the Westport District, Nelson', remains a key reference for those interested in this unique New Zealand wetland type, found mostly in the Tai Poutini West Coast and Whakatū Nelson areas of the South Island. (Other reference studies of pakihi and their soils were subsequently published by DSIR soil scientist George Mew.) Hughes would eventually become the first Parliamentary Commissioner for the Environment in 1987.

Dr Colin Burrows (1931–2014) was a botanist and palaeoecologist at the University of Canterbury with an interest in wetlands, particularly in Canterbury. Like Alan Mark, he had an active involvement with the famous Save Manapouri campaign. Burrows's chapter 'Bog and Flush Vegetation' in the second edition of *The Natural History of Canterbury* (1969) is still often referred to by many wetland scientists. He was renowned for his in-depth knowledge of the indigenous flora and vegetation of New Zealand, and for his research into alpine grasslands, wetlands, forests and glacial phenomena. He helped edit the third revision of *The Natural History of Canterbury*, published in 2008, and was invited to write a chapter on heathlands in the *Ecosystems of the World* book series mentioned above.

Another of Burrows' University of Canterbury colleagues who had an interest in wetlands was Dr Andrew Dobson, who may have been the first New Zealand ecologist to have offered a wetland classification, in his 1979 conference paper 'Mire Types of New Zealand'.[3]

The 'two Lucys', Dr Lucy May Cranwell and Lucy Moore, en route to the sixth International Plant Congress in Amsterdam, 1935.

Another wetland scientist of note among those born prior to the Second World War is Keith Thompson. An ecologist, he has made a huge contribution to our knowledge of New Zealand peatlands, particularly under the leadership and support of Professor Jack McCraw at the University of Waikato at a time when peatlands were under increased pressure from agricultural and horticultural development. Thompson arrived in New Zealand in the 1970s, having already gained a wide knowledge of wetlands in the United Kingdom and Europe, in tropical areas and across all continents. He taught for 28 years at the University of Waikato, and today continues to work as a consultant and serves as a trustee of the National Wetland Trust. He also served on the Scientific and Technical Review Panel of the Ramsar Convention. Thompson took part in national peatland surveys, and one of his key legacies is the *Annotated Bibliography of New Zealand Peat and Peatlands*, published in 1987, at the end of a decade that saw the publication of several other important New Zealand wetland reports.

One of those reports was written not by an academic, but by someone who played a special role in New Zealand conservation, Gordon Stephenson (1924–2015), whose list of contributions and awards is extraordinary. Another immigrant from England, Stephenson moved to the Waikato region with his family in 1958. He was a farmer but had a particular interest in science, and in 2013 he was awarded an honorary doctorate from the University of Waikato in recognition of his conservation and environmental work, which aimed at encouraging environmentally friendly farming practices. Gordon and his wife Celia founded the QEII National Trust in 1977, which provides a legal mechanism for protecting privately owned land in perpetuity.* In 1979, the native forest on the Stephensons' farm became the first QEII covenant declared in New Zealand, and many privately owned wetlands have since been covenanted under the scheme. Gordon was also the driving force behind the creation of the National Wetland Trust in 1999 (a millennium project), along with David Lawrie, another wetland conservationist and an ornithologist, who is still involved with the trust today.

COASTAL AND MARINE wetlands became better known and appreciated thanks to the work of several renowned early scientists.

One name that stands out is Professor Valentine Chapman (1910–1980), who was appointed the first professor of botany at Auckland University in 1945 following his arrival from England, where he had completed a PhD on the East Anglian saltmarshes. In addition to the taxonomy of marine algae, he had a lifelong interest in mangroves and was much concerned with conservation and resource management of coastal ecosystems and beyond. With zoologist John Morton, a fellow professor at the University of Auckland, he took a keen interest in the protection of kauri forests. Both professors were instrumental in the passing of the Marine Reserves Act in 1971.

---

* Covenanting natural features via the QEII National Trust is a legally binding protection agreement that is registered on the title of the land. It is voluntary, but once in place it binds current and all subsequent landowners. See https://qeiinationaltrust.org.nz.

The value of New Zealand's estuarine and coastal wetlands, including Christchurch's Ihutai Avon–Heathcote Estuary, were also widely promoted by Canterbury zoologist Professor George Knox (1919–2008). Knox recognised the value of all branches of natural history and became the general editor of the second edition of *The Natural History of Canterbury* (1969).

Other zoologists also deserve a mention — especially those involved in the field of ornithology, a discipline that is hard to separate from wetlands! Robert Falla (1901–1979) was widely recognised for his national and international work on birds, and was the founding president of the Ornithological Society of New Zealand and author of *Birds of New Zealand* (1966).

Dr Kaj Ejvind Westerskov (1919–2003) was one of three European ornithologists, along with Friedrich-Carl Kinsky (1911–1999) and Kazimierz Wodzicki (1900–1987), to have a lasting influence in their adopted home and inspire many wetland conservation colleagues interested in birds. Born in Denmark, Westerskov migrated to New Zealand with his family in 1952. He joined the New Zealand Wildlife Service as a research scientist and became head of the research section from 1960 to 1964. He then shifted to Ōtepoti Dunedin, where he created the one-year post-graduate Diploma in Wildlife Management at the University of Otago — a first for Australasia and still highly regarded today (several theses by students on the course have focused on wetland topics). Westerskov remained active outside the lecture theatre and continued to be involved in ornithology after his retirement in 1985, serving on several executive boards both locally and nationally.

Because many important New Zealand wetlands are associated with lake and river ecosystems, I should mention the important role played by two men involved in the field of freshwater ecology. Dr Edward Percival (1893–1969) was appointed professor at the University of Canterbury in 1929, where he would stay until his death in 1969, while Brian John Marples (1907–1997) was professor at the University of Otago from 1937 to 1967, where he was instrumental in helping set up the Diploma in Wildlife Management programme. Marples was also a co-founder of the New Zealand Ornithological Society and, like Percival, was said to be an inspirational teacher. Not the least among his achievements, in 1962 he published *New Zealand Freshwater Life*, the first book on the topic and an inspiration to many students with an interest in freshwater ecology. Some of those students went on to establish the New Zealand Freshwater Sciences Society (NZFSS) in 1968, known initially as the New Zealand Limnological Society.

Among the freshwater scientists inspired by Percival and Marples were several notable women. Dr Violet Hilary Jolly (1906–1975) carried out pioneering studies of lakes in Australia and New Zealand. She played a leading role in the foundation of the Australian Society for Limnology, appointed president from 1967 to 1968, and she became the main author, with Dr John Brown, of the first book on New Zealand lakes in 1975.

Professor Vida Stout (1930–2012) carried out research on the biology and chemistry of South Island lakes, including the nature and changes in zooplankton communities over time. She became the first female dean of a science department at a New Zealand university, in 1984 at the University of Canterbury.

Dr Ann Chapman (1937–2009) was one of the first New Zealand women scientists to visit Antarctica, and the first woman to lead a scientific expedition to the frozen continent. A specialist in crustaceans, she inspired a whole new generation of freshwater specialists.

Another important figure in the history of the NZFSS is Professor Carolyn Burns, a limnologist of high repute who specialises in *Daphnia*. She was awarded a personal chair in zoology in 1993, as well as being head of the University of Otago's Department of Zoology between 1987 and 1989. On her retirement in 2010, Burns was conferred the title of Professor Emeritus by the university. She was very engaged in the conservation of wetlands through her membership of the Nature Conservation Council from 1975 to 1990, including five years as its chair between 1978 and 1983, and was also a member of the National Parks and Reserves Authority from 1981 to 1990. In addition, Burns was active in the International Union for Conservation of Nature (IUCN), serving on its Commission on Education from 1979 to 1988, Commission on Ecology from 1988 to 1990, and Commission on National Parks and Protected Areas (1988–). She was a regional councillor of the IUCN from 1984 to 1990, and chaired the New Zealand committee of IUCN members from 1986 to 1990.

Other scientific societies, including the New Zealand Ecological Society and the New Zealand Hydrological Society (founded in the early 1950s and in 1961, respectively, before the NZFSS), would no doubt have had the odd session around wetland fish or wetland birds. Ironically, the New Zealand Hydrological Society would probably have been more concerned in those days with how to improve the drainage of wetlands than their conservation!

An ichthyologist and active member of the NZFSS was Dr Robert (Bob) McDowall (1939–2011). His work on whitebait in south Westland contributed greatly to a better understanding of lowland wetlands and their importance in the conservation of these iconic New Zealand species. Two of his colleagues from the Fisheries Division of the Ministry of Agriculture and Fisheries, Dr Don Jellyman (an eel specialist) and Tony Eldon (a mudfish specialist), have also been ardent advocates for the protection of wetlands.

New Zealand's ratification of the Ramsar Convention in 1976 may have triggered a renewed interest in wetlands among a new generation of scientists, who were no doubt also inspired by those who pioneered study in the area. The report *Wetlands: A Diminishing Resource*, published in 1983, was co-authored by a panel of several wetland conservationists and scientists comprising Gordon Stephenson, Bernard Card, Alan Mark, Dr Roger McLean, Keith Thompson and Bob Priest. Prepared by the Environmental Council of New Zealand for the Minister for the Environment, this report has undoubtedly played a key role in generating further knowledge on wetlands in New Zealand.

The creation of the Resource Management Act (RMA) in 1991 brought wetlands to the fore, especially Section 6 of the Act, which states that:

> In achieving the purpose of this Act, all persons exercising functions and powers under it, in relation to managing the use, development, and protection of natural and physical resources, shall recognise and provide for the following matters of national importance: (a) the preservation of the natural character of the coastal environment (including the coastal marine area), wetlands, and lakes and rivers and their margins, and the protection of them from inappropriate subdivision, use, and development.

In the 1990s, during the development of environmental freshwater indicators by the Ministry for the Environment (MfE) to support the implementation of the new RMA, it became clear that there was very little information on appropriate indicators for monitoring wetlands. To help fill this knowledge gap, Dr Jonet Ward and James Lambie from Lincoln Environmental led a series of workshops conducted under the MfE Sustainable Management Fund. A wide range of scientists and practitioners from the newly created Crown Research Institutes, universities and government agencies attended. Outputs from the project included two handbooks, *Wetland Types in New Zealand* (by Peter Johnson and Philippe Gerbeaux) and *Handbook for Monitoring Wetland Condition* (by Dr Beverley Clarkson, Brian Sorrell, Paula Reeves, Paul Champion, Trevor Partridge and Bruce Clarkson), and *Māori Environmental Performance Indicators for Wetland Condition and Trend* (by Dr Garth Harmsworth). These publications are still very much used by wetland practitioners today.

The development of GIS tools and improved remote sensing techniques brought new ways of mapping wetlands. I would like to mention in particular the important contribution made in this respect by Dr Anne-Gaelle Ausseil from Manaaki Whenua — Landcare Research in helping to assemble, for the first time, credible, quantified information on the actual loss of wetlands in New Zealand and an assessment of the condition of those remaining, as part of the 2008 Wetlands of National Importance project funded by the Department of Conservation.[4]

WETLAND SCIENCE in New Zealand today is mostly undertaken by Crown Research Institutes, NIWA (National Institute of Water and Atmospheric Research Taihoro Nukurangi) and Manaaki Whenua — Landcare Research, universities and wānanga, all recognising and including mātauranga Māori as an essential component of research. Additional work, particularly applied research, is conducted by a range of other agencies, including the Department of Conservation, regional councils, consultants and private institutions or foundations.

The very successful restoration symposia organised every two years by the National Wetland Trust will continue to be important fora that create a safe and comfortable space for young scientists, iwi, landowners, community volunteers and citizen scientists to share their learnings.

Shortly after the thirtieth anniversary of the Ramsar Convention, in 2003, I wrote a review of wetland management in New Zealand. In that article, I documented a chronology of wetland destruction in Aotearoa and the first attempts, through science and policy decisions, to reverse the trend and provide tools to conserve and protect areas of wetlands that remained.

The release in 2020 of sweeping new wetland policies and rules by the Minister for the Environment was a significant step forward, but more needs to be done to attract young students towards wetland science to ensure their future implementation. While this book celebrates present and past wetland scientists, we are keenly aware of a looming shortage of such experts. The creation of dedicated wetland training courses at tertiary level would be a very positive step in that direction.

The National Wetland Trust has built a wetland discovery trail with educational pontoon panels at Rotopiko to inspire a future generation of scientists.

Appendix 1

Supporting Wetland Science

This book was produced in association with the National Wetland Trust of New Zealand to showcase our current understanding of wetlands and some of the people — both current and from days past — whose observations and research have revealed the secrets of our shallows.

In Aotearoa New Zealand, formal wetland research is conducted and supported by a number of institutes, universities, agencies, iwi groups and private consultants. Many are represented in the chapters of this book. The organisations listed here have substantially contributed to the production of this book.

About the National Wetland Trust

The National Wetland Trust (NWT) aims to get people 'into' wetlands, in both senses of the word. The brainchild of the late Gordon Stephenson ONZM, and David Lawrie MNZM, the trust was founded in 1999 to increase public understanding and awareness of wetland values, improve our understanding of wetlands, and encourage their protection, creation and restoration. The NWT advocates for stronger wetland policies, engages the public via education materials and events, takes a lead promoting World Wetlands Day and brings the community and professionals together at biannual conferences.

At Rotopiko near Hamilton Airport, the trust is developing an education facility, with a showcase restoration project, interpretation panels and a hands-on wetland discovery trail. Here, families, schools and other groups can learn about wetlands while having fun, engaging in such activities as mudfish scrabble, eels and ladders, and wildlife detection. People best appreciate what they experience, but wetlands are hard to access. So the trust has also created wetland trails and a directory of wetlands to visit nationwide, to bring people and wetlands together in a safe, enjoyable way.

Wetlands have long been the forgotten ecosystem — unappreciated, destroyed in greater proportions than any other New Zealand ecosystem, and underprotected. Finally, the tide is turning for wetlands, and their recognition, protection and community-led restoration efforts are on the rise. But we need to combine cultural knowledge and grassroots experience with robust up-to-date science if we want to create and restore healthy, functioning, representative wetlands. The NWT has been encouraging students to study wetlands, and rewarding them through its Golden Plover Wetland Research Award. By championing *Life in the Shallows*, the trust aims to inspire and engage aspiring new scientists and showcase the latest research into Gordon Stephenson's shy places.

www.wetlandtrust.org.nz

WETLAND TRUST
NATIONAL WETLAND TRUST OF NEW ZEALAND

GOLDEN PLOVER WETLAND RESEARCH AWARD

The Golden Plover Wetland Research Award supports students who are conducting original research into any aspect of the ecology or management of wetlands at honours or master's level at a New Zealand university. The award was established in 2001 by Dr Tony Reiger and Dr Steven Messerschmidt, in conjunction with the NWT and the University of Waikato. It is currently partly funded by the trust and the New Zealand Freshwater Sciences Society, with annual contributions from David and Lynne Lawrie.

The Pacific golden plover is one of many migratory birds that make their way from Alaska to New Zealand. The award's name symbolises this connection between the two countries, as the key donor was originally from Alaska but took up residence in New Zealand, where he was instrumental in restoring and protecting Southland's Big Lagoon.

RECIPIENTS OF THE GOLDEN PLOVER AWARD 2007–2021

2020
No recipient

2019
No recipient

2018
Clara Wilson
Recovery of *Empodisma robustum*-dominated bog vegetation from fire.

2017
Katrina Brown
Eco-hydrological characterisation of Opuatia Wetland.

2016
Chantal Whitby
Overcoming barriers and opportunities to restoring ecological function within wetlands on privately owned agricultural landscapes in New Zealand.

2015
Adam Purcell
Trait-based predictions of wetland plant composition along a hydrological gradient.

2014
Erin Doyle
Assessing the use of molecular techniques to determine terrestrial invertebrate diversity in New Zealand mangroves and detection of whio (blue ducks) in New Zealand streams.

2013
Alexandra Keyte Beattie
Functional role of *Empodisma robustum* in ecosystem exchanges of carbon and water in Kopuatai bog.

2012
Olivia Burge
Vegetation succession within the Awarua–Waituna Wetland.

2011
Bridget Johnson
Processes influencing the establishment and survival of native vegetation at Wairio Wetlands, Wairarapa.

2010
James Blythe
Eco-hydrological characteristics of Whangamarino Wetland.

2010
Yvonne Taura
Effects of willows and willow control on aquatic invertebrate communities in the South Taupō Wetland.

2009
Craig Allen
Effects of climate change and bore extraction on the hydrology of Te Hapua Wetlands near Foxton.

2008
No recipient

2007
Kieran Whelan
Using native plants in constructed wetlands to rehabilitate mines.

2007
Amy McDonald
Increasing the success of moving black mudfish, in relation to establishing new populations.

About the New Zealand Freshwater Sciences Society

The Limnological Society of New Zealand, now the New Zealand Freshwater Sciences Society (NZFSS), was founded in 1968. At the time, there was increasing community concern about freshwater issues, including the Manapōuri Dam proposal in Fiordland National Park and the effects of introduced aquatic weeds on the Rotorua Lakes. The main drive for the society's formation came from academics based at the country's universities, particularly pioneering women in freshwater science. This proved important to the society's inclusive ethos, with many of the early students going on to become leaders of the society and the science.

The NZFSS now has a welcome mix of more than 500 practising scientists, including academics, freshwater managers, consultants, Crown researchers, tangata whenua scientists and resource managers, and students. The NZFSS aims:

- to establish effective liaison between all persons interested in any aspect of fresh and brackish water research and management in Aotearoa New Zealand, and to encourage and promote these interests, new knowledge and research
- to embrace the rich heritage of Māori culture in Aotearoa New Zealand and acknowledge the rights of Māori under the Treaty of Waitangi and the principles of partnership, participation and protection
- to be fair and inclusive, and committed to equity and diversity in all aspects of its operations.

Around 10 per cent of the society's membership also belong to Te Wai Māori — NZFSS Rōpū Māori, which is supported by the wider society. The whakakitenga (vision) of Rōpū Māori is 'the successful participation of Māori in freshwater sciences, management and research where Māori principles, values and interests are identified and valued within the NZFSS'.

The NZFSS holds popular annual conferences across Aotearoa, from Shantytown to Waiheke Island. It often joins with the Australian Freshwater Sciences Society, the New Zealand Hydrological Society, the New Zealand Rivers Group and the New Zealand Marine Sciences Society to share multi-disciplinary ideas and research.

The society has published books on a range of freshwater topics, from taxonomy to management, and produces an annual newsletter for members.

https://freshwater.science.org.nz

Established 1968
New Zealand Freshwater Sciences Society

About Manaaki Whenua — Landcare Research

Manaaki Whenua — Landcare Research is a Crown Research Institute (a government-owned science research business). Established under the Crown Research Institutes Act 1992, its primary role is to drive innovation in New Zealand's management of terrestrial (land-based) biodiversity and land resources, in order to both protect and enhance the terrestrial environment and grow New Zealand's prosperity. One of its 10 key active programmes on restoration and functioning of wetland ecosystems was developed and led by Dr Beverley Clarkson.

The institute's research aims to assist landowners, managers and policy makers in the protection and restoration of wetlands by providing scientifically based tools and guidelines for identifying, managing and restoring wetlands. The research will benefit Aotearoa New Zealand by helping prevent further loss and degradation of wetlands, with a long-term outcome of improved wetland ecological condition and resilience by:

- understanding species–environment patterns in lowland wetlands (vegetation, nutrients, invertebrates) to inform limits for maintaining wetland ecological integrity in intact and modified wetlands throughout New Zealand
- field restoration experiments that can be upscaled to large wetland areas, including control of invasive grey willow and re-establishment of native kahikatea forest at the Ramsar-listed Whangamarino Wetland with the Department of Conservation's Arawai Kākāriki Programme
- development of New Zealand wetland delineation tools based on vegetation (completed 2014), soils (completed 2018) and hydrology (completed 2021)
- best-practice restoration techniques, summarised in *Wetland Restoration: A Handbook for New Zealand Freshwater Systems* (published 2010, and reprinted 2012)
- *Te Reo o te Repo: The Voice of the Wetland*, led by Māori researchers, focusing on the restoration of cultural resources in wetlands (volume I published 2017, volume II published 2021).

www.landcareresearch.co.nz

Manaaki Whenua
Landcare Research

About the Department of Conservation Te Papa Atawhai

The Department of Conservation (DOC) is the government agency charged with conserving New Zealand's natural and historic heritage. Its outcomes are Te ora o Papatūānuku | Healthy nature, Te ora o te Hapori | Thriving communities and Te hunga Atawhai | People who care. DOC engages in research to underpin its conservation activities and measure outcomes.

Arawai Kākāriki, translated as 'green waterways', is DOC's flagship wetland conservation and science programme. It was launched in 2007 to provide leadership in freshwater research and restoration at nationally important sites.

Arawai Kākāriki covers a diverse range of wetland, lake, estuary and river ecosystems from Northland to Southland, across five of New Zealand's most outstanding wetland systems.

DOC is working with scientists, iwi, other partners and the community to learn about and restore the health of these vital natural environments. Research, collaboration and public involvement are intrinsic elements of Arawai Kākāriki.

The Department selected three core sites for restoration based on their significant ecological values: Whangamarino Wetland in the Waikato region, Ō Tū Wharekai in Canterbury and Awarua-Waituna Wetland in Southland.

www.doc.govt.nz

Department of Conservation
Te Papa Atawhai

About NIWA (National Institute of Water and Atmospheric Research Taihoro Nukurangi)

NIWA is New Zealand's pre-eminent provider of environmental science for climate, freshwater, coasts and oceans, undertaking world-leading science on some of the most important issues facing the globe.

The institute's role is to enhance the sustainable management and economic value of New Zealand's marine and freshwater resources, increase understanding of our changing climate, and improve our resilience to weather and climate hazards. NIWA works with other Crown Research Institutes, universities, industry and iwi to fulfil its purpose of ensuring its science benefits all New Zealanders. Its research also supports government decisions and policies, and the institute works in partnership with Māori to support their goals and aspirations.

NIWA maintains four heritage assets – the National Climate Database, the Water Resources Database, the New Zealand Freshwater Fish Database and the NIWA Marine Invertebrate Collection. The National Climate Database is a record of weather and climate monitoring dating back more than 100 years with information from 7,500 climate stations covering Aotearoa New Zealand, the Southwest Pacific and Antarctica, and the Hydrology Database has information from a nationwide network of water and soil monitoring stations.

NIWA also houses the New Zealand High Performance Computing Facility, which includes the nation's largest supercomputer, has the Northland Marine Research Centre, New Zealand's leading facility for finfish aquaculture, and runs a fleet of state-of-the-art marine research vessels.

https://niwa.co.nz

NIWA
Taihoro Nukurangi

The Mirror Lakes reflect Mount Eglington on the Te Anau–Milford Road.

Appendix 2

Species List

Plant names are sourced from Ngā Tipu o Aotearoa New Zealand Plants, https://nzflora.landcareresearch.co.nz; bird names from New Zealand Birds Online, https://nzbirdsonline.org.nz.

Each species is listed alphabetically by both Māori and English names. The authors acknowledge that many more regional Māori names may exist than those presented here. Additional Māori names (not recorded in Ngā Tipu o Aotearoa) were extracted from the chapter submitted by Garth Harmsworth. Duplicate listings in the scientific name column indicate where multiple species share a common name.

Common name(s)	Scientific name(s)
alligator weed	*Alternanthera philoxeroides*
alligator weed flea beetle	*Agasicles hygrophila*
Argentine ant	*Linepithema humile*
aruhe, rārahu, rarauhe, rahurahu, tākaka / bracken	*Pteridium esculentum*
Australasian bittern / matuku	*Botaurus poiciloptilus*
Australasian crested grebe	*Podiceps cristatus australis*
Australasian shoveler / kuruwhengi	*Anas rhynchotis*
bachelor's button	*Cotula coronopifolia*
bamboo spike sedge/ kūkuta, kuta, kutakuta	*Eleocharis sphacelata*
banded dotterel / piopio, pohowera, tūturiwhatu	*Charadrius bicinctus*
banded kōkopu	*Galaxias fasciatus*
banded rail / mioweka	*Gallirallus philippensis assimilis*
bar-tailed godwit	*Limosa lapponica*
black mudfish / waikaka	*Neochanna diversus*
black shag / kawau	*Phalacrocorax carbo*
black stilt / kakī	*Himantopus novaezelandiae*
black swan / kakīānau	*Cygnus atratus*
black-backed gull / karoro	*Larus dominicanus*
black-billed gull / tārapunga	*Larus bulleri*
black-fronted tern / tara pirohe	*Chlidonias albostriatus*
blue duck / whio	*Hymenolaimus malacorhynchos*
bog clubmoss	*Lycopodiella serpentina*
bog pine	*Halocarpus bidwillii*
bracken / aruhe, rārahu, rarauhe, rahurahu, tākaka	*Pteridium esculentum*
brown mudfish / waikaka	*Neochanna apoda*
brown teal/ pāteke	*Anas chlorotis*
brushtail possum	*Trichosurus vulpecula*
bulrush / raupō	*Typha orientalis*
cabbage tree / tī kōuka, tī rakau, whanake	*Cordyline australis*
Canada goose	*Branta canadensis*
Caspian tern / taranui	*Hydropogne rufopectus*
common bully / tīpokopoko	*Gobiomorphus cotidianus*
common tiger beetle / pāpapa	*Neocicindela tuberculata*

Common name(s)	Scientific name(s)
crack willow	*Salix × fragilis*
creeping bent	*Agrostis stolonifera*
dwarf mistletoe	*Korthalsella salicornioides*
eastern curlew	*Numenius madagascariensis*
eel grass	*Vallisneria australis*
fairy tern / tara iti	*Sterna nereis*
fantail / pīwakawaka	*Rhipidura fuliginosa*
fernbird / mātātā	*Bowdleria punctata*
ferret	*Mustela furo*
flax / harakeke	*Phormium tenax*
forked sundew	*Drosera binata*
Fred the thread	*Houdinia flexilissima*
freshwater mussel / kākahi	*Hyridella menziesi*
geothermal kānuka	*Kunzea tenuicaulis*
giant cane rush (Chatham Island)	*Sporadanthus traversii*
giant cane rush (North Island)	*Sporadanthus ferrugineus*
giant kōkopu	*Galaxias argenteus*
giant umbrella sedge / ūpoko-a-tangata	*Cyperus ustulatus*
glasswort / ureure	*Sarcocornia quinqueflora*
golden sand sedge / pingao	*Ficinia spiralis*
gorse	*Ulex europaeus*
great bindweed	*Calystegia silvatica*
grey duck / pārera	*Anas superciliosa*
grey teal / tētē moroiti	*Anas gracilis*
grey willow	*Salix cinerea*
Haast's eagle / pouākai, te hōkioi	*Hieraaetus moorei*
Hall's tōtara	*Podocarpus laetus*
harakeke / flax	*Phormium tenax*
hauhau, kōwaro, waikaka / mudfish	*Neochanna diversus / Neochanna apoda*
hedgehog	*Erinaceus europaeus*
hīnau	*Elaeocarpus dentatus*
hornwort	*Ceratophyllum demersum*
horse's mane	*Ruppia megacarpa / Ruppia polycarpa*
houhere / lacebark, ribbonwood	*Hoheria species*

Common name(s)	Scientific name(s)
Hutton's shearwater	*Puffinus huttoni*
īnanga	*Galaxias maculatus*
jointed twig rush / mokuautoto	*Machaerina articulata*
jointed wire rush / oioi	*Apodasmia similis*
kahikatea / white pine	*Dacrycarpus dacrydioides*
kāhu / swamp harrier	*Circus approximans*
kaikawaka	*Libocedrus bidwillii*
kākahi / freshwater mussel	*Hyridella menziesi*
kākāpō	*Strigops habroptilus*
kakī / black stilt	*Himantopus novaezelandiae*
kakīānau / black swan	*Cygnus atratus*
kāmahi	*Pterophylla racemosa*
kanakana, pirahau / lamprey	*Geotria australis*
kārearea / New Zealand falcon	*Falco novaeseelandiae*
karoro / black-backed gull	*Larus dominicanus*
katipō spider	*Latrodectus katipo*
kawau / black shag	*Phalacrocorax carbo*
kea	*Nestor notabilis*
kekeno / New Zealand fur seal	*Arctocephalus forsteri*
kererū / wood pigeon	*Hemiphaga novaeseelandiae*
kēwai / freshwater crayfish	*Paranephrops planifrons / Paranephrops zealandicus*
kiore / Pacific rat	*Rattus exulans*
kiwi	*Apteryx species*
knobby clubrush / wīwī	*Ficinia nodosa*
kōaro	*Galaxias brevipinnis*
koi carp	*Cyprinus carpio*
koitareke / marsh crake	*Porzana pusilla*
kōkopu	*Galaxias species*
kōpūngāwhā, kuawa, kūwāwā, wāwā / lake clubrush	*Schoenoplectus species*
kōtare / sacred kingfisher	*Todiramphus sanctus*
kōtuku / white heron	*Ardea modesta*
kōtuku ngutupapa / royal spoonbill	*Platalea regia*
kōura / freshwater crayfish	*Paranephrops planifrons / Paranephrops zealandicus*
kōwaro, hauhau, waikaka / mudfish	*Neochanna diversus / Neochanna apoda*
kōwhitiwhiti / water cress	*Nasturtium microphyllum / Nasturtium officinale*
kuawa, kūwāwā, kōpūngāwhā, wāwā / lake clubrush	*Schoenoplectus species*
kūkuta, kuta, kutakuta / bamboo spike sedge	*Eleocharis sphacelata*
kuruwhengi /Australasian shoveler	*Anas rhynchotis*
lacebark / houhere	*Hoheria species*
lagarosiphon	*Lagarosiphon major*
lake clubrush / kōpūngāwhā, kuawa, kūwāwā, wāwā	*Schoenoplectus species*

Common name(s)	Scientific name(s)
lamprey / kanakana, piharau	*Geotria australis*
laughing owl, / whēkau	*Sceloglaux albifacies*
longfin eel / tuna	*Anguilla dieffenbachii*
maire tawake / swamp maire	*Syzygium maire*
mākaka / marsh ribbonwood	*Plagianthus divaricatus*
mako / wineberry	*Aristotelia serrata*
mallard duck	*Anas platyrhynchos*
mānawa / mangrove	*Avicennia marina subsp. australasica*
mangrove / mānawa	*Avicennia marina subsp. australasica*
manoao / silver pine	*Manoao colensoi*
mānuka / tea-tree	*Leptospermum scoparium*
mānuka beetle	*Pyronota festiva*
Māori musk	*Thyridia repens*
mārohi, aruhe, rārahu, rarauhe, rahurahu, tākaka / bracken	*Pteridium esculentum*
marsh crake / koitareke	*Porzana pusilla*
marsh cress / poniu	*Rorippa palustris*
marsh ribbonwood / mākaka	*Plagianthus divaricatus*
marsh clubrush / purua	*Bolboschoenus fluviatilis*
mataī	*Prumnopitys taxifolia*
mātātā / fernbird	*Bowdleria punctata*
matuku / Australasian bittern	*Botaurus poiciloptilus*
matuku moana / white-faced heron	*Egretta novaehollandiae*
mauve sun orchid	*Thelymitra malvina*
mercer grass	*Paspalum distichum*
mingimingi	*Coprosma propinqua*
mioweka / banded rail	*Gallirallus philippensis assimilis*
miro	*Prumnopitys ferruginea*
mokuautoto / Jointed twig rush	*Macherina articulata*
morepork / ruru	*Ninox novaeseelandiae*
mountain toatoa	*Phyllocladus alpinus*
mud buttercup	*Ranunculus limosella*
mudwort	*Limosella lineata*
nana, rimurehia / seagrass	*Zostera muelleri*
narrow-leaved houhere	*Houheria angustifolia*
New Zealand dabchick / weweia	*Poliocephalus rufopectus*
New Zealand falcon / kārearea	*Falco novaeseelandiae*
New Zealand fur seal / kekeno	*Arctocephalus forsteri*
New Zealand scaup / pāpango	*Aythya novaeseelandiae*
ngāwhā, kūkuta, kuta, kutakuta / bamboo spike sedge	*Eleocharis sphacelata*
ngutu parore /wrybill	*Anarhynchus frontalis*
nīkau palm	*Rhopalostylis sapida*
North Island freshwater crayfish / kēwai, kōura	*Paranephrops planifrons*

Common name(s)	Scientific name(s)
Norway rat	*Rattus norvegicus*
oioi / jointed wire rush	*Apodasmia similis*
oval sedge	*Carex leporina*
Pacific golden plover	*Pluvialis fulva*
Pacific rat / kiore	*Rattus exulans*
pakihi rush	*Machaerina teretifolia*
pākura, pūkeko / purple swamp hen	*Porphyrio melanotus*
panoko, papangoko / torrent fish	*Cheimarrichthys fosteri*
paopao /clubrush and spike sedge	*Schoenoplectus tabernaemontani / Eleocharis sphacelata*
pāpango / New Zealand scaup	*Aythya novaeseelandiae*
pāpapa / common tiger beetle	*Neocicindela tuberculata*
paradise shelduck / pūtangitangi	*Tadorna variegata*
paratya shrimp	*Paratya curvirostris*
pārera / grey duck	*Anas superciliosa*
parrot's feather	*Myriophyllum aquaticum*
pāteke / brown teal	*Anas chlorotis*
pied stilt / poaka	*Himantopus himantopus*
piharau, kanakana / lamprey	*Geotria australis*
pīngao / golden sand sedge	*Ficinia spiralis*
pink bindweed	*Calystegia sepium subsp. roseata*
pink pine	*Halocarpus biformis*
piopio / banded dotterel	*Charadrius bicinctus*
pīpīwharauroa / shining cuckoo	*Chrysococcyx lucidus*
pīwakawaka / fantail	*Rhipidura fuliginosa*
poaka / pied stilt	*Himantopus himantopus*
pohowera / banded dotterel	*Charadrius bicinctus*
pōkākā	*Elaeocarpus hookerianus*
poniu / marsh cress	*Rorippa palustris*
pouākai, te hōkioi / Haast's eagle	*Hieraaetus moorei*
pūhā / sowthistle	*Sonchus species*
pukatea	*Laurelia novae-zelandiae*
pūkeko, pākura / purple swamp hen	*Porphyrio melanotus*
purei / pukio	*Carex secta / Carex virgata*
pūriri	*Vitex lucens*
purple swamp hen / pākura, pūkeko	*Porphyrio melanotus*
purua / marsh clubrush	*Bolboschoenus fluviatilis*
pūtangitangi / paradise shelduck	*Tadorna variegata*
pūweto / spotless crake	*Porzana tabuensis tabuensis*
pygmy pine	*Lepidothamnus laxifolius*
rainbow trout	*Oncorhynchus mykiss*
rārahu, rarauhe, rahurahu / bracken	*Pteridium esculentum*
raupō / bulrush	*Typha orientalis*

Common name(s)	Scientific name(s)
red knot	*Calidris canutus*
remuremu	*Selliera radicans*
rifleman/ tītipounamu	*Acanthisitta chloris*
rimu	*Dacrydium cupressinum*
rimurehia, nana / seagrass	*Zostera muelleri*
rove beetle	*Physobryaxis inflata*
royal fern	*Osmunda regalis*
royal spoonbill / kōtuku ngutupapa	*Platalea regia*
ruru / morepork	*Ninox novaeseelandiae*
sacred kingfisher / kōtare	*Todiramphus sanctus*
sea rush	*Juncus kraussii subsp. australiensis*
seagrass / nana, rimurehia	*Zostera muelleri*
sebaea	*Sebaea ovata*
shining cuckoo / Pīpīwharauroa	*Chrysococcyx lucidus*
ship rat	*Rattus rattus*
shore plover / tūturuatu	*Thinornis novaeseelandiae*
shore primrose	*Samolus repens var. repens*
shortfin eel / tuna	*Anguilla australis*
shortjaw kōkopu	*Galaxias postvectis*
silver pine / manoao	*Manoao colensoi*
silvereye / tauhou	*Zosterops lateralis*
small-leaved hoheria	*Hoheria angustifolia*
soft twig rush	*Machaerina rubiginosa*
South Island freshwater crayfish / kēwai, kōura	*Paranephrops zealandicus*
South Island pied oystercatcher / tōrea	*Haematopus finschi*
sowthistle / pūhā	*Sonchus species*
spartina, cord grass	*Spartina anglica*
sphagnum moss	*Sphagnum cristatum / Sphagnum perichaetiale*
spotless crake / pūweto	*Porzana tabuensis tabuensis*
stoat	*Mustela erminea*
stout water milfoil	*Myriophyllum robustum*
swamp harrier / kāhu	*Circus approximans*
swamp helmet orchid	*Corybas carsei*
swamp īnaka, swamp īnanga	*Dracophyllum oliveri*
swamp maire / maire tawake	*Syzygium maire*
swamp millet	*Isachne globosa*
tākaka / bracken	*Pteridium esculentum*
tamingi	*Epacris pauciflora*
tānekaha	*Phyllocladus trichomanoides*
tangle fern	*Gleichenia dicarpa*
tara iti / fairy tern	*Sterna nereis*
tara pirohe / black-fronted tern	*Chlidonias albostriatus*
taranui / Caspian tern	*Hydropogne rufopectus*
tārapunga / black-billed gull	*Larus bulleri*
tarn speedwell	*Veronica lilliputiana*
tauhou / silvereye	*Zosterops lateralis*
te hōkioi, pouākai / Haast's eagle	*Hieraaetus moorei*

Common name(s)	Scientific name(s)
tea-tree / mānuka	*Leptospermum scoparium*
tētē moroiti / grey teal	*Anas gracilis*
thermal ladder fern	*Nephrolepis flexuosa*
tī kōuka, tī rakau / cabbage tree	*Cordyline australis*
tīpokopoko / common bully	*Gobiomorphus cotidianus*
tītipounamu/ rifleman	*Acanthisitta chloris*
tōrea / South Island pied oystercatcher	*Haematopus finschi*
torea-pango / variable oystercatcher	*Haematopus unicolor*
toro	*Myrsine salicina*
torrent fish / panoko, papangoko	*Cheimarrichthys fosteri*
tōtara	*Podocarpus totara*
tūī	*Prosthemadera novaeseelandiae*
tuna / native freshwater eel	*Anguilla dieffenbachii / Anguilla australis*
tūpākihi, tutu	*Coriaria species*
tūturiwhatu / banded dotterel	*Charadrius bicinctus*
tūturuatu / shore plover	*Thinornis novaeseelandiae*
ūpoko-a-tangata / giant umbrella sedge	*Cyperus ustulatus*
ureure / glasswort	*Sarcocornia quinqueflora*
variable oystercatcher / torea-pango	*Haematopus unicolor*
waikaka, kōwaro, hauhau / mudfish	*Neochanna diversus / Neochanna apoda*
wātakirihi, kōwhitiwhiti / water cress	*Nasturtium microphyllum / Nasturtium officinale*

Common name(s)	Scientific name(s)
water brome	*Amphibromus fluitans*
water hyacinth	*Eichhornia crassipes*
watercress / kōwhitiwhiti, wātakirihi	*Nasturtium microphyllum / Nasturtium officinale*
wāwā, kōpūngāwhā, kuawa, kūwāwā / lake clubrush	*Schoenoplectus species*
weasel	*Mustela erminea*
weweia / New Zealand dabchick	*Poliocephalus rufopectus*
whanake / cabbage tree	*Cordyline australis*
whēkau / laughing owl	*Sceloglaux albifacies*
whio / blue duck	*Hymenolaimus malacorhynchos*
white heron / kōtuku	*Ardea modesta*
white pine / kahikatea	*Dacrycarpus dacrydioides*
whitebait	*Galaxias species and Retropinna retropinna (juveniles)*
white-faced heron / matuku moana	*Egretta novaehollandiae*
willow	*Salix species*
wineberry / mako	*Aristotelia serrata*
wire rush (northern species)	*Empodisma robustum*
wire rush (southern species)	*Empodisma minus*
wīwī / knobby clubrush/ sea rush	*Ficinia nodosa / Juncus kraussii subsp. australiensis*
wood pigeon / kererū	*Hemiphaga novaeseelandiae*
wrybill / ngutu parore	*Anarhynchus frontalis*
yellow flag iris	*Iris pseudacorus*
zig-zag bog-rush	*Schoenus brevifolius*

Appendix 3
Featured Wetlands and Ramsar Sites

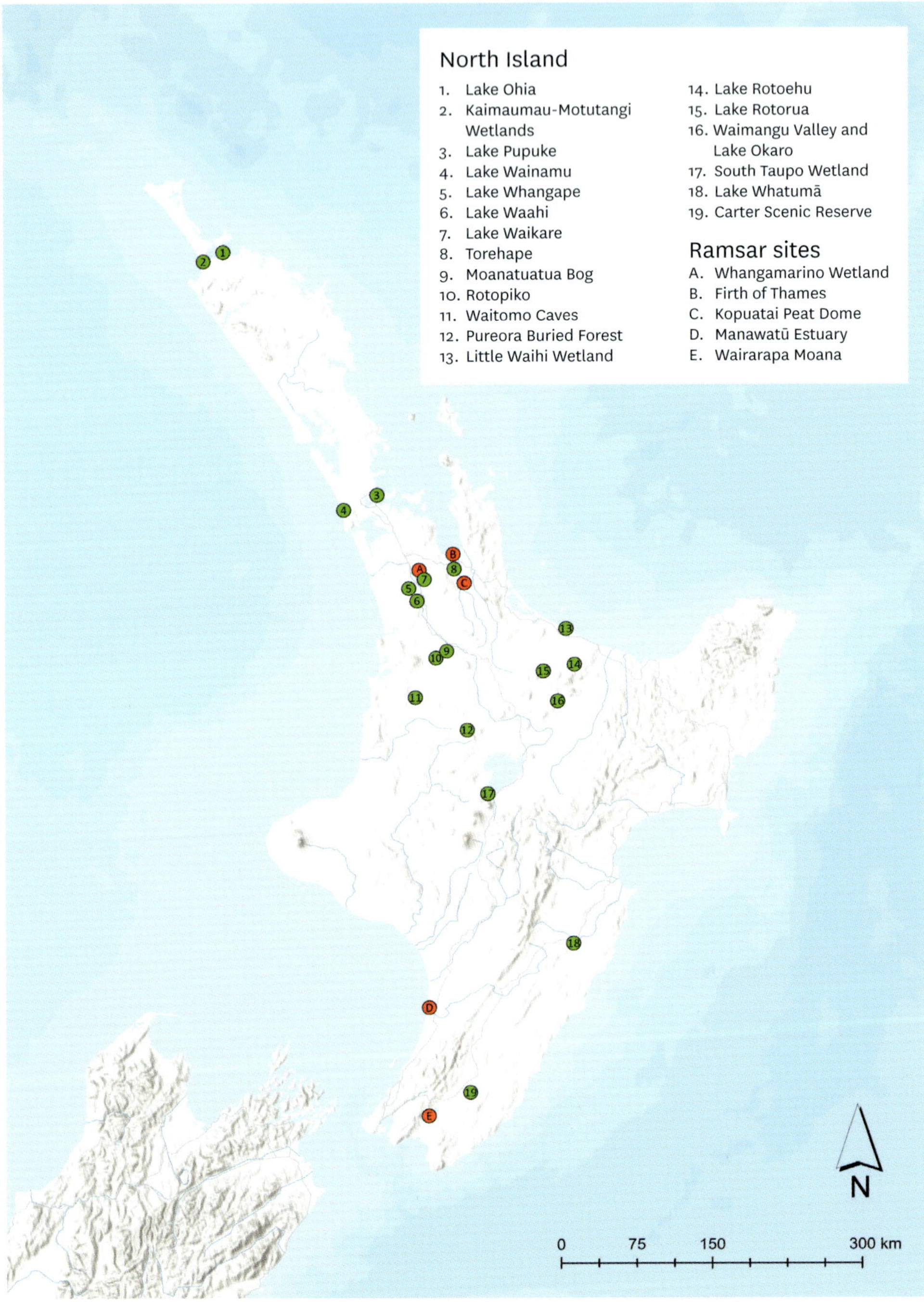

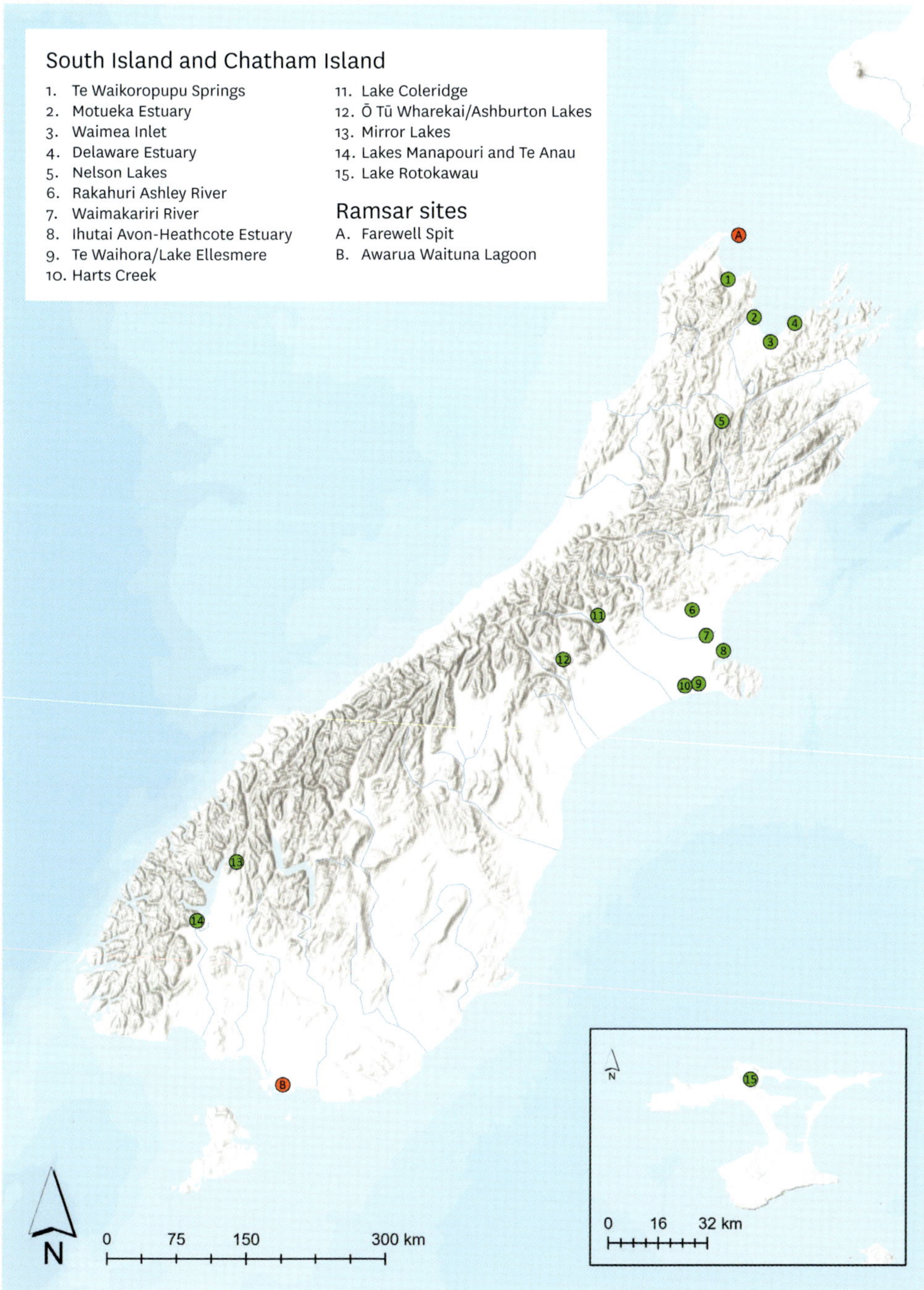
South Island and Chatham Island
1. Te Waikoropupu Springs
2. Motueka Estuary
3. Waimea Inlet
4. Delaware Estuary
5. Nelson Lakes
6. Rakahuri Ashley River
7. Waimakariri River
8. Ihutai Avon-Heathcote Estuary
9. Te Waihora/Lake Ellesmere
10. Harts Creek
11. Lake Coleridge
12. Ō Tū Wharekai/Ashburton Lakes
13. Mirror Lakes
14. Lakes Manapouri and Te Anau
15. Lake Rotokawau
Ramsar sites
A. Farewell Spit
B. Awarua Waituna Lagoon
N
0 75 150 300 km
0 16 32 km

Notes

Getting into Wetlands

1 M. Peters, The Ecology of Community Environmental Groups: Integrating Restoration, Partnerships and Citizen Science, PhD thesis, University of Waikato, 2016, https://researchcommons.waikato.ac.nz/handle/10289/10618

2 M. Peters and B. Clarkson (eds.), *Wetland Restoration: A Handbook for New Zealand Freshwater Systems* (Lincoln: Manaaki Whenua Press, 2010), www.mwpress.co.nz/wetlands/wetland-restoration-a-handbook-for-new-zealand-freshwater-systems

3 'Restiad Wetlands', NZ Landcare Trust, accessed 15 September 2021, www.landcare.org.nz/completed-project-item/restiad-wetlands

4 Y. Taura, C. Van Shravendijk-Goodman and B. Clarkson (eds.), *Te Reo o Te Repo: The Voice of the Wetland* (Hamilton: Manaaki Whenua — Landcare Research, 2017), www.landcareresearch.co.nz/publications/te-reo-o-te-repo

5 'WETMAK', NZ Landcare Trust, accessed 15 September 2012, www.landcare.org.nz/resource-item/wetmak

6 https://knowledgeauckland.org.nz/publications/auckland-community-ecological-monitoring-guide-a-framework-for-selecting-monitoring-methods

7 Ministry for the Environment and Stats NZ, *New Zealand's Environmental Reporting Series: Our Freshwater 2020* (Wellington: Ministry for the Environment and Stats NZ, 2020), https://environment.govt.nz/assets/Publications/Files/our-freshwater-2020.pdf

8 Dymond et al., 'Revised Extent of Wetlands'.

9 Based on analysis of table 4 (extant freshwater and coastal species, excluding colonisers) in C. M. Miskelly, J. E. Dowding, G. P. Elliott, R. A. Hitchmough, R. G. Powlesland, H. A. Robertson, P. M. Sagar, R. P. Scofield and G. A. Taylor, 'Conservation Status of New Zealand Birds', *Notornis* 55, no. 3 (2008): 117–135, https://notornis.osnz.org.nz/system/files/Notornis_55_3_117.pdf

10 J. M. Goodman, N. R. Dunn, P. J. Ravenscroft, R. M. Allibone, J. A. T. Boubee, B. O. David, M. Griffiths, N. Ling, R. A. Hitchmough and J. R. Rolfe, *Conservation Status of New Zealand Freshwater Fish*, New Zealand Threat Classification Series 7 (Wellington: Department of Conservation, 2014).

11 A. Suren and B. Sorrell, *Aquatic Invertebrate Communities of Lowland Wetlands in New Zealand.* Characterising Spatial, Temporal and Geographic Distribution Patterns, Science for Conservation 305 (Wellington: Department of Conservation, 2010).

12 B. Clarkson, C. Watts, B. Sorrell, S. Bartlam, D. Thornburrow, N. Fitzgerald, C. Chague-Goff, K. Bodmin and P. Champion, *Biotic Composition of New Zealand Lowland Wetlands: I Vegetation and II Invertebrates*, Landcare Research Contract Report LC0708/142 (Hamilton: Manaaki Whenua — Landcare Research, 2008).

13 R. Costanza, R. d'Arge, R. de Groot et al., 'The Value of the World's Ecosystem Services and Natural Capital', *Nature* 387 (1997): 253–26, https://doi.org/10.1038/387253a0

14 J. P. Goodrich, D. I. Campbell and L. A. Schipper, 'Southern Hemisphere Bog Persists as a Strong Carbon Sink During Droughts', *Biogeosciences*, 14 (2017): 4563–4576, https://doi.org/10.5194/bg-14-4563-2017

15 David I. Campbell, Georgie L. Glover-Clark, Jordan P. Goodrich, Christopher P. Morcom, Louis A. Schipper, Aaron M. Wall, 'Large Differences in CO_2 Emissions from Two Dairy Farms on a Drained Peatland Driven by Contrasting Respiration Rates During Seasonal Dry Conditions', *Science of the total environment* 760 (2021): 143410.

16 Interview with Karen Denyer and Monica Peters, 3 March 2021.

The Wetlands of Aotearoa New Zealand

1 J. Dymond, M. Sabetizade, P. F. Newsome, G. R. Harmsworth and A.-G. Ausseil, 'Revised Extent of Wetlands in New Zealand', *New Zealand Journal of Ecology* 45, no. 2 (2021): 3444, https://newzealandecology.org/nzje/3444.pdf

2 P. Johnson and P. Gerbeaux, *Wetland Types in New Zealand* (Wellington: Department of Conservation, 2004).

3 P. Cromarty and D. A. Scott (eds), *A Directory of Wetlands in New Zealand* (Wellington: Department of Conservation, 1995), www.doc.govt.nz/upload/documents/science-and-technical/nzwetlands00.pdf

4 M. C. McGlone, 'Postglacial History of New Zealand Wetlands and Implications for their Conservation', *New Zealand Journal of Ecology* 33 (2009): 1–23.

5 Ibid.

6 A. Ausseil, L. Chadderton, P. Gerbeaux, R. T. Stephens and J. R. Leathwick, 'Applying Systematic Conservation Planning Principles to Palustrine and Inland Saline Wetlands of New Zealand', *Freshwater Biology* 56 (2011): 142–161.

17 P. N. Johnson and P. Gerbeaux, *Wetland Types in New Zealand* (Wellington: Department of Conservation, 2004), www.doc.govt.nz/documents/science-and-technical/WetlandsBW.pdf
18 'Types of Wetlands', National Wetland Trust, accessed 17 September 2021, www.wetlandtrust.org.nz/what-we-do/resources/types-of-wetlands

Mātauranga Māori o ngā Repo/Māori Knowledge of Wetlands

1 G. R. Harmsworth and G. Tipa, *Māori Environmental Monitoring in New Zealand: Progress, Concepts, and Future Direction*, report for the ICM website (2006), http://icm.landcareresearch.co.nz/knowledgebase/publications/public/2006_Maorienvmonit%20paper.pdf
2 I. Shiklomanov, 'World Fresh Water Resources, in *Water in Crisis: A Guide to the World's Fresh Water Resources*, edited by Peter H. Gleick (New York: Oxford University Press, 1993).
3 A.-G. E. Ausseil, P. Gerbeaux, W. L. Chadderton, R. T. T. Stephens, D. Brown and J. R. Leathwick, *Wetland Ecosystems of National Importance for Biodiversity: Criteria, Methods and Candidate List of Nationally Important Inland Wetlands*, Landcare Research Contract Report LC0708/158 prepared for the Department of Conservation (Lincoln: Manaaki Whenua — Landcare Research, 2008); B. R. Clarkson, A.-G. Ausseil and P. Gerbeaux, 'Wetland Ecosystem Services', in *Ecosystem Services in New Zealand — Conditions and Trends*, edited by J. R. Dymond (Lincoln: Manaaki Whenua Press, 2013), 192–202; Y. Taura, C. van Schravendijk-Goodman and B. Clarkson (eds), *Te Reo o Te Repo: The Voice of the Wetland. Connections, Understandings and Learnings for the Restoration of Our Wetlands* (Hamilton: Manaaki Whenua — Landcare Research and Waikato Raupatu River Trust, 2017), www.landcareresearch.co.nz/publications/te-reo-o-te-repo; Y. Taura, C. van Schravendijk-Goodman and B. Clarkson (eds), *Te Reo o Te Repo — Kei Konei Tonu Au: The Voice of the Wetland — I Am Still Here* (Hamilton: Manaaki Whenua — Landcare Research, 2021); J. R. Dymond, M. Sabetizade, P. F. Newsome, G. R. Harmsworth and A. Gaelle-Ausseil, 'Revised Extent of Wetlands in New Zealand', *New Zealand Journal of Ecology* 45, no. 2 (2021): 34–44, https://newzealandecology.org/nzje/j45_2
4 E. Best, *The Māori As He Was*, 2nd ed. (Wellington: Dominion Museum, 1942); E. Best, *Māori Religion and Mythology*, Dominion Museum Bulletin 10 (Wellington: Board of Māori Ethnological Research and Dominion Museum, 1924); P. Buck Te Rangi Hiro, *The Coming of the Māori* (Wellington: Māori Purposes Fund Board, 1950); M. Marsden and M. King, *God, Man and Universe: A Māori View Te Ao Hurihuri: The World Moves On* (Wellington: Hicks Smith & Sons, 1975); M. Marsden, *The Natural World and Natural Resources. Māori Value Systems and Perspectives*, Resource Management Law Reform Working paper no. 29, Part A (Wellington: Ministry for the Environment, 1988); M. Marsden, *Resource Management Law Reform: Part A, the Natural World and Natural Resources, Māori Value Systems and Perspectives; Part B, Water Resources and the Kai Tahu Claim* (Wellington: Ministry for the Environment, 1989); H. M. Mead and N. Grove, *Ngā Pepeha a ngā Tipuna* (Wellington: Victoria University Press, 2001); H. M. Mead, *Tikanga Māori: Living by Māori Values* (Wellington: Huia Publishers, 2003); T. Black (ed.), *Enhancing Mātauranga Māori and Global Indigenous Knowledge* (Wellington: New Zealand Qualifications Authority, 2014), www.nzqa.govt.nz/assets/Māori/Te-Rautaki-Māori/Publications/Enhancing-Mtauranga-Mori-and-Global-Indigenous-Knowledge.pdf; D. Hikuroa, 'Mātauranga Māori — the Ūkaipō of Knowledge in New Zealand', *Journal of the Royal Society of New Zealand* 47, no. 1 (2017): 5–10, doi:10.1080/03036758.2016.1252407; O. R. Mercier, 'Mātauranga and Science', *New Zealand Science Review* 74, no. 4 (2018): 83–90.
5 M. Durie, *Whaiora: Māori Health Development*, 2nd ed. (Auckland: Oxford University Press: 1998); G. R. Harmsworth, *Māori Environmental Performance Indicators for Wetland Condition and Trend. Coordinated Monitoring of New Zealand Wetlands, Phase 2, Goal 2, A Ministry for the Environment SMF Project 5105*, Landcare Research Report LC 0102/099 (Palmerston North: Manaaki Whenua — Landcare Research, 2002); B. Raskin, 'The Adoption and Use of Technology in Food Production: Māori Experiences and Perspectives', summer studentship project report, Institute of Environmental Science and Research Ltd, 2009, www.esr.cri.nz/assets/Future-Foods/Reports/Maori-experiences-and-perspectives.pdf; M. Roskruge, A. Grimes, P. McCann and J. Poot, 'Social Capital and Regional Social Infrastructure Investment: Evidence from New Zealand', *International Regional Science Review* 35, no. 1 (2012): 3–25; G. R. Harmsworth and S. Awatere,

'Indigenous Māori Knowledge and Perspectives of Ecosystems', in *Ecosystem Services in New Zealand — Conditions and Trends*, edited by J. R. Dymond (Lincoln: Manaaki Whenua Press, 2013), 274–286; S. Awatere and G. Harmsworth, *Ngā Aroturukitanga Tika Mō ngā Kaitiaki: Summary Review of Mātauranga Māori Frameworks, Approaches, and Culturally Appropriate Monitoring Tools for Management of Mahinga Kai*, Landcare Research Contract Report LC1774, prepared for the MBIE programme Ngā Tohu o te Taiao: Sustaining and Enhancing Mahinga Kai and Wai Māori, (Hamilton: Manaaki Whenua — Landcare Research, 2014); S. Awatere, M. Robb, Y. Taura, K. Reihana, G. Harmsworth, J. Te Maru and E. Watene-Rawiri, *Wai Ora Wai Maori — a Kaupapa Maori Assessment Tool*, Policy Brief No. 19 (Hamilton: Manaaki Whenua — Landcare Research, 2017); O. R. Mercier, 'Mātauranga and Science', *New Zealand Science Review* 74, no. 4 (2018): 83–90; New Zealand Association of Scientists (NZAS), 'Mātauranga Māori and Science', *New Zealand Science Review* 75, no. 4, Special Issue (2019); New Zealand Association of Scientists (NZAS), 'Mātauranga Māori and Science — Part 2', *New Zealand Science Review* 76, nos 1–2, Special Issue (2020); Environmental Protection Authority (EPA) — Te Mana Rauhī Taiao, *Partnership in Action: The EPA's Mātauranga Framework* (Wellington: Environmental Protection Authority, 2020).

6 Best, *The Māori as He Was*; Buck, *The Coming of the Māori*; Marsden and King, *God, Man and Universe*; Mead and Grove, *Ngā Pepeha a ngā Tipuna*; Mead, *Tikanga Māori*; New Zealand Qualifications Authority Mana Tohu Mātuaranga o Aotearoa, *Conversations on Mātauranga Māori* (Wellington: New Zealand Qualifications Authority Mana Tohu Mātuaranga o Aotearoa, 2012), www.nzqa.govt.nz/assets/Maori/ConversationsMMv6AW-web.pdf

7 Waitangi Tribunal, *Report of the Waitangi Tribunal on Indigenous Flora and Fauna and Cultural Intellectual Property Claim (Wai 262)* (Wellington: Waitangi Tribunal, 2011).

8 H. Whaanga and T. Roa, 'Te Reo o Te Repo — The Language of the Swamp', in Taura et al., *Te Reo o Te Repo — Kei Konei Tonu Au*, 65–76.

9 Ibid.

10 Whaanga and Roa, 'Te Reo o Te Repo - The Language of the Swamp'.

11 M. Henare, *Ngā Tikanga me ngā Ritenga o Te Ao Māori: Standards and Foundations of Māori Society* (Wellington: Royal Commission on Social Policy, 1988); M. Marsden, *Resource Management Law Reform*; C. Barlow, *Tikanga Whakaaro: Key Concepts in Māori Culture* (Auckland: Oxford University Press, 1991); G.R. Harmsworth, 'Māori Values for Land-use Planning', *Broadsheet: Newsletter of the New Zealand Association of Resource Management*, February 1997, 37–52; Mead, *Tikanga Māori*.

12 Marsden, *The Natural World and Natural Resources*; Marsden, *Resource Management Law Reform*.

13 Henare, *Ngā Tikanga me ngā Ritenga o Te Ao Māori*; Marsden, *The Natural World and Natural Resources*; C. Barlow, *Tikanga Whakaaro: Key Concepts in Māori Culture* (Auckland: Oxford University Press, 1993); Durie, *Whaiora*; M. Kawharu, 'Kaitiakitanga: A Māori Anthropological Perspective of the Māori Socio-Environmental Ethic of Resource Management', *Journal of the Polynesian Society* 109, no. 4 (2000): 349–370; G. R. Harmsworth, T. Warmenhoven, P. Pohatu and M. Page, *Waiapu Catchment Report: Māori Community Goals for Enhancing Ecosystem Health*, Landcare Research Report LC0102/100 for Te Whare Wananga o Ngāti Porou (Palmerston North: Manaaki Whenua — Landcare Research, 2002); Mead, *Tikanga Māori*.

14 Harmsworth and Awatere, 'Indigenous Māori Knowledge and Perspectives of Ecosystems'.

15 Durie, *Whaiora*.

16 E. M. K. Douglas, *Waiora, Wai Māori, Waitai, Waikino, Waimate. Māori Perceptions of Water and the Environment*, Occasional Paper 27 (Hamilton: Centre for Māori Studies and Research, University of Waikato, 1984); Harmsworth, *Māori Environmental Performance Indicators*; G. Harmsworth, 'Māori and Public Health: Working in Partnership to Manage Freshwater Resources', paper presented to the master class as part of the Managing Fresh Water Resources for Public Health and Sustainability. Connecting Communities, Policy, and Science: Proceedings of the New Zealand Population Health Congress, Auckland, 8 October 2014, 145–159, www.researchgate.net/publication/266971241_Maori_and_public_health_working_in_partnership_to_manage_freshwater_resources; J. C. Kitson, A. M. Cain, M. N. T. H. Johnstone, R. Anglem, J. Davis, M. Grey, A. Kaio, S. Blair and D. Whaanga, 'Murihiku Cultural Water Classification System: Enduring Partnerships Between People, Disciplines and Knowledge Systems', *New Zealand Journal of Marine and Freshwater Research* 52, no. 4 (2018): 511–525.

17 Douglas, *Waiora, Wai Māori*; Harmsworth, *Māori Environmental Performance Indicators for Wetland Condition and Trend*; Harmsworth, 'Māori and Public Health'.

18 Ibid.

19 B. R. Clarkson, B. K. Sorrell, P. N. Reeves, P. D. Champion, T. R. Partridge and B. D. Clarkson, *Handbook for Monitoring Wetland Condition* (Wellington: Ministry for the Environment, 2003), https://datastore.landcareresearch.co.nz/dataset/handbook-for-monitoring-wetland-condition; P. Johnson and P. Gerbeaux, *Wetland Types in New Zealand* (Wellington: Department of Conservation and Ministry for the Environment, 2004); G. R. Harmsworth, *Māori Needs for Vegetation Mapping in Aotearoa-New Zealand*, Landcare Research Contract Report LC3850 (Landcare Research — Manaaki Whenua, 2020).

20 Johnson and Gerbeaux, *Wetland Types in New Zealand.*

21 Harmsworth, *Māori Needs for Vegetation Mapping.*

22 Johnson and Gerbeaux, *Wetland Types in New Zealand.*

23 Harmsworth and Roskruge, 'Chapter 9 — Beliefs, and Concepts of Soils, the Environment and Land'; Harmsworth and Roskruge, 'Chapter 20 — Māori Use and Knowledge of Soils over Time'.

24 Ibid.

25 Waitangi Tribunal, Wai 262 claim.

26 'Cultural Indicators', Science Learning Lab Pokapū Akoranga Pūtaiao video, 1:46, 18 November 2020, www.sciencelearn.org.nz/videos/2018-cultural-indicators

27 Ibid.

28 G. R. Harmsworth, R. G. Young, D. Walker, J. E. Clapcott and T. James, 'Linkages Between Cultural and Scientific Indicators of River and Stream Health', *New Zealand Journal of Marine and Freshwater Research* Special Issue 45, no. 3 (2011): 423–436.

29 H. Rainforth and G. Harmsworth, *Kaupapa Māori freshwater Assessments: A Summary of Iwi and Hapū-based Tools, Frameworks, and Methods for Assessing Freshwater Environments* (Perception Planning, 2019), www.mahaki.com/uploads/1/0/9/7/109751215/kaupapa_maori_freshwater_assessments.pdf

30 G. Harmsworth, S. Awatere and M. Robb, 'Indigenous Māori Values and Perspectives to Inform Freshwater Management in Aotearoa-New Zealand', *Ecology and Society* 21, no. 4 (2016): 9, www.ecologyandsociety.org/vol21/iss4/art9

31 C. Pauling, B. Mattingley and J. Aitken, *Te Ahuatanga o te Waiau — Cultural Health Baseline Report 2005*, Report for Te Rūnanga o Ngāi Tahu and the Ministry for the Environment (Wellington: Ministry for the Environment, 2005).

32 G. Tipa, 'Exploring Indigenous Understandings of River Dynamics and River Flows: A Case from New Zealand', *Environmental Communication* 3 (2009): 95–120, http://dx.doi. org/10.1080/17524030802707818; G. Tipa, 'Environmental Flow Assessments: A Participatory Process Enabling Māori Cultural Values to Inform Flow Regime Setting', in *Water, Cultural Diversity and Global Environmental Change: Emerging Trends, Sustainable Futures?*, eds. B. R. Johnston, L. Hiwasaki, I. J. Klaver, A. R. Castillo and V. Strang, 467–491 (Dordrecht: Springer, 2021), http://dx.doi. org/10.1007/978-94-007-1774-9_32; G. Tipa and C. Severne, *Including Mātauranga Māori in Environmental Flow Setting Decisions*, NIWA Report HAM 2010-030 for the Ministry for the Environment (Wellington: NIWA, 2010); G. Tipa and K. Nelson, 'Identifying Cultural Flow Preferences: Kakaunui River Case Study', *Journal of Water Resources Planning and Management* 138, no. 6 (2012): 660–670; Tipa & Associates, *Cultural Values, Flow and Water Management Issues for the Waikirikiri/Selwyn-Te Waihora Catchments. Data Collected by the Taumutu/Tuahuriri COMAR Team and Horomaka COMAR Team*, report prepared by Tipa & Associates (Christchurch: Environment Canterbury Regional Council, 2013), http://files.ecan.govt.nz/public/lwrp/variation1/cultural-values-flow-water-managementissues-waikirikiri-selwyn-te-waihora-catchments.pdf; H. Rainforth, 'A Difficult Convert? Combining Cultural Assessments and the IFIM to Determine Flow Setting', paper presented at the joint conference for the New Zealand Hydrological Society, New Zealand Freshwater Sciences Society and IPE New Zealand Rivers Group, 2014 Water Symposium Integration: The Final Frontier, 24–28 November 2014.

33 G. R. Harmsworth, 'Māori Values and GIS: The New Zealand Experience', GIS Asia/Pacific April (1997): 40–43; G. R. Harmsworth, 'Indigenous Values and GIS: A Method and Framework', *Indigenous Knowledge and Development Monitor* 6, no. 3 (1998): 3–7; G. T. Tipa, 'Bringing the Past into Our Future — Using Historic Data to Inform Contemporary Freshwater Management',

Kotuitui: New Zealand Journal of Social Sciences 8 (2013): 40–63, http://dx.doi. org/10.1080/1177 083x.2013.837080
34 'Kaitiaki Tools', NIWA (National Institute of Water and Atmospheric Research Taihoro Nukurangi), accessed 14 September 2021, www.niwa.co.nz/ freshwater/management-tools/water-quality-tools/kaitiaki-tools
35 G. Stewart, 'Kaupapa Māori, Philosophy, and Schools', *Educational Philosophy and Theory* 46, no. 11 (2014): 1270–1275, doi :10.1080/00131857.2013.771448; K. Maxwell and W. Penetito, 'How the Use of Rāhui for Protecting Taonga Has Evolved Over Time', *MAI Review* 2: 1–15.
36 T. K. Morgan, 'Waiora and Cultural Identity: Water Quality Assessment Using the Mauri Model', *AlterNative* 3, no. 1 (2007): 44–69; T. K. Morgan and R. Manuel, 'Western Science and Indigenous Wisdom: Is Integration Possible, Practical, Plausible?', *New Zealand Science Review* 76, nos 1–2, Special Issue — Mātauranga Māori and Science — Part 2 (2020): 6–12.
37 R. Jefferies and N. Kennedy, *Māori Outcome Evaluation: A Kaupapa Māori Outcomes and Indicators Framework and Methodology* (Hamilton: International Global Change Institute (IGCI), University of Waikato, 2009).
38 I. Ruru, I. 'The Mauri Compass', Te Runanga o Turanganui a Kiwa Internal Memo to Kahui Kaumatua, 14 August 2014, www.mauricompass. com; I. Ruru, *The Mauri Compass. A Concept Paper Showing the Mauri Compass as an Evaluation Tool in a RMA Freshwater Context* (Gisborne: Te Rūnanga o Turanganui a Kiwa, 2015).
39 I. A. Kusabs, J. M. Quinn and D. P. Hamilton, 'Effects of Benthic Substrate, Nutrient Enrichment and Predatory Fish on Freshwater Crayfish (Kōura, *Paranephrops planifrons*) Population Characteristics in Seven Te Arawa (Rotorua) Lakes, North Island, New Zealand', *Marine and Freshwater Research* 66, no. 7 (2015): 631–643, doi:10.1071/MF14148; I. A. Kusabs, B. J. Hicks, J. M. Quinn and D. P. Hamilton, 'Sustainable Management of Freshwater Crayfish (Kōura, *Paranephrops planifrons*) in Te Arawa (Rotorua) Lakes, North Island, New Zealand', *Fisheries Research* 168 (2015): 35–46.
40 E. Williams, J. Boubée, J. Jameson, P. Horton, J. Potangaroa, C. Baker, M. Stewart, A. Zernack and S. Crow, *Glass Eel Recruitment, Species Separation Trials and Juvenile Eel Surveys. Rangitaane Tuna Aquaculture — Year 1 Review of Activities*, NIWA Client Report WLG2014-83, prepared for Te Ohu Tiaki o Rangitaane Te Ika a Maui Trust (Wellington: NIWA, 2014).
41 H. J. Rainforth, 'Tiakina Kia Ora — Protecting Our Freshwater Mussels', master's thesis, Victoria University of Wellington, 2008, http://researcharchive.vuw.ac.nz/ xmlui/ handle/10063/839
42 Te Ao Marama Incorporated and Waikawa Whānau, *Waikawa Kanakana Research 2009: Monitoring Observations and recommendations for Future Monitoring and Research*, report for Te Putea Whakakaha Mahika Kai (Invercargill: Te Ao Marama, 2010); J. Kitson, V. Leith, D. Whaanga, J. Hay, A. Quarterman, S. Ledington and C. Pauling, *Kanakana Harvest Mātauranga: Potential Tools to Monitor Population Trends on the Waikawa River, Southland/Murihiku (a Scoping Project)*, final technical teport for Ngā Pae o Te Māramatanga (Auckland: Ngā Pae o Te Māramatanga, Auckland University, 2010).
43 B. Morris, C. van Schravendijk-Goodman, J. Williams and G. Ormsby, *Identifying Traditional Whitebait Stands in the Lower Waikato River — A Joint Spatial Analysis Project*, Waikato Regional Council Technical Report 2013/18 (Hamilton: Waikato Regional Council, 2013), www. waikatoregion.govt.nz/TR201318
44 M. M. Kapa and B. D. Clarkson, 'Biological Flora of New Zealand 11. *Eleocharis sphacelata*, Kuta, Paopao, Bamboo Spike Sedge', *New Zealand Journal of Botany* 47, no. 1 (2009): 43–52, www.tandfonline.com/doi/ pdf/10.1080/00288250909509792
45 D. Rickard and A. Swales, 'Field Trials of Ngā Waihotanga Iho', *Water and Atmosphere* 17, no. 1 (2009): 9; D. Rickard and A. Swales, 'Ngā Waihotanga Iho — The Estuarine Monitoring Toolkit for Iwi', *New Zealand Coastal Society Coastal News* 40 (2009): 1–3.
46 B. Williamson, J. Quinn, E. Williams and C. Van Schravendijk-Goodman, *2016 Pilot Waikato River Report Card: Methods and Technical Summary*, report prepared for the Waikato River Authority (Hamilton: Waikato River Authority, 2016).
47 M. Stewart, G. Tipa, E. Williams, M. Home, G. Olsen and C. Hickey, *Impacts of Bioaccumulative Contaminants in the Te Waihora Catchment on Mahinga Kai Gatherers: Data Report and Risk Assessment*, NIWA Client Report HAM2014-012 for Te Waihora Management Board and Environment Canterbury Regional Council (Christchurch: Te Waihora Management Board and Environment Canterbury Regional Council, 2014).

48 G. R. Harmsworth, R. G. Young, D. Walker, J. E. Clapcott and T. James, 'Linkages Between Cultural and Scientific Indicators of River and Stream Health', *New Zealand Journal of Marine and Freshwater Research* 45, no. 3 (2011): 423–436.

49 G. Tipa, 'Consideration of a Significance Assessment Method for Tangata Whenua River Values', in *The River Values Assessment System, Volume 2: Application to Cultural, Production and Environmental Values*, LEaP Report No. 24B, eds K. F. D. Hughey and A. J. M. Baker, 1–36 (Christchurch: Lincoln University, 2010), www.lincoln.ac.nz/ Documents/LEaP/Chap-8-Tangata-whenua-Southland.pdf

50 Ministry for the Environment Manatū Mō Te Taiao, *Using the Cultural Health Index: How to Assess the Health of Streams and Waterways* (Wellington: Ministry for the Environment Manatū Mō Te Taiao, 2006), www.mfe.govt.nz/sites/default/files/chi-for-streams-and-waterways-feb06-full-colour.pdf

51 G. Tipa, *Māori Environmental Indicators: Taieri River Case Study*, Technical Paper No. 58 (Wellington: Ministry for the Environment, 1999); G. Tipa and L. Teirney, *A Cultural Health Index for Streams and Waterways: Indicators for Recognising and Expressing Māori Values* (Wellington: Ministry for the Environment, Wellington, 2003); G. Tipa and L. Teirney, *Using the Cultural Health Index: How to Assess the Health of Streams and Waterways* (Wellington: Ministry for Environment, 2006), www.mfe.govt.nz/publications/cultural-health-index-streams-andwaterways-feb06; G. Tipa and L. Teirney, A Cultural Health Index for Streams and Waterways: A Tool for Nationwide Use, final technical report ME701 (Wellington: Ministry for the Environment, 2006), www.mfe.govt.nz/publications/ cultural-health-index-streams-and-waterways-tech-report-apr06; C. R. Townsend, G. Tipa, L. D. Teirney and D. K. Niyogi, 'Development of a Tool to Facilitate Participation of Māori in the Management of Stream and River Health', *EcoHealth* 1, no. 2 (2004): 184–195, http://dx.doi.org/10.1007/s10393-004-0006-9; K. Nelson and G. Tipa, *Cultural Indicators, Monitoring Frameworks and Assessment Tools* (Christchurch: Tipa & Associates, 2012); Tipa & Associates, *Cultural Values*; Tipa and Nelson, 'Identifying Cultural Flow Preferences'.

52 Townsend et al., 'Development of a Tool'; Taranaki District Council, *A Cultural Health Index for Reservoir Creek. Indicators for Recognising and Expressing Tangata Whenua ki Whakatu Values* (Stratford, Taranaki District Council, 2007); K. F. D. Hughey and K. J. W. Taylor (eds.), *Te Waihora/ Lake Ellesmere: State of the Lake and Future Management* (Christchurch: EOS Ecology, 2009), www. wet.org.nz/wp-content/uploads/2009/10/Ch08-Cultural-health-ofthe-lake.pdf

53 B. Mattingley and C. Pauling, *State of the Takiwa: Cultural Monitoring and Reporting on the Health of Our Environment. Development of the Takiwa Database* (Christchurch: Te Rūnanga ō Ngāi Tahu, 2005); C. Pauling, T. M. Lenihan, M. Rupene, N. Tirikatene-Nash and R. Couch, *State of the Takiwā: Te Āhuatanga o Te Ihutai. Cultural Health Assessment of the Avon-Heathcote Estuary and Catchment* (Christchurch: Te Rūnanga o Ngāi Tahu, 2007), www.ngaaho.Maori.nz/documents/ resources/2010091421 2427StateoftheTakiwa.pdf; Te Rūnanga ō Ngāi Tahu, *Ki Uta ki Ta— Mountains to the Sea natural Resource Management: A Scoping Document for Developing Mountains to the Sea Natural Resource Management Tools for Ngāi Tahu*, draft report (April 2003) prepared by Kaupapa Taiao for Ngā Papatipu Runanga (Christchurch: Te Rūnanga o Ngāi Tahu, 2003); Te Rūnanga ō Ngāi Tahu, 'Te Waipounamu Freshwater Report 2007: Cultural Health Assessment of South Island Waterways', in *State of the Takiwa: Ngā Wai Pounamau*, ed. C. Pauling (Christchurch, Te Rūnanga ō Ngāi Tahu, 2007); C. Pauling, 'Making Mātauranga Matter: The Role of Cultural Monitoring and Health Assessment in Water Management', paper presented at the Critical and Sensitive Research Issues Symposium (CSRI) 2010, Tangaroa Ki Uta, Tangaroa Ki Tai: Water, Our Future, hosted by Ngā Pae o te Māramatanga, 15 November 2010, Christchurch, New Zealand.

54 G. R. Harmsworth, *Māori Environmental Performance Indicators for Wetland Condition and Trend. Coordinated Monitoring of New Zealand Wetlands, Phase 2, Goal 2. Ministry for the Environment SMF Project 5105*, Landcare Research Report LC 0102/099 (Hamilton: Manaaki Whenua — Landcare Research, 2002).

55 M. Robb, 'When Two Worlds Collide: Mātauranga Māori, Science and Health of the Toreparu Wetland', MSc thesis, University of Waikato, Hamilton, 2014.

56 Harmsworth, *Māori Environmental Performance Indicators*.

57 Adapted from Harmsworth et al., 'Indigenous Māori Values and Perspectives'.

Fire and Water

Any quotes not referenced are taken from an interview with Beverley Clarkson on 19 January 2021.

1 C. J. N. Wilson and G. P. L. Walker, 'The Taupo Eruption, New Zealand. I. General Aspects', *Philosophical Transactions of the Royal Society of London* A314 (1985): 199–228; R. J. Sparks, W. H. Melhuish, J. W. A. McKee, J. Ogden and J. G. Palmer, 'C Calibration in the Southern Hemisphere and the Date of the Last Taupo Eruption: Evidence from Tree-ring Sequences', *Radiocarbon* 37, no. 2 (1995): 155–163.

2 G. P. L. Walker, 'The Taupo Pumice: Product of the Most Powerful Known (Ultraplinian) Eruption?', *Journal of Volcanology and Geothermal Research* 8, no. 1, (1980): 69–94, www.sciencedirect.com/science/article/abs/pii/0377027380900086

3 B. R. Clarkson, R. N. Patel and B. D. Clarkson, 'Composition and Structure of Forest Overwhelmed at Pureora, Central North Island, New Zealand, During the Taupo Eruption (*c.* AD 130)', *Journal of the Royal Society of New Zealand* 18 (1988): 417–436, www.tandfonline.com/doi/pdf/10.1080/03036758.1988.10426466

4 Ibid.

5 B. R. Clarkson, L. A. Schipper and B. D. Clarkson, 'Vegetation and Peat Characteristics of Restiad Bogs on Chatham Island (Rēkohu), New Zealand', *New Zealand Journal of Botany* 42, no. 2 (2004): 293–312.

6 P. J. de Lange, P. B. Heenan, B. D. Clarkson and B. R. Clarkson, 'Taxonomy, Ecology, and Conservation of *Sporadanthus* (Restionaceae) in New Zealand', *New Zealand Journal of Botany* 37 (1999): 413–431.

7 S. J. Wagstaff and B. R. Clarkson, 'Systematics and Ecology of the Australasian Genus *Empodisma* (Restionaceae) and Description of a New Species from Peatlands in Northern New Zealand', *Phytokeys* 13 (2012): 39–79.

8 Beverley R. Clarkson, 'Vegetation Recovery Following Fire in Two Waikato Peatlands at Whangamarino and Moanatuatua, New Zealand', *New Zealand Journal of Botany* 35, no. 2 (1997): 167–179.

9 G. L. W. Perry, J. M. Wilmshurst and M. S. McGlone, 'Ecology and Long-term History of Fire in New Zealand', *New Zealand Journal of Ecology* 38, no. 2 (2014): 157–176, https://newzealandecology.org/nzje/3198.pdf

10 Nancy Swarbrick, 'Waikato Region — Plants and Animals', *Te Ara — the Encyclopedia of New Zealand*, accessed 22 August 2021, www.TeAra.govt.nz/en/video/27060/peat-fire

11 Clarkson, 'Vegetation Recovery Following Fire'.

12 Ibid.

13 Ibid.

14 Ibid.

15 Ibid.

16 Department of Conservation, 'Record Numbers of Rare Whangamarino Orchid Allows New Research', *Conservation Blog: A Behind-the-scenes Look at DOC's Conservation Work* (blog), 26 May 2021, https://blog.doc.govt.nz/tag/corybas-carsei

17 J. Horsman, 'New Plant Records', *Auckland Botanical Society Newsletter* 23, no. 3 (1966): 22.

18 Department of Conservation, *Arawai Kākāriki: 2020 Programme Update* (Wellington: Department of Conservation, 2021), www.doc.govt.nz/globalassets/documents/conservation/land-and-freshwater/wetlands/ak-programme-update-2020.pdf

19 P. J. de Lange, '*Corybas carsei*', New Zealand Plant Conservation Network, accessed 17 February 2021, www.nzpcn.org.nz/flora/species/corybas-carsei

20 L. A. Schipper, B. R. Clarkson, M. Vojvodic-Vukovic and R. Webster, 'Restoring Cut-over Peat Bogs: A Factorial Experiment of Nutrients, Seeds and Cultivation', *Ecological Engineering* 19 (2002): 29–44.

21 Clarkson et al., 'Composition and Structure of Forest Overwhelmed at Pureora'.

Fish Hotels and 'Carpuccinos'

Any quotes not referenced are taken from an interview with Bruno David on 3 March 2021.

1 Bruno O. David, Mark P. Hamer and Kevin J. Collier, 'Mussel Spat Ropes Provide Passage for Banded Kokopu (*Galaxias fasciatus*) in Laboratory Trials', *New Zealand Journal of Marine and Freshwater Research* 43, no. 4 (2009): 883–888, doi: 10.1080/00288330909510046

2 B. O. David, J. D. Tonkin, K. W. T. Taipet and H. T. Hokianga, 'Learning the Ropes: Mussel Spat Ropes Improve Fish and Shrimp Passage Through Culverts', *Journal of Applied Ecology* 51 (2014): 214–223, doi: 10.1111/1365-2664.12178

3 B. O. David, D. Özkundakci, M. Pingram, D. Bergin and M. Bergin, '"The Carp-N Neutral Project" — Removal, Processing and Re-use of Invasive Fish in Local Terrestrial Conservation Projects', *Journal of Applied Ecology* 55 (2018): 1567–1574, www.coastalrestorationtrust.org.nz/site/assets/files/1178/1_-_journal_article_-_the_carp-n_neutral_project-_journal_of_app_ecology.pdf

4 S. Lowe, M. Browne, S. Boudjelas and M. De Poorter, *100 of the World's Worst Invasive Alien Species: A selection from the Global Invasive*

Species Database (Auckland: Invasive Species Specialist Group, 2000), www.iucngisd.org/gisd/pdf/100English.pdf
5 P. J. Bellingham, D. R. Towns, E. K. Cameron, J. J. Davis, D. A. Wardle, J. M. Wilmshurst and C. P. H. Mulder, 'New Zealand Island Restoration: Seabirds, Predators, and the importance of History', *New Zealand Journal of Ecology* 34, no. 1 (2010): 115–136, https://newzealandecology.org/nzje/2905.pdf
6 Deed of Settlement in Relation to the Waikato River (2009), www.govt.nz/assets/Documents/OTS/Waikato-Tainui/Waikato-Tainui-Deed-of-Settlement-in-relation-to-the-Waikato-River-17-Dec-2009.pdf

Filters and Floaters

Any quotes not referenced are taken from an interview with Chris Tanner on 28 July 2021.
1 'Wetlands', Department of Conservation, accessed 15 September 2021, www.doc.govt.nz/nature/habitats/wetlands
2 Ministry for the Environment and Stats NZ, *Aotearoa 2019: New Zealand's Environmental Reporting Series* (Wellington: Ministry for the Environment and Stats NZ, 2019), www.stats.govt.nz/information-releases/new-zealands-environmental-reporting-series-environment-aotearoa-2019
3 'Septic Tanks as Point Source of Pollutants and their Impact on Water Quality', James Hutton Institute, accessed 15 September 2021, www.hutton.ac.uk/research/projects/septic-tanks-point-source-pollutants-and-their-impact-water-quality
4 'Road Runoff and Environmental Pollution', Eniscuola, 22 March 2017, www.eniscuola.net/en/2017/03/22/road-runoff-environmental-pollution
5 Chris Tanner, James Sukias and Ben Woodward, *Provisional Guidelines for Constructed Wetland Treatment of Pastoral Farm Run-off*, NIWA client report, prepared for Dairy NZ (Hamilton: NIWA, 2020), https://niwa.co.nz/sites/niwa.co.nz/files/Provisional%20CW%20Guidelines-Final%2025.4.20%2Bnote.pdf
6 'Plant Use Details', Māori Plant Use, accessed 15 September 2021, https://maoriplantuse.landcareresearch.co.nz/WebForms/PeoplePlantsDetails.aspx?firstcome=firstcome&PKey=B39B557C-ADFD-4CF5-95DB-70D1318EC46D
7 'Water and Wastewater Challenges for Marae', NIWA Taihoro Nukurangi, 16 April 2014, https://niwa.co.nz/te-k%C5%ABwaha/publications/Aue-te-piro/Issue-1/water-and-wastewater-challenges-for-marae; 'Kokiri Centre, Whāingaroa', NIWA Taihoro Nukurangi, 12 September 2014, https://niwa.co.nz/te-k%C5%ABwaha/publications/issue-2-september-2014/kokiri-centre-wh%C4%81ingaroa
8 'News: Floating Wetlands May Help Sink Algae — Young Scientists Converge on NIWA', NIWA Taihoro Nukurangi, 1 June 2007, https://niwa.co.nz/publications/wa/vol15-no2-june-2007/news-floating-wetlands-may-help-sink-algae-young-scientists-converge-on-niwa
9 S. Ladislas, C. Gérente, F. Chazarenc, J. Brisson and Y. Andrès, 'Floating Treatment Wetlands for Heavy Metal Removal in Highway Stormwater Ponds', *Ecological Engineering* 20 (2015): 85–91, https://doi.org/10.1016/j.ecoleng.2014.09.115
10 T. R. Headley and C. Tanner, *Application of Floating Wetland for Enhanced Stormwater Treatment: A Review*, NIWA client report prepared for Auckland Regional Council (Hamilton: NIWA, 2006), www.aucklandcity.govt.nz/council/documents/technicalpublications/TP324%20-%20Floating%20Wetland%20Review-Final.pdf
11 'Restoring Water Quality in Lake Rotoehu', Rural Delivery, October 2012, www.ruraldelivery.net.nz/stories/Restoring-Water-Quality-in-Lake-Rotoehu-2017-04-04-01-27-26Z
12 J. P. S. Sukias, C. R. Yates and C. Tanner, *Assessment of Floating Treatment Wetlands for Remediation of Eutrophic Lake Waters — Maero Stream (Lake Rotoehu)*, NIWA client report (Hamilton: NIWA, 2010), https://cdn.boprc.govt.nz/media/415907/assessment-of-floating-treatment-wetlands-for-remediation-of-eutrophic-lake-waters-maero-stream-lake-rotoehu-.pdf
13 'Wetlands Can Clean Our Water, but Location is Key', NIWA Taihoro Nukurangi, 20 September 2011, https://niwa.co.nz/news/wetlands-can-clean-our-water-but-location-is-key
14 'Restoring Water Quality in Lake Rotoehu'.
15 'Floating Wetland Helps Clean Up Lake Rotorua', Beehive.govt.nz, 28 September 2012, www.beehive.govt.nz/release/floating-wetland-helps-clean-lake-rotorua
16 Ibid.
17 Alice Guy, 'Rotorua's Floating Wetland in "Quite a State"', *New Zealand Herald*, 5 September 2018, www.nzherald.co.nz/rotorua-daily-post/news/lake-rotoruas-floating-wetland-in-quite-a-state/RET5NDA5B23Q7BSCZ3OYSMK3LA
18 'Floating Wetland to be Removed from Lake Rotorua', *Rotorua Daily Post*, 24 August 2020, www.nzherald.co.nz/rotorua-daily-post/news/floating-wetland-to-be-removed-from-lake-rotorua/6M73POOQNHFHRATMJAS5I4LMGY

19 C. Tanner, J. Sukias and J. Woodward, *Technical Guidelines for Constructed Wetland Treatment of Pastoral Farm Run-off*, NIWA client report prepared for DairyNZ (Hamilton: NIWA, 2021); C. C. Tanner, J. P. S. Sukias and C. R. Yates, *New Zealand Guidelines: Constructed Wetland Treatment of Tile Drainage*, NIWA Information Series 75 (Hamilton: NIWA, 2010), https://niwa.co.nz/sites/niwa.co.nz/files/import/attachments/NZCW-guide4press_small.pdf

20 'Waitangi Park — An Urban Wetland Recreated', Envirohistory NZ, 12 December 2010, https://envirohistorynz.com/2010/12/12/waitangi-park-an-urban-wetland-recreated

21 'Waitangi Park', Wraight + Associates, accessed 15 September 2021, www.waal.co.nz/wsud/waitangi-park

Wetlands on Ice

Any quotes not referenced are taken from an interview with Clive Howard-Williams on 22 July 2021.

1 I. Hawes and C. Howard-Williams, 'Pond Life on the McMurdo Ice Shelf, One of the World's Strangest Ecosystems', *Water and Atmosphere* 11 (2003): 18–19.

2 Ibid.

3 'The Antarctic Treaty', Secretariat of the Antarctic Treaty, www.ats.aq/index_e.html

4 N. Roberts, 'Antarctica and New Zealand - The Ross Dependency', *Te Ara - the Encyclopedia of New Zealand*, accessed 3 September 2021, www.TeAra.govt.nz/en/antarctica-and-new-zealand

5 C. M. Prieto-Barajasa, E. Valencia-Canterob and G. Santoyo, 'Microbial Mat Ecosystems: Structure Types, Functional Diversity, and Biotechnological Application', *Electronic Journal of Biotechnology* 31 (January 2018): 48–56.

6 Hawes and Howard-Williams, 'Pond Life on the McMurdo Ice Shelf'.

7 D. Theobald, 'A Formal Test of the Theory of Universal Common Ancestry'. *Nature* 465 (2010): 219–222, https://doi.org/10.1038/nature09014

8 P. Sánchez-Baracaldo, G. Bianchini, J. D. Wilson and A. Knoll 'Cyanobacteria and Biogeochemical Cycles Through Earth History', *Trends in Microbiology* (4 July 2021), https://doi.org/10.1016/j.tim.2021.05.008

9 Heinrich D. Holland, 'The Oxygenation of the Atmosphere and Oceans', *Philosophical Transactions of the Royal Society B* 361, no. 1470 (2006): 903–915.

10 W. F. Vincent and C. Howard-Williams, 'Life on Snowball Earth', *Science* 287 (2000): 2421.

11 C. Howard-Williams, I. Hawes and N. Gilbert, 'Why Do So Few Antarctic Specially Protected Areas Protect Inland Waters?', *Antarctic Science* 33, no. 3 (2021): 231–232, doi:10.1017/S0954102021000201

12 E. H. Shackleton, *The Heart of the Antarctic* (London: William Heineman, 1909).

13 L. Shmakova, S. Malavin, N. Iakovenko, T. Vishnivetskaya, D. Shain, M. Plewka and E. Rivkina, 'A Living Bdelloid Rotifer from 24,000-year-old Arctic Permafrost', *Current Biology* 31, no. 11 (2021): R712–R713, doi: 10.1016/j.cub.2021.04.077; K. E. D. Coan, '24,000 Years in Permafrost Still Didn't Kill These Ancient Critters', Wired, 27 June 2021, www.wired.com/story/scientists-find-24000-years-in-siberian-permafrost-still-didnt-kill-the-ancient-rotifer

Feathers versus Fur

Any quotes not referenced are taken from an interview with Colin O'Donnell on 19 July 2021.

1 R. Holdaway, 'New Zealand's Pre-human Avifauna and its Vulnerability. *New Zealand Journal of Ecology* 12, Supplement (1989), https://newzealandecology.org/nzje/1856.pdf

2 Ibid.

3 Richard Holdaway, 'Extinctions — New Zealand Extinctions Since Human Arrival', *Te Ara — the Encyclopedia of New Zealand*, accessed 16 August 2021, www.TeAra.govt.nz/en/extinctions

4 C. F. J. O'Donnell, K. B. Clapperton and J. M. Monks, 'Impacts of Introduced Mammalian Predators on Indigenous Birds of Freshwater Wetlands in New Zealand', *New Zealand Journal of Ecology* 39, no. 1 (2015): 19–33.

5 'Rat Facts and Control Tips', Predator Free NZ, accessed 10 September 2021, https://predatorfreenz.org/resources/introduced-predator-facts/rat-facts

6 C. M. King, A. Veale, B. Patty and L. Hayward, 'Swimming Capabilities of Stoats and the Threat to Inshore Sanctuaries', *Biological Invasions* 16 (2014): 987–995.

7 Duncan Grey and Jon S. Harding, *Braided River Ecology: A Literature Review of Physical Habitats and Aquatic Invertebrate Communities*, Science for Conservation 279 (Wellington: Department of Conservation, 2007), www.doc.govt.nz/globalassets/documents/science-and-technical/sfc279entire.pdf

8 Ibid.

9 C. F. J. O'Donnell, M. Sanders, C. Woolmore and R. F. Malone, *Management and Research Priorities for Conserving Biodiversity on New Zealand's Braided Rivers* (Wellington:

Department of Conservation, 2016), https://braidedrivers.org/wp-content/uploads/ODonnell-et-al.-2016-Management-and-research-priorities-for-conserving-biodiversity-on-New-Zealands-braided-rivers.pdf
10 'Banded Dotterel', New Zealand Birds Online, accessed 10 September 2021, https://nzbirdsonline.org.nz/species/banded-dotterel
11 'Haast's Eagle', New Zealand Birds Online, accessed 10 September 2021, https://nzbirdsonline.org.nz/?q=node/299
12 'Hutton's Shearwater', New Zealand Birds Online, accessed 10 September 2021, https://nzbirdsonline.org.nz/species/huttons-shearwater
13 P. Dilks, C. O'Donnell and G. Elliott, *Predator Trapping in the Eglinton Valley, Fiordland 1990–91. The Effect of Bait Type, Tunnel Design and Trap Position*, Science & Research Internal Report No. 114 (Wellington: Department of Conservation, 1992), www.doc.govt.nz/Documents/science-and-technical/SRIR114.pdf
14 C. F. J. O'Donnell and J. M. Hoare, 'Meta-analysis of Status and Trends in Breeding Populations of Black-fronted Terns (*Chlidonias albostriatus*) 1962–2008', *New Zealand Journal of Ecology* 35 (2011): 30–43.
15 'Black Stilt/Kakī', Department of Conservation, accessed 10 September 2021, www.doc.govt.nz/nature/native-animals/birds/birds-a-z/black-stilt-kaki
16 'Kakī Released After Great Breeding Season', Department of Conservation, 2 September 2021, www.doc.govt.nz/news/media-releases/2021-media-releases/kaki-released-after-great-breeding-season
17 C. F. J. O'Donnell and J. M. Monks, 'Distribution, Long Term Population Trends and Conservation Status of Banded Dotterels (*Charadrius bicinctus bicinctus*) on Braided Rivers in New Zealand', *Notornis* 67, no. 4 (2020): 733–753, www.notornis.osnz.org.nz/node/4529
18 'Fake Clues: Using Misinformation About Odour to Protect Rare Bird Species', Manaaki Whenua — Landcare Research, 11 March 2021, www.landcareresearch.co.nz/news/fake-clues-using-misinformation-about-odour-to-protect-rare-bird-species
19 'The History of Predator Free 2050', Department of Conservation, accessed 10 September 2021, www.doc.govt.nz/nature/pests-and-threats/predator-free-2050/history

The Tiny Majority

Any quotes not referenced are taken from an interview with Corinne Watts on 16 February 2021.

1 M. A. Peters, D. Hamilton and C. Eames, 'Action on the Ground: A Review of Community Environmental Groups' Restoration Objectives, Activities and Partnerships in New Zealand', *New Zealand Journal of Ecology* 29, no. 2 (2015): 179–189.
2 R. H. Hilderbrand, A. C. Watts, and A. M. Randle, 'The Myths of Restoration Ecology', *Ecology and Society* 10, no. 1 (2005): 19, www.ecologyandsociety.org/vol10/iss1/art19
3 Ministry for the Environment Manatū Mō Te Taiao, *The State of New Zealand's Environment* (Wellington: GP Publications, 1997): table 9.20, https://environment.govt.nz/publications/the-state-of-new-zealands-environment-1997
4 C. H. Watts and N. Mason, 'If We Build It — They Mostly Come: Partial Functional Recovery but Persistent Compositional Differences in Wetland Beetle Community Restoration', *Restoration Ecology* 23 (2015): 555–565.
5 L. A. Schipper, B. R. Clarkson, M. Vojvodic-Vukovic and R. Webster, 'Restoring Cut-over Peat Bogs: A Factorial Experiment of Nutrients, Seeds and Cultivation', *Ecological Engineering* 19 (2002): 29–44.
6 Watts and Mason, 'If We Build It'.
7 T. J. Walker (ed.), 'Chapter 39: Fastest Runner', in *University of Florida Book of Insect Records* (Gainesville, FL, University of Florida, 2009), https://entnemdept.ufl.edu/walker/ufbir/chapters/chapter_39.shtml
8 C. H. Watts and R. K. Didham, 'Influences of Habitat Isolation on Invertebrate Colonization of *Sporadanthus ferrugineus* in a Mined Peat Bog', *Restoration Ecology* 14, no. 3 (2006): 412–419, https://doi.org/10.1111/j.1526-100X.2006.00149.x
9 University of Waikato, 'Fred the Thread', Science Learning Hub Pokapū Akoranga Pūtaiao video, 2:14, 14 April 2009, www.sciencelearn.org.nz/videos/752-fred-the-thread

Estuaries and eDNA

Any quotes not referenced are taken from an interview with Dana Clark in July 2021.

1 G. R. Daborn and A. M. Redden, 'Estuaries', in *The Wetland Book*, edited by C. Finlayson, G. Milton, R. Prentice and N. Davidson (Dordrecht: Springer, 2016), https://doi.org/10.1007/978-94-007-6173-5_281-1
2 T. Hume, P. Gerbeaux, D. Hart, H. Kettles and D. Neale, (2016). *A Classification of New Zealand's Coastal Hydrosystems* (Hamilton:

NIWA Taihoro Nukurangi, prepared for Ministry for the Environment, 2016).
3 Ministry for the Environment and Stats NZ, 'Issue 2: Our Activities on Land Are Polluting Our Marine Environment', in *New Zealand's Environmental Reporting Series: Our Marine Environment 2019* (Wellington: Ministry for the Environment and Stats NZ, 2019), 34–42.
4 David R. Plew, John R. Zeldis, Bruce D. Dudley, Amy L. Whitehead, Leigh M. Stevens, Barry M. Robertson and Ben P. Robertson, 'Assessing the Eutrophic Susceptibility of New Zealand Estuaries', *Estuaries and Coasts* 43 (2020): 2015–2033, https://shiny.niwa.co.nz/Estuaries-Screening-Tool-1/pdfs/Plew_et_al_2020.pdf
5 W. M. Kemp, W. Boynton, J. Adolf et al., 'Eutrophication of Chesapeake Bay: Historical Trends and Ecological Interactions', *Marine Ecology Progress Series* 303 (2005): 1–29, https://doi.org/10.3354/meps303001
6 Chesapeake Bay Foundation, *State of the Bay Report* (Chesapeake Bay Foundation, 2020), www.cbf.org/document-library/cbf-reports/2020-state-of-the-bay-report.pdf
7 R. Costanza, R. S. de Groot, P. Sutton, S. van der Ploeg, S. J. Anderson, I. Kubiszewski, S. Farber and R. K. Turner, 'Changes in the Global Value of Ecosystem Services', *Global Environmental Change* 26 (2014): 152–158, https://doi.org/10.1016/j.gloenvcha.2014.04.002
8 'Seagrass and Seagrass Beds', Ocean: Find Your Blue, accessed 6 September 2021, https://ocean.si.edu/ocean-life/plants-algae/seagrass-and-seagrass-beds

Salty and Sweet

Any quotes not referenced are taken from an interview with Don Jellyman on 26 July 2021.

1 'A Taonga for Māori', Parliamentary Commissioner for the Environment Te Kaitiaki Taiao a Te Whare Pāremata, accessed 1 September 2021, www.pce.parliament.nz/explore/longfin-eel/a-taonga-for-m%C4%81ori
2 Parliamentary Commissioner for the Environment, *On a Pathway to Extinction? An Investigation Into the Status and Management of the Longfin Eel* (Wellington: Parliamentary Commissioner for the Environment, 2013), www.pce.parliament.nz/media/1239/pce-eels-final2.pdf
3 P. R. Todd, 'Morphometric Changes, Gonad Histology, and Fecundity Estimates in Migrating New Zealand Freshwater Eels (*Anguilla* spp.)', *New Zealand Journal of Marine Freshwater Resources* 15 (1981): 155–170.
4 V. J. van Ginneken and G. E. van den Thillart, 'Eel Fat Stores Are Enough to Reach the Sargasso', *Nature* 403, no. 6766 (2000): 156–157, doi: 10.1038/35003110
5 M. J. Miller, T. Otake, J. Aoyama, S. Wouthuyzen, S. Suharti, H. Y. Sugeha and K. Tsukamoto, (2011) Observations of Gut Contents of Leptocephali in the North Equatorial Current and Tomini Bay, Indonesia', *Coastal Marine Science* 35 (2011): 277–288; K. Tsukamoto and M. J. Miller, 'The Mysterious Feeding Ecology of Leptocephali: A Unique Strategy of Consuming Marine Snow Materials', *Fisheries Science* 87 (2021): 11–29, https://doi.org/10.1007/s12562-020-01477-3
6 J. D. McCleave and D. J. Jellyman, 'Male Dominance in the New Zealand Longfin Eel Population of a New Zealand River: Probable Causes and Implications for Management', *North American Journal of Fisheries Management* 24, no. 2 (2004): 490–505, doi: 10.1577/M03-045.1
7 'Tuna — Maturation and Identifying Sex', NIWA Taihoro Nukurangi, accessed 1 September 2021, https://niwa.co.nz/te-k%C5%ABwaha/tuna-information-resource/biology-and-ecology/maturation-and-identifying-sex
8 D. Jellyman, 'An Enigma: How Can Freshwater Eels (*Anguilla* spp.) Be Such a Successful Genus Yet Be Universally Threatened?', *Reviews in Fish Biology and Fisheries* (2021): 1–18, doi: 10.1007/s11160-021-09658
9 Ibid.
10 D. Kelly and D. Jellyman, 'Shortfin Eels in Te Waihora', NIWA Taihoro Nukurangi, 1 March 2004, https://niwa.co.nz/publications/wa/vol12-no1-march-2004/shortfin-eels-in-te-waihora
11 D. Jellyman, *Fish Recruitment into Te Waihora/Lake Ellesmere. A Consideration of the Requirements of Key Species*, NIWA client report prepared for Environment Canterbury (Christchurch: NIWA, 2012), www.wet.org.nz/wp-content/uploads/2013/10/Jellyman-D.-2012-Fish-Recruitment-into-Te-Waihora.pdf
12 D. Jellyman, J. Walsh, M. de Winton and D. Sutherland, *A Review of the Potential to Re-establish Macrophyte Beds in Te Waihora (Lake Ellesmere)*, NIWA client report prepared for Environment Canterbury (Christchurch: NIWA, 2009).
13 D. J. Jellyman and P. R. Todd, 'Why Are Migrating Male Shortfinned Eels (*Anguilla australis*) in Lake Ellesmere, New Zealand, Getting Smaller But Not Younger?', *Bulletin Français de la Pêche et de la Pisciculture* 349 (1998): 141–152.
14 'Blocking Nature to Nurture a Lake Ecosystem?', NIWA Taihoro Nukurangi, 8 November 2016,

https://niwa.co.nz/freshwater-and-estuaries/freshwater-and-estuaries-update/freshwater-update-71-november-2016/blocking-nature-to-nurture-a-lake
15 D. Jellyman, *The Status of Longfin Eels in New Zealand — An Overview of Stocks and Harvest*, NIWA client report prepared for Parliamentary Commissioner for the Environment (Christchurch: NIWA, January 2012), www.pce.parliament.nz/media/1237/jellyman-report-final2.pdf
16 Ibid.
17 '150 Years of Brown Trout', Fish & Game New Zealand, accessed 7 September 2021, https://fishandgame.org.nz/freshwater-fishing-in-new-zealand/fishing-news-and-events/150-years-of-brown-trout
18 A. M. R. Burnet, *A Study of the Relationships Between Brown Trout and Eels in a New Zealand Stream*, Fisheries Technical Report 26 (Wellington: New Zealand Marine Department, 1968).
19 Parliamentary Commissioner for the Environment, *On a Pathway to Extinction? An Investigation Into the Status and Management of the Longfin Eel* (Wellington: Parliamentary Commissioner for the Environment, 2013), www.pce.parliament.nz/media/1239/pce-eels-final2.pdf
20 Ibid.
21 J. Montgomerie, 'Eeler Apologises for Killing 1.4m Longfin Eel in Opihi Lagoon', Stuff, 8 April 2015, www.stuff.co.nz/timaru-herald/news/67633815/eeler-apologises-for-killing-14m-longfin-eel-in-opihi-lagoon
22 A. James, *Ecology of the New Zealand Lamprey (*Geotria australis*): A Literature Review* (Whanganui: Department of Conservation, 2008).
23 I. C. Potter, H. S. Gill, B. B. Renaud and D. Haoucher, 'The Taxonomy, Phylogeny, and Distribution of Lampreys', in *Lampreys: Biology, Conservation and Control*, Vol. 1, edited by M. F. Docker (Dordrecht: Springer, 2015), 35–73.
24 C. F. Baker, D. J. Jellyman, S. Crow, M. Stewart, T. Buchinger and W. Li, 'First Observations of Spawning Nests in the Pouched Lamprey (*Geotria australis*)', *Canadian Journal of Fisheries and Aquatic Sciences* 4 (2017): 1603–1611, https://doi.org/10.1139/cjfas-2016-0292
25 'Scientists Help Lamprey to Spread the Love', NIWA Taihoro Nukurangi, 30 September 2019, https://niwa.co.nz/news/scientists-help-lamprey-to-spread-the-love
26 'Visiting the Lake', Waihora Ellesmere Trust, accessed 7 September 2021, https://wet.org.nz/about-the-lake/visiting-the-lake

Mysterious Matuku

Any quotes not referenced are taken from an interview with Emma Williams on 22 July 2021.

1 C. F. J. O'Donnell and H. Robertson, 'Changes in the Status and Distribution of Australasian Bittern (*Botaurus poiciloptilus*) in New Zealand, 1800s–2011', *Notornis* 63 (2016): 152–166.
2 Basil Keane, 'Kurī — Polynesian Dogs', *Te Ara — the Encyclopedia of New Zealand*, accessed 9 August 2021, www.TeAra.govt.nz/en/kuri-polynesian-dogs
3 'Conservation Dogs Programme', Department of Conservation, accessed 9 September 2021, www.doc.govt.nz/our-work/conservation-dog-programme
4 Robin Ormerod, 'Henry, Richard Treacy', *Dictionary of New Zealand Biography*, first published 1993, updated July 2013, *Te Ara — the Encyclopedia of New Zealand*, accessed 6 August 2021, https://teara.govt.nz/en/biographies/2h30/henry-richard-treacy
5 Bob Brockie, 'Native Plants and Animals — Overview — Species Unique to New Zealand', *Te Ara — the Encyclopedia of New Zealand*, accessed 6 August 2021, www.TeAra.govt.nz/en/native-plants-and-animals-overview/page-1
6 'Matuku Bittern', Bird of the Year, accessed 9 September 2021, www.birdoftheyear.org.nz/bittern
7 Caroline Williams, 'Bird of the Year 2020: 1500 Fraudulent Votes Placed for Little Spotted Kiwi', *Stuff*, 10 November 2020, www.stuff.co.nz/environment/300154684/bird-of-the-year-2020-1500-fraudulent-votes-placed-for-little-spotted-kiwi
8 Nick Evershed, 'New Zealand Bird of the Year Leaderboard: Check the Pecking Order', *The Guardian*, 21 October 2017, www.theguardian.com/environment/ng-interactive/2017/oct/21/new-zealand-bird-of-the-year-leaderboard-check-the-pecking-order

Tipping Points

Any quotes not referenced are taken from an interview with Hugh Robertson on 21 July 2021.

1 Ngāi Tahu Claims Settlement Act 1998, Schedule 73: Statutory Acknowledgement for Waituna Wetland, New Zealand Government, accessed 8 September 2021, www.legislation.govt.nz/act/public/1998/0097/latest/DLM431306.html?search=sw_096be8ed81a807fb_waituna_25_se&p=1&sr=1

2 J. E. Cavanagh, K. L. Hogsden and J. S. Harding, *Effects of Suspended Sediment on Freshwater Fish*, Manaaki Whenua — Landcare Research contract report prepared for West Coast Regional Council (Lincoln: Manaaki Whenua — Landcare Research, 2014), https://envirolink.govt.nz/assets/Envirolink/1445-WCRC129-Effects-of-suspended-sediment-on-freshwater-fish.pdf

Drones on Duty

Any quotes not referenced are taken from an interview with Marinus Boon on 1 March 2021.

1 D. C. Kotze, D. M. Macfarlane, R. J. Edwards and B. Madikizela, 'WET-EcoServices Version 2: A Revised Ecosystem Services Assessment Technique, and Its Application to Selected Wetland and Riparian Areas', *Water SA* 46, no. 4 (2020): 679–688.
2 B. R. Clarkson, B. K. Sorrell, P. N. Reeves, P. D. Champion, T. R. Partridge and B. D. Clarkson, *Handbook for monitoring Wetland Condition: Coordinated Monitoring of New Zealand Wetlands*, (Hamilton: A Ministry for the Environment Sustainable Management Fund Project, 2004).

Looking Back to See the Future

Any quotes not referenced are taken from an interview with Olivia Burge on 12 March 2021.

1 'Arawai Kākāriki Wetland Restoration Programme', Department of Conservation, accessed 7 September 2021, www.doc.govt.nz/our-work/freshwater-restoration/arawai-kakariki-wetland-restoration
2 B. Rance, W. Cooper and Jan Riddell, *Awarua Ramsar Wetland Proposal: Background Information* (Invercargill: Department of Conservation Southland Conservancy, 2000).
3 'About Awarua-Waituna Wetlands', Department of Conservation, accessed 7 September 2021, www.doc.govt.nz/our-work/freshwater-restoration/arawai-kakariki-wetland-restoration/sites/awarua-waituna-wetlands/about-awarua-waituna-wetlands
4 P. Johnson, 'Vegetation Recovery After Fire on a Southern New Zealand Peatland', *New Zealand Journal of Botany* 39, no. 2 (2001): 251–267, doi: 10.1080/0028825X.2001.9512736
5 L. M. Cranwell and L. von Post, 'Post-Pleistocene Pollen Diagrams from the Southern Hemisphere, 1: New Zealand', *Geografska Annaler* 18 (1936): 308–347.
6 B. Rance, W. Cooper and Jan Riddell, *Awarua Ramsar Wetland Proposal: Background Information* (Invercargill: Department of Conservation Southland Conservancy, 2000).
7 Ibid.
8 Boffa Miskell Ltd and Urtica Inc., *Awarua/Waituna Wetlands. Hydrology and Vegetation Mapping Methodology, Vegetation Descriptions and Recommendations*, report prepared for the Department of Conservation (Boffa Miskell Ltd and Urtica Inc., 2010).
9 Johnson, 'Vegetation Recovery After Fire'.
10 Ibid.
11 Ibid.
12 Boffa Miskell Ltd. and Urtica Inc., 'Awarua/Waituna Wetlands'.
13 Colin J. Burrows, The Seeds Always Know Best', New Zealand Journal of Botany 32, no. 3 (1994): 349–363, doi: 10.1080/0028825X.1994.10410478
14 Dr Hugh Robertson, personal communication, 2021.

Green Invaders

Any quotes not referenced are taken from an interview with Paul Champion on 27 July 2021.

1 B. R. Clarkson, N.B. Fitzgerald, P. D. Champion, L. Forester and B. D. Rance, *New Zealand Wetland Plant List 2021* (Manaaki Whenua — Landcare Research Contract Report LC3975, prepared for Hawke's Bay Regional Council, 2021).
2 D. R. Given, *Rare and Endangered Plants of New Zealand* (Wellington: A. H. & A. W. Reed, 1981).
3 C. C. Ogle, '*Sebaea ovata* and Its Habitat Near Wanganui', *Wellington Botanical Society Bulletin* 45 (1989): 92–99.
4 P. D. Champion, D. E. Hofstra, M. E. Auger and C. E. C. Gemmill, C.E.C, *Growth and Habitat of* Sebaea ovata *(Gentianaceae) in New Zealand and Australia*, Science for Conservation 229 (Wellington: Department of Conservation, 2003).
5 'National Interest Pest Responses Programme', Ministry for Primary Industries Manatū Ahu Matua, accessed 9 August 20021, www.mpi.govt.nz/biosecurity/long-term-biosecurity-management-programmes/national-interest-pest-responses-programme
6 P. D. Champion and J. S. Clayton, *Border Control for Potential Aquatic Weeds. Stage 1: Weed Risk Model*, Science for Conservation 141 (Wellington: Department of Conservation, 2000).
7 'Pest Water Weed Eradicated from South Island', Ministry for Primary Industries Manatū Ahu Matua, accessed 9 August 20021, www.mpi.govt.nz/news/media-releases/pest-water-weed-eradicated-from-south-island
8 Ibid.

9 D. E. Hofstra and P. D. Champion, *Assessment of Environmental Effects for the Use of Endothall in Four Hydrilla Affected Lakes in Hawke's Bay*, NIWA client report prepared for MAF Biosecurity New Zealand (Hamilton: NIWA, 2008), www.mpi.govt.nz/dmsdocument/16999/direct
10 P. D. Champion, 'Knowledge to Action on Aquatic Invasive Species: Island Biosecurity — the New Zealand and South Pacific Story', *Management of Biological Invasions* 9, no. 4 (2018): 383–394, https://doi.org/10.3391/mbi.2018.9.4.02
11 'New Weapon in Fight Against Invasive Weeds', NIWA Taihoro Nukurangi, accessed 20 July 2021, https://niwa.co.nz/news/new-weapon-in-fight-against-invasive-aquatic-weeds
12 Adriana Weber, 'Sniffer Dogs to Help Detect Pesky Weed', Radio New Zealand, 12 December 2017, www.rnz.co.nz/news/country/345998/sniffer-dogs-to-help-detect-pesky-weed
13 P. D. Champion, D. E. Hofstra and M. D. de Winton, *Best Management Practice for Aquatic Weed Control. Part One: The Framework*, NIWA client report prepared for Envirolink (Hamilton: NIWA, 2019), https://niwa.co.nz/sites/niwa.co.nz/files/Best%20Management%20Practice%20for%20Aquatic%20Weeds%20Framework%20May%202019.pdf
14 P. D. Champion, *Selective Control of Weeds in New Zealand Wetlands* (NIWA, Hamilton: NIWA, 2000).
15 Environmental Protection Authority Te Mana Rauhī Taiao, *Application for the Modified Reassessment of Aquatic Herbicides* (Wellington: New Zealand Government, 2012), www.epa.govt.nz/assets/FileAPI/hsno-ar/APP201365/d75a14b69a/APP201365-APP201365-EnR-report.pdf
16 Y. Taura, C. van Schravendijk-Goodman and B. Clarkson (eds), 2017. *Te Reo o Te Repo: The Voice of the Wetland: Connections, Understandings and Learnings for the Restoration of our Wetlands* (Hamilton: Manaaki Whenua — Landcare Research and Waikato Raupatu River Trust, 2017), www.landcareresearch.co.nz/publications/te-reo-o-te-repo
17 Ibid.
18 L. Hayes (comp.), *Biological Control Agents for Weeds in New Zealand: A Field Guide* (Lincoln: Manaaki Whenua — Landcare Research, 2005), www.landcareresearch.co.nz/uploads/public/Discover-Our-Research/Biosecurity/Biocontrol-ecology-of-weeds/weeds_field_guide.pdf
19 Simon Nathan, 'Lakes — New Zealand lakes', *Te Ara — the Encyclopedia of New Zealand*, accessed 16 August 2021, www.TeAra.govt.nz/en/lakes/page-1
20 'Lakes', Northland Regional Council, accessed 7 September 2021, www.nrc.govt.nz/environment/water/lakes
21 'Mapping Northland's Dune Lakes', NIWA Taihoro Nukurangi, 19 August 2015, https://niwa.co.nz/news/mapping-northlands-dune-lakes
22 Northland Regional Council, *Lake Ngatu Management Plan* (Northland Regional Council, n.d.), accessed 16 August 2021, www.nrc.govt.nz/media/gsqhwrns/lakengatumanagementplanwebsite.pdf
23 'Oxygen Weed Gone from Lake Ngakapua', *Northland Age*, 1 May 2018, www.nzherald.co.nz/northland-age/news/oxygen-weed-gone-from-lake-ngakapua/SSL37D7D37HBEGDW5PRIE2ZYHE/ Accessed 9 August 2021
24 Northland Regional Council, *Lake Ngatu Management Plan*.
25 Ibid.

Hands-and-knees Botany

Any quotes not referenced are taken from an interview with Peter Johnson on 1 March 2021.

1 National Party, 'Dancing Cossacks – 1975 National Party', NZ On Screen Iwi Whitiāhua video, 2:58, 1975, www.nzonscreen.com/title/dancing-cossacks-national-party-ad-1975
2 G. Williams, 'Powerful Exploits of Engineering', *Otago Daily Times*, 28 September 2019, www.odt.co.nz/regions/southland/powerful-exploits-engineering
3 P. N. Johnson, 'Lakes Manapouri and Te Anau: Applied Ecological Studies of Shoreline Vegetation at Lakes Manapouri and Te Anau, Fiordland. Part 1: Vegetation of Lake Manapouri Shoreline', *Proceedings of the New Zealand Ecological Society* 19 (1972): 102–119.
4 P. N. Johnson and P. Gerbeaux, *Wetland Types in New Zealand* (Wellington: Department of Conservation, 2004), www.researchgate.net/publication/240311238_Wetland_Types_in_New_Zealand

Life in Lakes

Any quotes not referenced are taken from interviews with Susie Wood in July 2021.

1 'Lake Water Quality', Stats NZ, accessed 12 August 2021, www.stats.govt.nz/indicators/lake-water-quality
2 Ramsar Sites Information Service, *Wairarapa Moana Wetland*, Ramsar information sheet (Ramsar Sites Information Service, 2020), https://rsis.ramsar.org/RISapp/files/RISrep/NZ2432RIS_2008_en.pdf

Working Wetlands

Any quotes not referenced are taken from an interview with Suzanne Lambie on 25 May 2021.

1 B. R. Clarkson, A. E. Ausseil and P. Gerbeaux, 'Wetland Ecosystem Services', in Ecosystem Services in New Zealand — Conditions and Trends, edited by J. R. Dymond (Lincoln: Manaaki Whenua Press, 2013), www.mwpress.co.nz/__data/assets/pdf_file/0020/77042/1_14_Clarkson.pdf

2 M. B. Cunha-Santino, A. L. Bitar and I. Bianchini Jr, 'Gas Emission from Anaerobic Decomposition of Plant Resources' *Acta Limnologica Brasiliensia* 28 (2016): e30, www.scielo.br/j/alb/a/yQv3hYqrjjDYXSnmRH5bVYM/?lang=en

3 Parliamentary Commissioner for the Environment, *Growing for Good: Intensive Farming, Sustainability and New Zealand's Environment* (Wellington: Parliamentary Commissioner for the Environment, 2004), www.pce.parliament.nz/publications/archive/1997-2006/growing-for-good-intensive-farming-sustainability-and-new-zealands-environment

4 Ministry for the Environment Manatū Mō Te Taiao, National Policy Statement for Freshwater Management 2020 (Wellington: New Zealand Government, 2020), https://environment.govt.nz/acts-and-regulations/national-policy-statements/national-policy-statement-freshwater-management

5 'Farm Environment Plans (FEPs)', NZ Landcare Trust, accessed 9 September 2021, www.landcare.org.nz/resource-item/farm-environment-plans-feps

Weaving Knowledge Systems

Any quotes not referenced are taken from an interview with Yvonne Taura on 24 February 2021.

1 Y. Taura, C. van Schravendijk-Goodman and B. Clarkson (eds.), *Te Reo o Te Repo: The Voice of the Wetland. Connections, Understandings and Learnings for the Restoration of Our Wetlands* (Hamilton, New Zealand: Manaaki Whenua — Landcare Research and Waikato Raupatu River Trust, 2017), www.landcareresearch.co.nz/publications/te-reo-o-te-repo

2 Y. Taura, C. van Schravendijk-Goodman and B. Clarkson (eds.), *Te Reo o Te Repo — Kei Konei Tonu Au: The Voice of the Wetland — I Am Still Here* (Hamilton, New Zealand: Manaaki Whenua — Landcare Research, 2021).

3 G. Harmsworth, 'Indicators for Cultural Resources' in Taura et al., *Te Reo o Te Repo: The Voice of the Wetland*, 51–55, www.landcareresearch.co.nz/uploads/public/Publications/Te-reo-o-te-repo/4_Indicators_for_Cultural_resources.pdf

Western Wetland Science Pioneers

1 M. McGlone, 'Postglacial History of New Zealand Wetlands and Implications for Their Conservation', *New Zealand Journal of Ecology* 33, no. 1 (2009): 1–23.

2 E. O. Campbell, EO. 1981. 'The Water Relations of Heathlands: Morphological Adaptations to Water Logging', in *Heathlands and Related Scrublands, Analytical Studies, Ecosystems of the World vol. 9B*, edited by R. L. Sprecht, 107–109 (Amsterdam: Elsevier, 1981); E. O. Campbell, 'Mires of Australasia', in *Mires: Swamp, Bog, Fen and Moor. Ecosystems of the World Vol. 4B*, edited by A. J. P. Gore, 153–180 (Oxford: Elsevier, 1983).

3 A. T. Dobson, 'Mire Types of New Zealand', in *Proceedings of the International Symposium on Classification of Peat and Peatlands, Hyytlala, Finland, 17–21 September 1979*, 82–95 (International Peat Society, 1979).

4 A.-G. Ausseil, P. Gerbeaux, W. Chadderton, T. Stephens, D. Brown and J. L. Leathwick, *Wetland Ecosystems of National Importance for Biodiversity: Criteria, methods and Candidate List of Nationally Important Wetlands*, Landcare Research Contract Report LC0708/158, prepared for the Department of Conservation (Palmerston North: Landcare Research, 2008).

Waihora Lagoon,
Pureora Forest Park.

Further Reading

General references

Candler, G., and N. Barraud. *Up the River: Explore & discover New Zealand's rivers, lakes & wetlands*. Nelson: Potton & Burton, 2017.

Clarkson, B. R. 'Swamps, Fens and Bogs'. In *Botany of Waikato*, edited by B. D. Clarkson, M. Merrett and T. Downs, 47–55. Hamilton: Waikato Botanical Society, 2000.

Clarkson, B. R., A. G. E. Ausseil and P. Gerbeaux. 'Wetland Ecosystem Services'. In *Ecosystem Services in New Zealand: Conditions and trends*, edited by J. R. Dymond, 192–202. Lincoln: Manaaki Whenua Press, 2013.

Cromarty, P., and D. A. Scott, eds. *A Directory of Wetlands in New Zealand*. Wellington: Department of Conservation, 1995. www.doc.govt.nz/upload/documents/science-andtechnical/nzwetlands00.pdf

Denyer, K., and M. Peters. 'The Root Causes of Wetland Loss in New Zealand: An analysis of public policies and processes.' Pukekohe: National Wetland Trust, 2020. www.wetlandtrust.org.nz/wp-content/uploads/2021/07/ROOT-CAUSES-OF-WETLAND-LOSS-IN-NZ_Jan-2021.pdf

Gallagher, K., and D. Pyle, directors. *Rohe Kōreporepo: The swamp, the sacred place*. Wick Candle, 2021.

Hunt, J. *Wetlands of New Zealand: A bitter-sweet story*. Auckland: Random House, 2007.

Johnson, P., and P. Gerbeaux. *Wetland Types in New Zealand*. Wellington: Department of Conservation, 2004.

McGlone, M. C. 'Postglacial History of New Zealand Wetlands and Implications for their Conservation'. *New Zealand Journal of Ecology 33 (2009): 1–23.*

National Wetland Trust. *'Our Wet & Wild Places'.* Pukekohe: National Wetland Trust, 2005.

Stephenson, G. *Wetlands: Discovering New Zealand's shy places*. Wellington: Government Printing Office, 1989.

Wetland management

Denyer, K., and M. Peters. *WETMAK: Wetlands monitoring and assessment kit*. Hamilton: NZ Landcare Trust, 2012.

NIWA. Freshwater and Estuaries Management Tools. https://niwa.co.nz/freshwater-and-estuaries/management-tools

Peters, M. and B. Clarkson, eds. *Wetland Restoration: A handbook for New Zealand freshwater systems*. Lincoln: Manaaki Whenua Press, 2010. www.mwpress.co.nz/wetlands/wetland-restoration-a-handbook-for-newzealand-freshwater-systems

Taura, Y., C. van Schravendijk-Goodman and B. Clarkson, eds. *Te Reo o Te Repo — Kei Konei Tonu Au: The voice of the wetland — I am still here*. Hamilton: Manaaki Whenua — Landcare Research, 2021.

Taura, Y., C. van Schravendijk-Goodman and B. Clarkson, eds. *Te Reo o Te Repo: The voice of the wetland: Connections, understandings and learnings for the restoration of our wetlands*. Hamilton: Manaaki Whenua — Landcare Research and Waikato Raupatu River Trust, 2017. www.landcareresearch.co.nz/publications/tereo-o-te-repo

Plants and animals

Hicks, B. J., and E. M. Watene-Rawiri. 'Ikawai: Freshwater fishes in Māori culture and economy'. *New Zealand Journal of Marine and Freshwater Research* 46, no. 3 (2012): 437–439. www.tandfonline.com/doi/pdf/10.1080/00288330.2012.680194

Johnson, P. N., and P.A. Brooke. *Wetland Plants in New Zealand*. Wellington: Department of Scientific and Industrial Research, 1989.

Marinov, M., and M. Ashbee. *Dragonflies and Damselflies of New Zealand*. Auckland: Auckland University Press, 2019.

McDowall, R. M. *Ikawai: Freshwater fishes in Māori culture and economy*. Christchurch: Canterbury University Press, 2011.

Moon, G. *New Zealand Wetland Birds and their World*. Auckland: New Holland, 2008.

New Zealand Landcare Trust. 'Hooked on Native Fish'. Factsheet series. www.landcare.org.nz/resource-item/hooked-on-native-fish

Ngā Tipu Whakaoranga Māori Plant Use Database: https://maoriplantuse.landcareresearch.co.nz/WebForms/default.aspx

NIWA. Fish Finder. https://niwa.co.nz/freshwater-and-estuaries/nzffd/NIWA-fish-atlas/fish-finder

NIWA. Key to freshwater fish families in New Zealand. https://niwa.co.nz/freshwater-and-estuaries/nzffd/NIWA-fish-atlas/key-to-families

O'Donnell, C. F. J., K. B. Clapperton and J. M. Monks. 'Impacts of Introduced Mammalian Predators on Indigenous Birds of Freshwater Wetlands in New Zealand'. *New Zealand Journal of Ecology* 39, no. 1 (2015): 19–33

Potangaroa, J. *Tuna Kuwharuwharu: The longfin eel, facts, threats and how to help*. Wellington: Rangitāne o Wairarapa, Greater Wellington Regional Council, Department of Conservation, 2010. www.longfineel.co.nz/wp-content/uploads/2011/06/Tuna-Kuwharuwharu-Longfin-Eel.pdf

Image Credits

p. 1: Shaun Lee, Rotokawa pools, Waitonga Falls Track, Tongariro National Park; p. 2: Karen Denyer, Nukuhou Saltmarsh boardwalk, Ohiwa Harbour, Bay of Plenty; p. 4: Neil Fitzgerald, kōtuku (white heron, *Ardea modesta*); p. 7: Karen Denyer; p. 9: Craig Cook; p. 13: courtesy of Monica Peters; p. 14: Karen Denyer; p. 19: Ruud Kleinpaste; p. 20: Neil Fitzgerald; p. 21: Veronika Meduna; p. 22 (from left): Alastair Suren, Franziska Eller; p. 24 (clockwise from top left): Peter Johnson, Jacqui Geaux, P. Gerbeaux, Karen Denyer; p. 26: Karen Denyer, He Matapuna Paru, Whakatāne, courtesy of Te Rūnanga o Ngāti Awa; p. 29: Susie Elcock; p. 33: Karen Denyer; p. 41: Karen Denyer; p. 43 (from top): Ruud Kleinpaste, Stefan Marks; p. 44 (clockwise from left): Karen Denyer, Ruud Kleinpaste, Stefan Marks; p. 45: Karen Denyer; p. 47 (from top): Jim Denyer, Karen Denyer; p. 50: Karen Denyer; p. 54: Abraham Witana, Kaimaumau fire; p. 55: Vera Luthardt; p. 57: Courtesy of Beverley Clarkson, first published in B. R. Clarkson , B. D. Clarkson and R. N. Patel, 'The Pre-Taupo Eruption (c. AD 130) Forest of the Benneydale-Pureora District, Central North Island, New Zealand', *Journal of the Royal Society of New Zealand* 22, no. 2 (1992); p. 58: Courtesy of Beverley Clarkson, first published in B. Clarkson, 'Swamp, Fens and Bogs', in B. Clarkson, M. Merrett, and T. Downs (eds.), *Botany of the Waikato* (Hamilton: Waikato Botanical Society, 2002); p. 61 (from top): Karen Denyer, Dave Campbell; p. 62: Catherine Beard; p. 63: Karen Denyer; p. 65: Dave Campbell; p. 67: Mary Beech, Department of Conservation; p. 68: Karen Denyer, koi carp; p. 69: Waikato Regional Council; p. 70: Shaun Lee; pp. 71, 73, 77 & 80: Bruno David; p. 83: Courtesy of Dr David G. Schmale III, Professor and Fulbright New Zealand Alumnus, Virginia Tech, Blacksburg, Virginia; p. 84: Karen Denyer; p. 85: Chris Tanner; pp. 87, 91 & 93: Karen Denyer; p. 94: Courtesy of Chris Tanner, first published in T. R. Headley and C. Tanner, 'Application of Floating Wetland for Enhanced Stormwater Treatment: A review', NIWA client report prepared for Auckland Regional Council (Hamilton: NIWA, 2006); p. 97: Thomas Devereaux; p. 98: Marc Schallenberg, McMurdo Ice Shelf; p. 99: Tim Higham; p. 103: Anne Jungblut; p. 104: Anne Jungblut; p. 105: Wiedehopf20, Wikimedia Commons, CC-4.0; p. 108: Jenny Webster Brown; p. 111: Courtesy of Waimangu Volcanic Valley, Rotorua; p. 112: Imogen Warren, tūturiwhatu (banded dotterel); p. 113: Don Geddes; p. 114: Steve Attwood; p. 116: Department of Conservation; p. 119: Steve Attwood; p. 121: Department of Conservation; p. 123: Liz Brown, Department of Conservation; p. 124: Luke Eberhart-Phillips; p. 127: Environment Canterbury; p. 128: Neil Fitzgerald, forked sundew (*Drosera binata*); p. 129: Manaaki Whenua — Landcare Research; p. 133: Danny Thornburrow; p. 135: Corinne Watts; p. 138: Birgit Rhode, Manaaki Whenua — Landcare Research; p. 141: Karen Denyer; p. 142: Shaun Lee, pied stilts, Tahuna Torea Nature Reserve; p. 143: Cawthron Institute; pp. 145 & 150: Dana Clark; p. 153: Emily Goldthorpe; p. 154: Auckland Zoo, freshwater eel (tuna); p. 155: NIWA; p. 156: Salvor Gissurardottir, Wikimedia Commons, CC-2.5; p. 158: Paul Yancey; p. 161: NIWA; p. 163: Stuart Mackay, NIWA; p. 164: Cindy Baker, NIWA; p. 165: Jane Kitson; p. 167: P. Gerbeaux; p. 168: Imogen Warren, matuku (Australasian bittern, *Botaurus poiciloptilus*); p. 169: Colin O'Donnell; p. 170: John Cheyne; p. 173: Lauren Buchholz; p. 174 (from left): Karen Denyer, Jono Gribble; pp. 177: Courtesy of Emma Williams; p. 180: Imogen Warren; p. 183: Brian Hatton; p. 184: Karen Denyer, Awarua tarn; p. 185, 187 & 193: Karen Denyer; p. 197: Neil Fitzgerald; pp. 198 & 199: Hanneke Boon; pp. 203 & 204: Marinus Boon; p. 207: Judi Lapsley Miller; p. 208: Karen Denyer, Awarua tarn; p. 209: Manaaki Whenua — Landcare Research; p. 211: Brian Rance, Department of Conservation; p. 213: Sally Chesterfield; p. 216: Bradley White, Manaaki Whenua — Landcare Research; p. 221: & 222: Karen Denyer; p. 223: Tracey Burton; p. 226: Rowan Wells, NIWA; p. 229: Rob Suisted; pp. 231, 232 & 234: Karen Denyer; p. 235 (clockwise from top left): Karen Denyer (four images), Monica Peters (two images); p. 237: Paul Champion, NIWA; p. 238: Peter Johnson, Wairepo; p. 239: Pru Johnson; p. 240: Peter Johnson; pp. 242, 247 & 248: Karen Denyer; p. 249: Xavier Pochon, Cawthron; p. 253: Karen Denyer; pp. 254 & 255: Courtesy of Susie Wood; p. 259: Hamish Kendal, Flightworks; p. 260: Hamish Kendal, Flightworks, sediment traps; p. 261: Manaaki Whenua — Landcare Research; p. 265: Ilka Pelzer, Ministry for the Environment; p. 267: Hamish Kendal, Flightworks; pp. 268 & 271: Karen Denyer; p. 272: Karen Denyer, raupō (bulrush, *Typha orientalis*); p. 273: Manaaki Whenua — Landcare Research; p. 275: Adam Campbell; p. 276 (from left): Trevor James, Karen Denyer; pp. 278 & 281: Courtesy of Yvonne Taura; p. 283 (from left): Crispin Middleton, NIWA, Pat Enright; p. 284: Karen Denyer; p. 287: Department of Conservation; p. 288: Arthur Uden, Mangakaware, Waikato; p. 293: Auckland War Memorial Museum Tāmaki Paenga Hira, PH-2008-6-4-7; p. 298: Margaret Bellingham, NIWA; p. 305: Karen Denyer; p327: Karen Denyer

About the Authors

KAREN DENYER, MSc, Dip Envt Mgt, is the Executive Officer of the National Wetland Trust (NWT) and a freelance ecologist. With a background in botany, she undertakes wetland and terrestrial assessment and field monitoring for a range of clients. However, she is equally happy leading projects to translate science into pragmatic policy, supplying resources to support community-led conservation initiatives and raising awareness of wetlands. In partnership with Monica Peters, she has developed ecosystem monitoring manuals for community groups, and conducted many training workshops and NWT national symposia. Karen is also leading a volunteer project to restore a peatland complex and develop a wetland education facility at Rotopiko near Hamilton Airport for the NWT. She is the New Zealand representative for the World Wetland Network and the NGO co-focal point for the Ramsar Convention (shared with Annabeth Cohen of Forest & Bird).

MONICA PETERS works freelance at the interface between science, conservation and the public. After studying fine arts, she completed a PhD on community conservation and citizen science. She is a founding member and co-chair of the Citizen Science Association of Aotearoa New Zealand (#CitSciNZ, established in 2019) and co-led a three-year project to investigate citizen science in the context of ecological restoration. In 2019 she was awarded a Winston Churchill Fellowship to research citizen science policy, programmes and applications in Asia and Europe. She is a former trustee of the National Wetland Trust (2006–2013), and in 2018 became a board member of Forest & Bird. Read Monica's blog at www.monicalogues.com

GARTH HARMSWORTH (Te Arawa, Ngāti Tūwharetoa, Tūhourangi, Ngāti Raukawa) is Principal Scientist, Toi Rangahau Māori, at Manaaki Whenua — Landcare Research, and a senior Māori environmental scientist with over 37 years' experience. He has pioneered much of the collaborative Māori research for Manaaki Whenua — Landcare Research and across the New Zealand science sector, has led over 40 kaupapa Māori and science projects, given more than 100 conference and workshop presentations, and produced to date over 100 refereed publications. Garth is a key researcher in the development of knowledge bases/information systems, bridging mātauranga Māori and Western science, working with Māori communities and organisations, and building Māori research capacity, and has extensive experience and skills in environmental planning and resource management.

YVONNE TAURA (Ngāi Te Rangi, Ngāti Ranginui, Ngāti Hauā, Ngāti Uenuku, Ngāti Tūwharetoa) is a Kairangahau Māori, at Manaaki Whenua — Landcare Research. Her full biography is presented in the chapter 'Weaving Knowledge Systems' (page 272).

PHILIPPE GERBEAUX is an expert in the field of wetland ecology and management. Following postgraduate studies on mapping submerged aquatic vegetation in subalpine lakes in France, he headed to Aotearoa New Zealand in 1984 to complete a PhD thesis on factors influencing the decline of aquatic macrophytes in Te Waihora Lake Ellesmere. He took up a three-year position at the Department of Scientific and Industrial Research's Hydrology Centre in Ōtautahi Christchurch as a periphyton ecologist, before returning to Europe, to the Station Biologique de la Tour du Valat in Camargue, France, to lead an international programme supporting the development of management plans for wetland reserves around the Mediterranean region. Since his return to New Zealand in 1995, he has worked for the Department of Conservation as a wetland ecologist, with an interlude of three years in Fiji as Chief Technical Advisor for IUCN Oceania. Philippe has also had a long association with the Ramsar Convention, first as the New Zealand National Focal Point on its Scientific and Technical Review Panel (STRP), and then as the Oceania representative on the STRP for two terms. In his spare time, he frequently visits the Pacific Islands to survey the freshwater fauna of streams and rivers, in collaboration with France's Muséum national d'Histoire naturelle.

Acknowledgements

The authors and the National Wetland Trust wish to thank all of the contributors to this book — it is a very extensive list! Thanks to the featured scientists who generously shared their experiences and career highlights with the authors via face to face or virtual interviews, and who patiently fact-checked drafts and supplied some of the images for their chapter. We are indebted to you for allowing us a glimpse into the day-to-day life of a scientist and explaining the intricacies of your work, so that the public can better understand what goes on in the field, in labs and in offices around the country. Thanks also to Garth Harmsworth and Yvonne Taura for weaving wider narratives around cultural and ecological values of Aotearoa New Zealand's wetlands into the 'Mātauranga Māori o te Repo' chapter, and to Dr Philippe Gerbeaux for comprehensively documenting a long line of early Western scientists. The combined contributions from all of the scientists enriched our own experiences, as we hope it will for readers.

No book is complete without images to bring the text to life. We are humbled at the generosity of so many excellent photographers for allowing us to feature their beautiful images — a full list is provided in the image credits. We aim to encourage readers to have a real-life experience of each scientist's work by including a relevant wetland to visit, and we extend our thanks to GIS experts Nic Ensor (PDP) for painstakingly creating maps of each site, and to Matt Moss for producing the broad location maps pinpointing all of the wetlands mentioned throughout the book.

But it takes more than text and images to bring a book like this to fruition — we are also grateful to the scientists' respective organisations for supporting individual scientists to participate in the creation of this book, with some also providing welcome sponsorship, or, in the case of the Cawthron Institute, chapter drafts of their two scientists written by Natalie Bird (Senior Communications Advisor, Cawthron Institute).

Finally, we extended our gratitude to the countless individuals, community groups, whānau, hapū, iwi and marae groups who freely give their time to help protect and restore wetlands or their species, many of whom utilise, contribute to or inspire the work our wetland scientists do in the shallows.

Index

Page numbers in **bold** refer to images.

MASSEY
UNIVERSITY
PRESS

First published in 2022 by Massey University Press
Private Bag 102904, North Shore Mail Centre
Auckland 0745, New Zealand
www.masseypress.ac.nz

Design by Alice Bell
Cover photograph by Shaun Lee

A catalogue record for this book is available from the National Library of New Zealand

Printed and bound in China by 1010 Printing Asia Limited

ISBN: 978-0-9951431-2-8